W9-BIT-245

AN ASSESSMENT OF THE SBIR PROGRAM

Committee for
Capitalizing on Science, Technology, and Innovation:
An Assessment of the Small Business Innovation Research Program

Policy and Global Affairs

Charles W. Wessner, Editor

NATIONAL RESEARCH COUNCIL
OF THE NATIONAL ACADEMIES

THE NATIONAL ACADEMIES PRESS
Washington, D.C.
www.nap.edu

THE NATIONAL ACADEMIES PRESS 500 Fifth Street, N.W. Washington, DC 20001

NOTICE: The project that is the subject of this report was approved by the Governing Board of the National Research Council, whose members are drawn from the Councils of the National Academy of Sciences, the National Academy of Engineering, and the Institute of Medicine. The members of the committee responsible for the report were chosen for their special competences and with regard for appropriate balance.

This study was supported by Contract/Grant No. DASW01-02-C-0039 between the National Academy of Sciences and the U.S. Department of Defense, NASW-03003 between the National Academy of Sciences and the National Aeronautics and Space Administration, DE-AC02-02ER12259 between the National Academy of Sciences and the U.S. Department of Energy, NSFDMI-0221736 between the National Academy of Sciences and the National Science Foundation, and N01-OD-4-2139 (Task Order #99) between the National Academy of Sciences and the U.S. Department of Health and Human Services. The content of this publication does not necessarily reflect the views or policies of the Department of Health and Human Services, nor does mention of trade names, commercial products, or organizations imply endorsement by the U.S. Government. Any opinions, findings, conclusions, or recommendations expressed in this publication are those of the author(s) and do not necessarily reflect the views of the organizations or agencies that provided support for the project.

International Standard Book Number-13: 978-0-309-11086-0
International Standard Book Number-10: 0-309-11086-6

Limited copies are available from the Policy and Global Affairs Division, National Research Council, 500 Fifth Street, N.W., Washington, DC 20001; 202-334-1529.

Additional copies of this report are available from the National Academies Press, 500 Fifth Street, N.W., Lockbox 285, Washington, DC 20055; (800) 624-6242 or (202) 334-3313 (in the Washington metropolitan area); Internet, http://www.nap.edu.

Printed in the United States of America

THE NATIONAL ACADEMIES

Advisers to the Nation on Science, Engineering, and Medicine

The **National Academy of Sciences** is a private, nonprofit, self-perpetuating society of distinguished scholars engaged in scientific and engineering research, dedicated to the furtherance of science and technology and to their use for the general welfare. Upon the authority of the charter granted to it by the Congress in 1863, the Academy has a mandate that requires it to advise the federal government on scientific and technical matters. Dr. Ralph J. Cicerone is president of the National Academy of Sciences.

The **National Academy of Engineering** was established in 1964, under the charter of the National Academy of Sciences, as a parallel organization of outstanding engineers. It is autonomous in its administration and in the selection of its members, sharing with the National Academy of Sciences the responsibility for advising the federal government. The National Academy of Engineering also sponsors engineering programs aimed at meeting national needs, encourages education and research, and recognizes the superior achievements of engineers. Dr. Charles M. Vest is president of the National Academy of Engineering.

The **Institute of Medicine** was established in 1970 by the National Academy of Sciences to secure the services of eminent members of appropriate professions in the examination of policy matters pertaining to the health of the public. The Institute acts under the responsibility given to the National Academy of Sciences by its congressional charter to be an adviser to the federal government and, upon its own initiative, to identify issues of medical care, research, and education. Dr. Harvey V. Fineberg is president of the Institute of Medicine.

The **National Research Council** was organized by the National Academy of Sciences in 1916 to associate the broad community of science and technology with the Academy's purposes of furthering knowledge and advising the federal government. Functioning in accordance with general policies determined by the Academy, the Council has become the principal operating agency of both the National Academy of Sciences and the National Academy of Engineering in providing services to the government, the public, and the scientific and engineering communities. The Council is administered jointly by both Academies and the Institute of Medicine. Dr. Ralph J. Cicerone and Dr. Charles M Vest are chair and vice chair, respectively, of the National Research Council.

www.national-academies.org

Committee for Capitalizing on Science, Technology, and Innovation: An Assessment of the Small Business Innovation Research Program

Chair

Jacques S. Gansler (NAE)
Roger C. Lipitz Chair in Public Policy and Private Enterprise
and Director of the Center for Public Policy and Private Enterprise
School of Public Policy
University of Maryland

David B. Audretsch
Distinguished Professor and
Ameritech Chair of Economic
Development
Director, Institute for Development
Strategies
Indiana University

Gene Banucci
Executive Chairman
ATMI, Inc.

Jon Baron
Executive Director
Coalition for Evidence-Based Policy

Michael Borrus
Founding General Partner
X/Seed Capital

Gail Cassell (IOM)
Vice President, Scientific Affairs and
Distinguished Lilly Research Scholar
for Infectious Diseases
Eli Lilly and Company

Elizabeth Downing
CEO
3D Technology Laboratories

M. Christina Gabriel
Director, Innovation Economy
The Heinz Endowments

Trevor O. Jones (NAE)
Founder and Chairman
Electrosonics Medical, Inc.

Charles E. Kolb
President
Aerodyne Research, Inc.

Henry Linsert, Jr.
CEO
Columbia Biosciences Corporation

W. Clark McFadden
Partner
Dewey & LeBoeuf, LLP

Duncan T. Moore (NAE)
Kingslake Professor of Optical
Engineering
University of Rochester

Kent Murphy
President and CEO
Luna Innovations

Linda F. Powers
Managing Director
Toucan Capital Corporation

Tyrone Taylor
President
Capitol Advisors
on Technology, LLC

Charles Trimble (NAE)
CEO, *retired*
Trimble Navigation

Patrick Windham
President
Windham Consulting

PROJECT STAFF

Charles W. Wessner
Study Director

Sujai J. Shivakumar
Senior Program Officer

McAlister T. Clabaugh
Program Associate

Adam H. Gertz
Program Associate

David E. Dierksheide
Program Officer

Jeffrey C. McCullough
Program Associate

RESEARCH TEAM

Zoltan Acs
University of Baltimore

Alan Anderson
Consultant

Philip A. Auerswald
George Mason University

Robert-Allen Baker
Vital Strategies, LLC

Robert Berger
Robert Berger Consulting, LLC

Grant Black
University of Indiana South Bend

Peter Cahill
BRTRC, Inc.

Dirk Czarnitzki
University of Leuven

Julie Ann Elston
Oregon State University

Irwin Feller
American Association for the Advancement of Science

David H. Finifter
The College of William and Mary

Michael Fogarty
University of Portland

Robin Gaster
North Atlantic Research

Albert N. Link
University of North Carolina

Rosalie Ruegg
TIA Consulting

Donald Siegel
University of California at Riverside

Paula E. Stephan
Georgia State University

Andrew Toole
Rutgers University

Nicholas Vonortas
George Washington University

POLICY AND GLOBAL AFFAIRS

Ad hoc Oversight Board for
Capitalizing on Science, Technology, and Innovation:
An Assessment of the Small Business Innovation Research Program

Robert M. White (NAE), Chair
University Professor Emeritus
Electrical and Computer Engineering
Carnegie Mellon University

Anita K. Jones (NAE)
Lawrence R. Quarles Professor of Engineering and Applied Science
School of Engineering and Applied Science
University of Virginia

Mark B. Myers
Senior Vice President, *retired*
Xerox Corporation

Reports in the Series

Capitalizing on Science, Technology, and Innovation: An Assessment of the Small Business Innovation Research Program

An Assessment of the Small Business Innovation Research Program—Project Methodology
Washington, DC: The National Academies Press, 2004

SBIR: Program Diversity and Assessment Challenges
Washington, DC: The National Academies Press, 2004

SBIR and the Phase III Challenge of Commercialization
Washington, DC: The National Academies Press, 2007

An Assessment of the SBIR Program at the National Science Foundation
Washington, DC: The National Academies Press, 2008*

An Assessment of the SBIR Program at the Department of Energy
Washington, DC: The National Academies Press, 2008*

An Assessment of the SBIR Program at the National Institutes of Health
Washington, DC: The National Academies Press, 2009*

An Assessment of the SBIR Program at the Department of Defense
Washington, DC: The National Academies Press, 2009*

An Assessment of the SBIR Program at the National Aeronautics and Space Administration
Washington, DC: The National Academies Press, 2009*

---

*The NSF report and this overview report were released in prepublication in July 2007. The NIH and DoD reports were released in prepublication in November 2007, and the DoE and NASA reports were released in prepublication in June 2008 and December 2008, respectively.

Contents

APPENDIXES

Preface

Today's knowledge economy is driven in large part by the nation's capacity to innovate. One of the defining features of the U.S. economy is a high level of entrepreneurial activity. Entrepreneurs in the United States see opportunities and are willing and able to take on risk to bring new welfare-enhancing, wealth-generating technologies to the market. Yet, while innovation in areas such as genomics, bioinformatics, and nanotechnology present new opportunities, converting these ideas into innovations for the market involves substantial challenges.[1] The American capacity for innovation can be strengthened by addressing the challenges faced by entrepreneurs. Public-private partnerships are one means to help entrepreneurs bring new ideas to market.[2]

The Small Business Innovation Research (SBIR) program is one of the largest examples of U.S. public-private partnerships. Founded in 1982, the SBIR program was designed to encourage small business to develop new processes and products and to provide quality research in support of the many missions of the U.S. government. By including qualified small businesses in the nation's R&D (research and development) effort, SBIR grants are intended to stimulate innovative new technologies to help agencies meet the specific research and development needs of the nation in many areas, including health, the environment, and national defense.

[1]See Lewis M. Branscomb, Kenneth P. Morse, Michael J. Roberts, Darin Boville, *Managing Technical Risk: Understanding Private Sector Decision Making on Early Stage Technology Based Projects*, Gaithersburg, MD: National Institute of Standards and Technology, 2000.

[2]For a summary analysis of best practice among U.S. public-private partnerships, see National Research Council, *Government-Industry Partnerships for the Development of New Technologies: Summary Report*, Charles W. Wessner, ed., Washington, DC: The National Academies Press, 2002.

As the SBIR program approached its twentieth year of operation, the U.S. Congress asked the National Research Council to conduct a "comprehensive study of how the SBIR program has stimulated technological innovation and used small businesses to meet federal research and development needs" and to make recommendations on still further improvements to the program.[3] To guide this study, the National Research Council (NRC) drew together an expert committee that included eminent economists, small businessmen and women, and venture capitalists. The membership of this committee is listed in the front matter of this volume. Given the extent of 'green-field research' required for this study, the Committee in turn drew on a distinguished team of researchers to, among other tasks, administer surveys and case studies, and develop statistical information about the program. The membership of this research team is also listed in the front matter of this volume.

This report is one of a series published by the National Academies in response to the Congressional request. The series includes reports on the Small Business Innovation Research Program at the Department of Defense, the Department of Energy, the National Aeronautics and Space Administration, the National Institutes of Health, and the National Science Foundation—the five agencies responsible for 96 percent of the program's operations. It includes, as well, an Overview Report that provides assessment of the program's operations across the federal government. Other reports in the series include a summary of the 2002 conference that launched the study, and a summary of the 2005 conference on *SBIR and the Phase III Challenge of Commercialization* that focused on the Department of Defense and NASA.

PROJECT ANTECEDENTS

The current assessment of the SBIR program follows directly from an earlier analysis of public-private partnerships by the National Research Council's Board on Science, Technology, and Economic Policy (STEP). Under the direction of Gordon Moore, Chairman Emeritus of Intel, the NRC Committee on Government-Industry Partnerships prepared eleven volumes reviewing—the drivers of cooperation among industry, universities, and government; operational assessments of current programs; emerging needs at the intersection of biotechnology and information technology; the current experience of foreign government partnerships and opportunities for international cooperation; and the changing roles of government laboratories, universities, and other research organizations in the national innovation system.[4]

This analysis of public-private partnerships included two published studies

[3]See the SBIR Reauthorization Act of 2000 (H.R. 5667—Section 108).

[4]For a summary of the topics covered and main lessons learned from this extensive study, see National Research Council, *Government-Industry Partnerships for the Development of New Technologies: Summary Report*, op. cit.

of the SBIR program. Drawing from expert knowledge at a 1998 workshop held at the National Academy of Sciences, the first report, *The Small Business Innovation Research Program: Challenges and Opportunities*, examined the origins of the program and identified some operational challenges critical to the program's future effectiveness.[5] The report also highlighted the relative paucity of research on this program.

Following this initial report, the Department of Defense (DoD) asked the NRC to assess the Department's Fast Track Initiative in comparison with the operation of its regular SBIR program. The resulting report, *The Small Business Innovation Research Program: An Assessment of the Department of Defense Fast Track Initiative*, was the first comprehensive, external assessment of the Department of Defense's program. The study, which involved substantial case study and survey research, found that the SBIR program was achieving its legislated goals. It also found that DoD's Fast Track Initiative was achieving its objective of greater commercialization and recommended that the program be continued and expanded where appropriate.[6] The report also recommended that the SBIR program overall would benefit from further research and analysis, a perspective adopted by the U.S. Congress.

SBIR REAUTHORIZATION AND CONGRESSIONAL REQUEST FOR REVIEW

As a part of the 2000 reauthorization of the SBIR program, Congress called for a review of the SBIR programs of the agencies that account collectively for 96 percent of program funding. As noted, the five agencies meeting this criterion, by size of program, are the Department of Defense, the National Institutes of Health, the National Aeronautics and Space Administration, the Department of Energy, and the National Science Foundation.

Congress directed the NRC, via H.R. 5667, to evaluate the quality of SBIR research and evaluate the SBIR program's value to the agency mission. It called for an assessment of the extent to which SBIR projects achieve some measure of commercialization, as well as an evaluation of the program's overall economic and non-economic benefits. It also called for additional analysis as required to support specific recommendations on areas such as measuring outcomes for

[5]See National Research Council, *The Small Business Innovation Research Program: Challenges and Opportunities*, Charles W. Wessner, ed., Washington, DC: National Academy Press, 1999.

[6]See National Research Council, *The Small Business Innovation Research Program: An Assessment of the Department of Defense Fast Track Initiative*, Charles W. Wessner, ed., Washington, DC: National Academy Press, 2000. Given that virtually no published analytical literature existed on SBIR, this Fast Track study pioneered research in this area, developing extensive case studies and newly developed surveys.

agency strategy and performance, increasing federal procurement of technologies produced by small business, and overall improvements to the SBIR program.[7]

ACKNOWLEDGMENTS

On behalf of the National Academies, we express our appreciation and recognition for the insights, experiences, and perspectives made available by the participants of the conferences and meetings, as well as by survey respondents and case study interviewees who participated over the course of this study. We are also very much in debt to officials from the leading departments and agencies. Among the many who provided assistance to this complex study, we are especially in debt to Kesh Narayanan, Joseph Hennessey, and Ritchie Coryell of the National Science Foundation; Ivory Fisher and later Michael Caccuitto of the Department of Defense; Robert Berger and later Larry James of the Department of Energy; Carl Ray and Paul Mexcur of NASA; and Jo Anne Goodnight and Kathleen Shino of the National Institutes of Health.

The Committee's research team deserves major recognition for their instrumental role in the preparation of this study. In particular, Dr. Robin Gaster deserves special recognition and thanks for his energy, commitment, and many insights. Without the research team's collective efforts, amidst many other competing priorities, it would not have been possible to prepare these reports.

NATIONAL RESEARCH COUNCIL REVIEW

This report has been reviewed in draft form by individuals chosen for their diverse perspectives and technical expertise, in accordance with procedures approved by the National Academies' Report Review Committee. The purpose of this independent review is to provide candid and critical comments that will assist the institution in making its published report as sound as possible and to ensure that the report meets institutional standards for objectivity, evidence, and responsiveness to the study charge. The review comments and draft manuscript remain confidential to protect the integrity of the process.

We wish to thank the following individuals for their review of this report: Robert Archibald, The College of William and Mary; Richard Bendis, Innovation Philadelphia; David Bodde, Clemson University; Anthony DeMaria, DeMaria Electro-Optics Systems; George Eads, CRA International; John Foster, TRW Defense and Space Sector (Retired); Fred Gault, Statistics Canada; Bronwyn Hall, University of California, Berkeley; Thomas Pelsoci, Delta Research Company;

[7]Chapter 3 of the Committee's Methodology Report describes how this legislative guidance was drawn out in operational terms. See National Research Council, *An Assessment of the Small Business Innovation Research Program—Project Methodology*, Washington, DC: The National Academies Press, 2004, accessed at *<http://www7.nationalacademies.org/sbir/SBIR_Methodology_Report.pdf>*.

Charles Phelps, University of Rochester; Michael Rodemeyer, Pew Initiative on Food and Biotechnology; Michael Squillante, Radiation Measurement Device, Inc.; Roland Tibbets, Search Corporation; and Richard Wright, National Institute of Standards and Technology (Retired).

Although the reviewers listed above have provided many constructive comments and suggestions, they were not asked to endorse the conclusions or recommendations, nor did they see the final draft of the report before its release. The review of this report was overseen by Robert Frosch, Harvard University, and Robert White, Carnegie Mellon University. Appointed by the National Academies, they were responsible for making certain that an independent examination of this report was carried out in accordance with institutional procedures and that all review comments were carefully considered. Responsibility for the final content of this report rests entirely with the authoring committee and the institution.

Jacques S. Gansler Charles W. Wessner

Summary

I. INTRODUCTION

The Small Business Innovation Research (SBIR) program was created in 1982 through the Small Business Innovation Development Act. In 2005, the 11 federal agencies administering the SBIR program disbursed over $1.85 billion dollars in competitive awards to innovative small firms. As the SBIR program approached its twentieth year of operation, the U.S. Congress requested the National Research Council (NRC) of the National Academies to "conduct a comprehensive study of how the SBIR program has stimulated technological innovation and used small businesses to meet Federal research and development needs" and to make recommendations with respect to the SBIR program. Mandated as a part of SBIR's reauthorization in late 2000, the NRC study has assessed the SBIR program as administered at the five federal agencies that together make up some 96 percent of SBIR program expenditures. The agencies, in order of program size, are the Department of Defense (DoD), the National Institutes of Health (NIH), the National Aeronautics and Space Administration (NASA), the Department of Energy (DoE), and the National Science Foundation (NSF).

Based on that legislation, and after extensive consultations with both Congress and agency officials, the NRC focused its study on two overarching questions.[1] First, how well do the agency SBIR programs meet four societal objectives

[1]Three primary documents condition and define the objectives for this study: These are the Legislation—H.R. 5667, the NAS-Agencies *Memorandum of Understanding*, and the NAS contracts accepted by the five agencies. These are reflected in the Statement of Task addressed to the Committee by the Academies' leadership. Based on these three documents, the NRC Committee developed a comprehensive and agreed-upon set of practical objectives to be reviewed. These are outlined in the Committee's formal Methodology Report, particularly Chapter 3: Clarifying Study Objec-

of interest to Congress? That is: (1) to stimulate technological innovation; (2) to increase private sector commercialization of innovations; (3) to use small business to meet federal research and development needs; and (4) to foster and encourage participation by minority and disadvantaged persons in technological innovation.[2] Second, can the management of agency SBIR programs be made more effective? Are there best practices in agency SBIR programs that may be extended to other agencies' SBIR programs?

To satisfy the congressional request for an external assessment of the program, the NRC analysis of the operations of the SBIR program involved multiple sources and methodologies. A large team of expert researchers carried out extensive NRC-commissioned surveys and case studies. In addition, agency-compiled program data, program documents, and the existing literature were reviewed. These were complemented by extensive interviews and discussions with program managers, program participants, agency "users" of the program, as well as program stakeholders.

The study as a whole sought to understand operational challenges and to measure program effectiveness, including the quality of the research projects being conducted under the SBIR program, the challenges and achievements in commercialization of the research, and the program's contribution to accomplishing agency missions. To the extent possible, the evaluation included estimates of the benefits (both economic and noneconomic) achieved by the SBIR program, as well as broader policy issues associated with public-private collaborations for technology development and government support for high technology innovation.

Taken together, this study is the most comprehensive assessment of SBIR to date. Its empirical, multifaceted approach to evaluation sheds new light on the operation of the SBIR program in the challenging area of early-stage finance. As with any assessment, particularly one across five quite different agencies and departments, there are methodological challenges. These are identified and discussed at several points in the text.[3] This important caveat notwithstanding, the scope and diversity of the report's research should contribute significantly to the understanding of the SBIR program's multiple objectives, measurement issues, operational challenges, and achievements.

---

tives. See National Research Council, *An Assessment of the Small Business Innovation Research Program—Project Methodology*, Washington, DC: The National Academies Press, 2004, accessed at <*http://books.nap.edu/catalog.php?record_id=11097#toc*>.

[2]These congressional objectives are found in the Small Business Innovation Development Act (PL 97-219). In reauthorizing the program in 1992 (PL 102-564), Congress expanded the purposes to "emphasize the program's goal of increasing private sector commercialization developed through Federal research and development and to improve the Federal government's dissemination of information concerning small business innovation, particularly with regard to woman-owned business concerns and by socially and economically disadvantaged small business concerns."

[3]See, for example, Box 4-1 in Chapter 4, which discusses the multiple sources of bias in innovation surveys.

II. SUMMARY OF KEY FINDINGS

The core finding of the study is that the SBIR program is sound in concept and effective in practice. It can also be improved. Currently, the program is delivering results that meet most of the congressional objectives. Specifically, the program is:

- **Stimulating Technological Innovation**[4]
  - **Generating Multiple Knowledge Outputs.** SBIR projects yield a variety of knowledge outputs. These contributions to knowledge are embodied in data, scientific and engineering publications, patents and licenses of patents, presentations, analytical models, algorithms, new research equipment, reference samples, prototypes products and processes, spin-off companies, and new "human capital" (enhanced know-how, expertise, and sharing of knowledge).[5]
  - **Linking Universities to the Public and Private Markets.** The SBIR program supports the transfer of research into the marketplace, as well as the general expansion of scientific and technical knowledge, through a wide variety of mechanisms. With regard to SBIR's role in linking universities to the market, three metrics from the NRC Phase II Survey and NRC Firm Survey reveal the relatively high level of university connections. Over a third of respondents to the NRC surveys reported university involvement in their SBIR project. Among those reporting university involvement,
    - More than two-thirds of companies reported that at least one founder was previously an academic;
    - About one-third of founders were most recently employed as academics before founding the company; and
    - Some 27 percent of projects had university faculty as contractors on the project, 17 percent used universities themselves as subcontractors, and 15 percent employed graduate students.

    These data underscore the significant level of involvement by universities in the program and highlight the program's contribution to the transition of university research to the marketplace.[6]

---

[4]See related Finding F in Chapter 2.

[5]Surveys commissioned by the NRC for this study indicate that 29.3 percent of projects received at least one patent related to their SBIR research (see Table 4-11). The NRC survey also determined that 45.4 percent of respondents reported publishing at least one related peer-reviewed scientific paper (see Table 4-12). These data fit well with case studies and interviews of firms, which suggest that SBIR companies are proud of the quality of their research. Further, as highlighted in some firm interviews, metrics like patents and publications do not tell the full story; there may be benefits from the development and diffusion of knowledge that are not reflected in any qualitative metric. See, for example, the case study of Language Weaver in Appendix C of this report.

[6]See Section 4.6.3.1 on university faculty and company formation. Also see Table 4-13 on university involvement in SBIR projects.

- **Increasing Private Sector Commercialization of Innovations**[7]
    - **A Commercial Enabler for Small Firms.** Small technology companies use SBIR awards to advance projects, develop firm-specific capabilities, and ultimately create and market new commercial products and services.
        - **Company Creation.** Just over 20 percent of companies responding to the NRC Firm Survey indicated that they were founded entirely or partly because of a prospective SBIR award.[8]
        - **The Decision to Initiate Research.** Companies responding to the NRC Phase II Survey reported that over two-thirds of SBIR projects would not have taken place without SBIR funding.[9]
        - **Providing Alternative Development Paths.** Companies often use SBIR to fund alternate development strategies, exploring technological options in parallel with other activities.
        - **Reaching the Market.** Although the data vary by agency, respondents to the NRC Phase II Survey indicate that just under half of the projects do reach the marketplace.[10] Given the very early stage of SBIR investments, and the high degree of technical risk involved (reflected in risk assessment scores developed during some agency selection procedures), the fact that a high proportion of projects reach the market place in some form is significant, even impressive.
    - **A Small Percentage of Projects Account for Most Successes.** As with investments made in early stage companies by angel investors or venture capitalists, SBIR awards result in sales numbers that are highly skewed.[11] A small percentage of projects will likely achieve large growth and significant sales revenues—i.e., become commercial "home runs." Meanwhile many small successes together will continue to meet agency research needs and

---

[7]See related Finding B in Chapter 2.

[8]See Chapter 4, Table 4-7.

[9]See Figure 4-19.

[10]There are many different sources of commercialization data, all of which suggest commercialization activity. Survey data from the NRC (see Appendix A) indicates that 47 percent of respondents for awards made 1992-2002 report some sales (see Section 4.2.3); resurveying respondents to their own survey from approximately the same time period, NIH found that 62.8 percent reported sales. (See National Institutes of Health, *Report on the Second of the 2005 Measures Updates: NIH SBIR Performance Outcomes Data System (PODS)*, September 26, 2005 p. 10.) DoD also maintains an outcomes database where data are submitted by firms each time they apply for another award. This database also shows that the percentage of projects grows as time elapses, and again generates comparable figures. For example, approximately 53 percent of Navy projects from 1999 have reached the market. (DoD Company Commercialization Report Database, provided to NRC by DoD in August 2005.) It therefore appears that about half of all funded projects reach the market; follow-on research similar to that conducted by NIH is likely to generate numbers that are both higher and better reflective of the long run commercialization from the program. See National Institutes of Health, *Report on the Second of the 2005 Measures Updates: NIH SBIR Performance Outcomes Data System (PODS)*, op. cit., p. 10.

[11]See Figure 4-2.

comprise a potentially important contribution to the nation's innovative capability.

- **SBIR Is an Input, Not a Panacea.** SBIR can be a key input to encourage small business commercialization, but most major commercialization successes require substantial post-SBIR research and funding from a variety of sources.[12] SBIR awards will have been in many cases a major, even decisive input—but only one of the many contributions needed for success.

- **Using Small Businesses to Meet Federal Research and Development Needs**[13]
  - **Flexible Adaptation to Agency Mission.** The effective alignment of the program with widely varying mission objectives, needs, and modes of operation is a central challenge for an award program that involves a large number of departments and agencies. The SBIR program has been adapted effectively by the management of the individual departments, services, and agencies, albeit with significant differences in mode of operation reflecting their distinct missions and operational cultures. This flexibility in program management and modes of operation is one of the great strengths of the program.
  - **Meeting Agency Procurement Needs.** The SBIR program helps to meet the procurement needs of diverse federal agencies. At the Department of Defense, the Navy has achieved significant success in improving the insertion of SBIR-funded technologies into the acquisition process. The commitment of upper management to the effective operation of the program appears to be a key element of this success. Teaming among the SBIR program managers, agency procurement managers, the SBIR awardees, and, increasingly, the prime contractors is important in the transition of technologies from projects to products to integration in systems. At DoD, the growing importance of the SBIR program within the defense acquisition system is reflected in the growing interest of prime contractors, who are seeking opportunities to be in support of SBIR projects—a key step toward acquisition.[14]

- **Providing Widely Distributed Support for Innovation Activity**[15]
  - **Large Number of Firms.** During the fourteen years between 1992 and

---

[12]See Table 4-2 for a list of sources of additional investments.

[13]See related Finding C in Chapter 2.

[14]The growing interest of Defense prime contractors is recorded in National Research Council, *SBIR and the Phase III Challenge of Commercialization*, Charles W. Wessner, ed., Washington, DC: The National Academies Press, 2007.

[15]See related Finding D in Chapter 2.

2005, inclusive, more than 14,800 firms received at least one Phase II award, according to the SBA Tech-Net database.[16]

- **Many New Participants.** Each year, over one third of the firms awarded SBIR funds participate in the program for the first time.[17] This steady infusion of new firms is a major strength of the program and suggests that SBIR is encouraging innovation across a broad spectrum of firms, creating additional competition among suppliers for the procurement agencies, and providing agencies new mission-oriented research and solutions.

- **Fostering Participation by Minority and Disadvantaged Persons in Technological Innovation**[18]
  - **A Mixed Record.** Woman- and minority-owned firms face substantial challenges in obtaining early-stage finance.[19] Recognizing these challenges, the legislation calls for fostering and encouraging the participation of women and minorities in SBIR. Given this objective, some current trends are troubling. Agencies do not have a uniformly positive record in collecting data and monitoring funding flows for research by woman- and minority-owned firms.
    - While support for woman-owned businesses is increasing, support for minority-owned firms has not increased. For example, at DoD, which accounts for over half the SBIR program funding, the share of Phase II awards going to woman-owned businesses increased from 8 percent at the time of the 1992 reauthorization (1992-1994) to 9.5 percent (in a program increasing in overall size) for the most recent years covered by the NRC Phase II Survey (1999-2001).[20]
    - The share of Phase I awards to minority-owned firms at DoD has declined quite substantially since the mid 1990s and fell below 10 percent for the first time in 2004 and 2005.[21] Data on Phase II awards suggest that the decline in Phase I award shares for minority-owned firms is reflected in Phase II.

---

[16]See U.S. Small Business Administration, Tech-Net Database, accessed at <*http://tech-net.sba.gov/index.cfm*>.

[17]See Section 4.5.3 for a discussion of the incidence of new entrants. See also Figures 4-21, 4-22, 4-23, and 4-24 for data on new entrants at NSF, NIH and DoD.

[18]See related Finding E in Chapter 2.

[19]Academics represent an important future pool of applicants, firm founders, principal investigators, and consultants. Recent research shows that owing to the low number of women in senior research positions in many leading academic science departments, few women have the chance to lead a spinout. "Under-representation of female academic staff in science research is the dominant (but not the only) factor to explain low entrepreneurial rates amongst female scientists." See Peter Rosa and Alison Dawson, "Gender and the commercialization of university science: academic founders of spinout companies," *Entrepreneurship & Regional Development*, 18(4):341-366, July 2006.

[20]See Figure 2-4 in National Research Council, *An Assessment of the SBIR Program at the Department of Defense*, Charles W. Wessner, ed., Washington, DC: The National Academies Press, 2009.

[21]This statistic is drawn from the DoD Awards database.

- Documenting and monitoring the participation by women and minorities is complex, given, *inter alia*, the variations in the demographics of the applicant pool. In some cases, agency efforts in this area have been inadequate. Agencies are encouraged to collect, analyze, and regularly report on this important element of the program.[22]

- **Support for Woman and Minority-Principal Investigators.** Beyond support for woman- and minority-owned firms, support for woman and minority principal investigators can be an important step, supporting the potential entrepreneurs of the future.

III. SUMMARY OF KEY RECOMMENDATIONS

The recommendations below are intended to improve the operation of an already effective program. They seek to maintain, and reinforce, positive features of program management, such as the flexibility in approach by different agencies. They also identify pressing needs, e.g., for better data collection and analysis and opportunities for improvements in program operations in areas such as award size, cycle time, and outreach to minorities.[23]

- **Retain Program Flexibility**[24]
  - **SBA and SBIR.** The SBA has oversight responsibility for the eleven SBIR programs underway across the federal government. The agency is to be commended for its flexibility in exercising its oversight responsibilities, which allows the agencies to adapt the program to fit their needs and methods of operation. This flexibility has proven fundamental to the program's success, and should be preserved.
  - **Encourage Program Innovation.**[25] As noted above, it is essential to retain and encourage the flexibility that enables SBIR program management to innovate towards an even more effective multiphase program.
  - **Preserve the Basic Program Structure.**[26] The three phase approach of the SBIR program should be maintained. Proposals to "bypass" Phase I are neither necessary nor appropriate. Permitting companies to apply directly to Phase II would have the potential to change the program, significantly reducing funds for Phase I. Such a shift does not seem necessary given the current flexibility in award size.

---

[22]This is a correction of the text in the prepublication version released on July 27, 2007.

[23]These recommendations are provided in substantially greater detail in Chapter 2.

[24]See related Recommendation A in Chapter 2.

[25]See related Recommendation H in Chapter 2.

[26]See related Recommendation G in Chapter 2.

- **Conduct Regular Evaluations.**[27] Regular, rigorous program evaluation is essential for quality program management and accountability, and improved program output. Accordingly, the SBIR program managers should give greater attention and resources to the systematic evaluation of the program supported by reliable data and should seek to make the program as responsive as possible to the needs of small company applicants.
  - **Annual Reports.** Top agency management should make a direct annual report to Congress on the state of the SBIR program at their agency. This report should include a statistical appendix, which would provide data on awards, processes, outcomes, and survey information.
  - **Internal Evaluation.** Agencies should be encouraged—and funded—to develop improved data collection technologies and evaluation procedures. Where possible, agencies should be encouraged to develop interoperable standards for data collection and dissemination.[28]
  - **External Evaluation.** Agencies should be directed to commission an external evaluation of their SBIR programs on a regular basis.

- **Improve Program Processes**
  - **Topic Definition.**[29] SBIR program managers should ensure that solicitation topics are broadly defined and that topics are defined from the "bottom-up" based on agency mission needs.
  - **Project Selection.**[30] Agencies should also ensure that project selection procedures are transparent and flexible and are attuned to the needs of small businesses.
  - **Cycle Time.**[31] The processing periods for awards vary substantially by agency, and appear to have significant effects on recipient companies.[32] Agencies should closely monitor and report on cycle times for each element of the SBIR program: topic development and publication, solicitation, application review, contracting, Phase II application and selection, and Phase III contracting. Agencies should also specifically report on initiatives to shorten decision cycles.
  - **Pilot Programs.**[33] The agencies should be strongly encouraged to develop pilot programs to address possible improvements to the SBIR program. Agencies should equally ensure that such program modifications are de-

---

[27]See related Recommendation B in Chapter 2.

[28]The agencies should consider providing data from a range of sources—including agency databases, agency surveys, the patent office, and bibliographic databases, along with data from award recipients themselves.

[29]See related Recommendation D in Chapter 2.

[30]Ibid.

[31]Ibid.

[32]Drawn from case study interviews and gap data from the NRC Phase II Survey.

[33]See related Recommendation H in Chapter 2.

signed, monitored and evaluated, so that positive and negative results can be effectively determined.

- **Readjust Award Sizes**[34]
  - **One-time Adjustment.** The real value of SBIR awards, last increased in 1995, has eroded due to inflation. Given that Congress did not indicate that the real value of awards should be allowed to decline, this erosion in the value of awards needs to be addressed. In order to restore the program to the approximate initial levels, adjusted for inflation, the Congress should consider making a one-time adjustment that would give the agencies latitude to increase the standard size of Phase I awards to $150,000, and to increase the standard size of Phase II awards to approximately $1,000,000.[35]
  - **Maintain Flexibility.** It should be stressed that recommendations are intended as *guidance* for standard award size. The SBA should continue to provide the maximum flexibility possible with regard to award size and the agencies should continue to exercise their judgment in applying the program standard. The diversity of agency and project needs does not permit a one-size-fits-all approach.

- **Continue to Focus on Increased Private-sector Commercialization**
  - **Encourage Continued Experimentation.**[36] The agencies should be strongly encouraged to develop programs that seek to improve the commercialization outcomes of the SBIR program. Some agencies have sought, with the approval of SBA, to experiment with SBIR funding beyond Phase II in order to improve the commercialization potential of SBIR funded technologies. NIH has substantially increased its use of supplementary awards—additional funding provided largely at the discretion of the program manager to help meet unexpected research costs. The NSF Phase IIB initiative and the NIH Competing Continuation Awards are positive examples that might well be adapted elsewhere.
  - **Mission Agencies Create a Phase III Pull.**[37] By working with prime contractors, create mechanisms (such as the Navy's Phase IIB SBIR or Phase III funding with program dollars) to help bridge the "Valley of Death" between Phase II and application funding.

---

[34]See related Recommendation I in Chapter 2.

[35]Recognizing that these values are not identical, the erosion by inflation of the amounts available for Phase I appear more constraining than for Phase II. One may argue that, for parity, the Phase II should be raised to approximately $1,125,000. The trade-offs involve attracting adequate numbers of quality proposals, and providing sufficient resources to enable firms to actually carry out the necessary work, while also minimizing the impact on the number of awards. While recognizing that there is necessarily an arbitrary element in fixing these amounts, the Committee is confident that a $1,000,000 Phase II award will maintain the program's attraction to innovative small businesses.

[36]See related Recommendation H in Chapter 2.

[37]Ibid.

- **Multiple Winners Should Be Judged on Output, Not Numbers of Awards.**[38] In the case of multiple award winners who qualify in terms of the selection criteria, the acceptance/rejection decision should be based on their performance on past grants in terms of commercialization success and addressing agency needs, rather than on the number of grants received. Firms able to provide quality solutions to solicitations should not be excluded, *a priori*, from the program except on clear and transparent criteria (e.g., quality of research and/or commercialization performance).

- **Improve Participation and Success by Women and Minorities**[39]
  - **Improve Data Collection and Analysis.** Agencies should arrange for an independent analysis of a sample of past proposals from woman- and minority-owned firms and from other firms (to serve as a control group). This will help identify specific factors accounting for the lower success rates of woman- and minority-owned firms, as compared with other firms, in having their Phase I proposals granted.
  - **Extend Outreach to Younger Women and Minority Students.** Agencies should be encouraged to solicit women and underrepresented minorities working at small firms to apply as principal investigators and senior co-investigators for SBIR awards, and should track their success rates.
  - **Encourage Participation.** Agencies should develop targeted outreach to improve the participation rates of woman- and minority-owned firms, and strategies to improve their success rates based on causal factors determined by analysis of past proposals and feedback from the affected groups.[40]

- **Increase Management Funding for SBIR**[41]
  - **Enhance Program Utilization.** To enhance program utilization, management, and evaluation, consideration should be given to the provision of additional program funds for management and evaluation. Additional funds might be allocated internally within the existing agency budgets, drawn from the existing set-aside for the program, or by modestly increasing the set-aside for the program, currently at 2.5 percent of external research budgets.
  - **Optimize the Return on Investment.** The key point is that a modest addition to funds for program management and evaluation are necessary to

[38]See related Recommendation J in Chapter 2.

[39]See related Recommendation E in Chapter 2.

[40]This recommendation should not be interpreted as lowering the bar for the acceptance of proposals from woman- and minority-owned companies, but rather as assisting them to become able to meet published criteria for grants at rates similar to other companies on the basis of merit, and to ensure that there are no negative evaluation factors in the review process that are biased against these groups.

[41]See related Recommendation C in Chapter 2.

optimize the nation's return on the substantial annual investment in the SBIR program.

- **Additional Resources Could be Used Effectively.** In summary, the program is proving effective in meeting congressional objectives. It is increasing innovation, encouraging participation by small companies in federal R&D, providing support for small firms owned by minorities and women, and resolving research questions for mission agencies in a cost-effective manner. Should the Congress wish to provide additional funds for the program in support of these objectives, those funds could be employed effectively by the nation's SBIR program.

1

Introduction

Small businesses are a major driver of high-technology innovation and economic growth in the United States, generating significant employment, new markets, and high-growth industries.[1] In this era of globalization, optimizing the ability of innovative small businesses to develop and commercialize new products is essential for U.S. competitiveness and national security. Developing better incentives to spur innovative ideas, technologies, and products—and ultimately to bring them to market—is thus a central policy challenge.

Created in 1982 through the Small Business Innovation Development Act, the Small Business Innovation Research (SBIR) is the nation's largest innovation program. SBIR offers competition-based awards to stimulate technological innovation among small private-sector businesses while providing government agencies new, cost-effective, technical and scientific solutions to meet their diverse mission needs. The program's goals are four-fold: "(1) to stimulate technological innovation; (2) to use small business to meet federal research and development needs; (3) to foster and encourage participation by minority and disadvantaged

[1]A growing body of evidence, starting in the late 1970s and accelerating in the 1980s indicated that small businesses were assuming an increasingly important role in both innovation and job creation. See, for example, J. O. Flender and R. S. Morse, *The Role of New Technical Enterprise in the U.S. Economy*, Cambridge, MA: MIT Development Foundation, 1975, and David L. Birch, "Who Creates Jobs?" *The Public Interest*, 65:3-14, 1981. Evidence about the role of small businesses in the U.S. economy gained new credibility with the empirical analysis by Zoltan Acs and David Audretsch of the U.S. Small Business Innovation Data Base, which confirmed the increased importance of small firms in generating technological innovations and their growing contribution to the U.S. economy. See Zoltan Acs and David Audretsch, "Innovation in Large and Small Firms: An Empirical Analysis," *The American Economic Review*, 78(4):678-690, Sept. 1988. See also Zoltan Acs and David Audretsch, *Innovation and Small Firms*, Cambridge, MA: The MIT Press, 1991.

persons in technological innovation; and (4) to increase private sector commercialization derived from federal research and development."[2]

A distinguishing feature of SBIR is that it embraces the multiple goals listed above, while maintaining an administrative flexibility that allows very different federal agencies to use the program to address their unique mission needs.

SBIR legislation currently requires federal agencies with extramural R&D budgets in excess of $100 million to set aside 2.5 percent of their extramural R&D funds for SBIR. In 2005, the 11 federal agencies administering the SBIR program disbursed over $1.85 billion dollars in innovation awards. Five agencies administer over 96 percent of the program's funds. They are the Department of Defense (DoD), the Department of Health and Human Services (particularly the National Institutes of Health [NIH]), the Department of Energy (DoE), the National Aeronautics and Space Administration (NASA), and the National Science Foundation (NSF). (See Figure 1-1.)

As the Small Business Innovation Research (SBIR) program approached its twentieth year of operation, the U.S. Congress asked the National Research Council (NRC) to carry out a "comprehensive study of how the SBIR program has stimulated technological innovation and used small businesses to meet federal research and development needs" and make recommendations on improvements to the program.[3] The NRC's charge is, thus, to assess the operation of the SBIR program and recommend how it can be improved.[4]

This report provides an overview of the NRC assessment. It is a complement to a set of five separate reports that describe and assess the SBIR programs at the Departments of Defense and Energy, the National Institutes of Health, the National Aeronautics and Space Administration, and the National Science Foundation.

The purpose of this introduction is to set out the broader context of the SBIR program. Section 1.1 provides an overview of the program's history and legislative reauthorizations. It also contrasts the common structure of the SBIR program with the diverse ways it is administered across the federal government. Section 1.2 describes the important role played by SBIR in the nation's innovation system, explaining that SBIR has no public or private sector substitute. Section 1.3 then lists the advantages and limitations of the SBIR concept, including benefits and challenges faced by entrepreneurs and agency officials. Section 1.4 summarizes some of the main challenges of the NRC study and opportunities for

---

[2]The Small Business Innovation Development Act (PL 97-219).

[3]See U.S. Congress, Public Law 106-554, Appendix I—H.R. 5667, Section 108.

[4]At the conference launching the NRC assessment, James Turner, Counsel to the House Science Committee, noted that the study is not expected to question whether the program should exist. "We're 20 years into the SBIR now," he said. "It is a proven entity; it's going to be with us." He suggested that the appropriate goals for the study would be to look ahead and craft a series of sound suggestions on how to improve the program and to give good advice to Congress on what legislative changes, if any, are necessary. See National Research Council, *SBIR: Program Diversity and Assessment Challenges*, Charles W. Wessner, ed., Washington, DC: The National Academies Press, 2004.

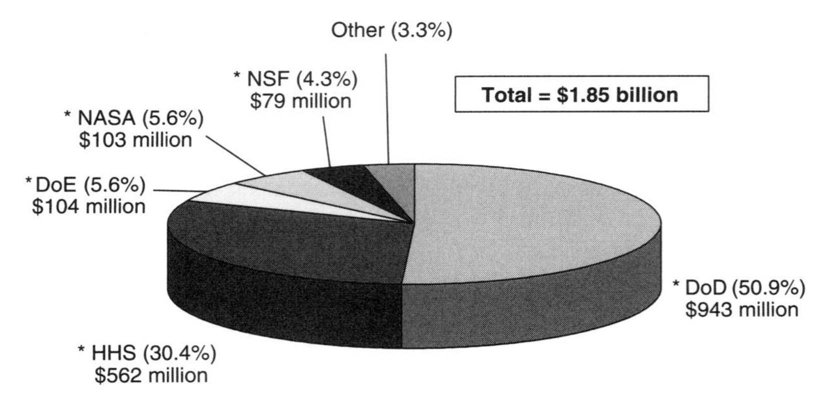

FIGURE 1-1 Dimensions of the SBIR program in 2005.
NOTE: Figures do not include STTR funds. Asterisks indicate those departments and agencies reviewed by the National Research Council.
SOURCE: U.S. Small Business Administration, Accessed at from <*http://tech-net.sba.gov*>, July 25, 2006.

improving SBIR. Finally, Section 1.5 looks at the changing perception of SBIR in the United States and the growing recognition of the SBIR concept around the world as an example of global best practice in innovation policy. The increasing adoption of SBIR-type programs in competitive Asian and European economies underlines the need, here at home, to improve upon and take advantage of this unique American innovation partnership program.

1.1 PROGRAM HISTORY AND STRUCTURE

In the 1980s, the country's slow pace in commercializing new technologies—compared with the global manufacturing and marketing success of Japanese firms in autos, steel, and semiconductors—led to serious concern in the United States about the nation's ability to compete economically. U.S. industrial competitiveness in the 1980s was frequently cast in terms of American industry's failure "to translate its research prowess into commercial advantage."[5] The pessimism of some was reinforced by evidence of slowing growth at corporate re-

[5]David C. Mowery, "America's Industrial Resurgence (?): An Overview," in David C. Mowery, ed., *U.S. Industry in 2000: Studies in Competitive Performance*, Washington, DC: National Academy Press, 1999, p. 1. Mowery examines eleven economic sectors, contrasting the improved performance of many industries in the late 1990s with the apparent decline that was subject to much scrutiny in the 1980s. Among the studies highlighting poor economic performance in the 1980s are Dertouzos, et al., *Made in America: The MIT Commission on Industrial Productivity*, Cambridge, MA: The MIT Press, 1989, and Otto Eckstein, *DRI Report on U.S. Manufacturing Industries*, New York: McGraw Hill, 1984.

search laboratories that had been leaders of American innovation in the postwar period and the apparent success of the cooperative model exemplified by some Japanese *kieretsu*.[6]

Yet, a growing body of evidence, starting in the late 1970s and accelerating in the 1980s, began to indicate that small businesses were assuming an increasingly important role in both innovation and job creation. David Birch, a pioneer in entrepreneurship and small business research, and others suggested that national policies should promote and build on the competitive strength offered by small businesses.[7]

Meanwhile, federal commissions from as early as the 1960s had recommended changing the direction of R&D funds toward innovative small businesses.[8] These recommendations were unsurprisingly opposed by traditional recipients of government R&D funding.[9] Although small businesses were beginning to be recognized by the late 1970s as a potentially fruitful source of innovation, some in government remained wary of funding small firms focused on high-risk technologies with commercial promise.

The concept of early-stage financial support for high-risk technologies with commercial promise was first advanced by Roland Tibbetts at the National Science Foundation. As early as 1976, Mr. Tibbetts advocated that the NSF should increase the share of its funds going to innovative, technology-based small businesses. When NSF adopted this initiative, small firms were enthused and proceeded to lobby other agencies to follow NSF's lead. When there was no immediate response to these efforts, small businesses took their case to Congress and to higher levels of the Executive branch.[10]

In response, the White House convened a conference on Small Business

---

[6]Richard Rosenbloom and William Spencer, *Engines of Innovation: U.S. Industrial Research at the End of an Era*. Boston, MA: Harvard Business Press, 1996.

[7]David L. Birch, "Who Creates Jobs?" *The Public Interest*, op. cit. Birch's work greatly influenced perceptions of the role of small firms. Over the last 20 years, it has been carefully scrutinized, leading to the discovery of some methodological flaws, namely making dynamic inferences from static comparisons, confusing gross and net job creation, and admitting biases from chosen regression techniques. See S. J. Davis, J. Haltiwanger, and S. Schuh, "Small Business and Job Creation: Dissecting the Myth and Reassessing the Facts, Working Paper No. 4492, Cambridge, MA: National Bureau of Economic Research, 1993. These methodological fallacies, however, "ha[ve]not had a major influence on the empirically based conclusion that small firms are over-represented in job creation," according to Per Davidsson. See Per Davidsson, "Methodological Concerns in the Estimation of Job Creation in Different Firm Size Classes," Working Paper, Jönköping International Business School, 1996.

[8]For an overview of the origins and history of the SBIR program, see George Brown and James Turner, "The Federal Role in Small Business Research," *Issues in Science and Technology*, Summer 1999, pp. 51-58.

[9]See Roland Tibbetts, "The Role of Small Firms in Developing and Commercializing New Scientific Instrumentation: Lessons from the U.S. Small Business Innovation Research Program," in *Equipping Science for the 21st Century*, John Irvine, Ben Martin, Dorothy Griffiths, and Roel Gathier, eds., Cheltenham UK: Edward Elgar Press, 1997. For a summary of some of the critiques of SBIR, see Section 1-3 of this Introduction.

[10]Ibid.

in January 1980 that recommended a program for small business innovation research. This recommendation was grounded in:

- Evidence that a declining share of federal R&D was going to small businesses;
- Broader difficulties among innovative small businesses in raising capital in a period of historically high interest rates; and
- Research suggesting that small businesses were fertile sources of job creation.

A widespread political appeal in seeing R&D dollars "spread a little more widely than they were being spread before" complemented these policy rationales. Congress responded, under the Reagan administration, with the passage of the Small Business Innovation Research Development Act of 1982, which established the SBIR program.[11]

1.1.1 The SBIR Development Act of 1982

The new SBIR program initially required agencies with R&D budgets in excess of $100 million to set aside 0.2 percent of their funds for SBIR. This amount totaled $45 million in 1983, the program's first year of operation. Over the next six years, the set-aside grew to 1.25 percent.[12]

The legislation authorizing SBIR had two broad goals[13]:

- "to more effectively meet R&D needs brought on by the utilization of small innovative firms (which have been consistently shown to be the most prolific sources of new technologies); and
- "to attract private capital to commercialize the results of federal research."

1.1.2 The SBIR Reauthorizations of 1992 and 2000

The SBIR program approached reauthorization in 1992 amidst continued worries about the U.S. economy's capacity to commercialize inventions. Finding that "U.S. technological performance is challenged less in the creation of new technologies than in their commercialization and adoption," the National Academy of Sciences at the time recommended an increase in SBIR funding

[11]Additional information regarding SBIR's legislative history can be accessed from the Library of Congress. See <*http://thomas.loc.gov/cgi-bin/bdquery/z?d097:SN00881:@@@L*>.

[12]The set-aside is currently 2.5 percent of an agency's extramural R&D budget.

[13]U.S. Congress, Senate, Committee on Small Business (1981), Senate Report 97-194, Small Business Research Act of 1981, September 25, 1981.

as a means to improve the economy's ability to adopt and commercialize new technologies.[14]

Following this report, the Small Business Research and Development Enhancement Act (P.L. 102-564), which reauthorized the SBIR program until September 30, 2000, doubled the set-aside rate to 2.5 percent. This increase in the percentage of R&D funds allocated to the program was accompanied by a stronger emphasis on encouraging the commercialization of SBIR-funded technologies.[15] Legislative language explicitly highlighted commercial potential as a criterion for awarding SBIR grants.[16]

The Small Business Reauthorization Act of 2000 (P.L. 106-554) extended SBIR until September 30, 2008. It also called for an assessment by the National Research Council of the broader impacts of the program, including those on employment, health, national security, and national competitiveness.[17]

1.1.3 Previous Research on SBIR

The current NRC assessment represents a significant opportunity to gain a better understanding of one of the largest of the nation's early-stage finance programs. Despite its size and 24-year history, the SBIR program has not previously been comprehensively examined. While there have been some previous studies, most notably by the General Accounting Office and the Small Business Administration, these have focused on specific aspects or components of the program.[18]

---

[14]See National Research Council, *The Government Role in Civilian Technology: Building a New Alliance*, Washington, DC: National Academy Press, 1992, p. 29.

[15]See Robert Archibald and David Finifter, "Evaluation of the Department of Defense Small Business Innovation Research Program and the Fast Track Initiative: A Balanced Approach," in National Research Council, *The Small Business Innovation Research Program: An Assessment of the Department of Defense Fast Track Initiative*, Charles W. Wessner, ed., Washington, DC: National Academy Press, 2000, pp. 211-250.

[16]In reauthorizing the program in 1992 (PL 102-564) Congress expanded the purposes to "emphasize the program's goal of increasing private sector commercialization developed through Federal research and development and to improve the federal government's dissemination of information concerning the small business innovation, particularly with regard to woman-owned business concerns and by socially and economically disadvantaged small business concerns."

[17]The current assessment is congruent with the Government Performance and Results Act (GPRA) of 1993: <*http://govinfo.library.unt.edu/npr/library/misc/s20.html*>. As characterized by the GAO, GPRA seeks to shift the focus of government decision making and accountability away from a preoccupation with the activities that are undertaken—such as grants dispensed or inspections made—to a focus on the results of those activities. See <*http://www.gao.gov/new.items/gpra/gpra.htm*>.

[18]An important step in the evaluation of SBIR has been to identify existing evaluations of SBIR. These include U.S. Government Accounting Office, *Federal Research: Small Business Innovation Research Shows Success But Can be Strengthened*, Washington, DC: U.S. General Accounting Office, 1992; and U.S. Government Accounting Office, "Evaluation of Small Business Innovation Can Be Strengthened," Washington, DC: U.S. General Accounting Office, 1999. There is also a 1999 unpublished SBA study on the commercialization of SBIR surveys Phase II awards from 1983 to 1993 among non-DoD agencies.

There have been few internal assessments of agency programs.[19] The academic literature on SBIR is also limited.[20]

Writing in the 1990s, Joshua Lerner of the Harvard Business School positively assessed the program, finding "that SBIR awardees grew significantly faster than a matched set of firms over a ten-year period." [21] Underscoring the importance of local infrastructure and cluster activity, Lerner's work also showed that the "positive effects of SBIR awards were confined to firms based in zip codes with substantial venture capital activity." These findings were consistent with both the corporate finance literature on capital constraints and the growth literature on the importance of localization effects.[22]

To help fill this assessment gap, and to learn about a large, relatively under-evaluated program, the National Academies' Committee for Government-Industry Partnerships for the Development of New Technologies was asked by the Department of Defense to convene a symposium to review the SBIR program as a whole, its operation, and current challenges. Under its chairman, Gordon Moore, Chairman Emeritus of Intel, the Committee convened government policymakers, academic researchers, and representatives of small business for the first comprehensive discussion of the SBIR program's history and rationale, review

---

[19]Agency reports include an unpublished 1997 DoD study on the commercialization of DoD SBIR technologies. NASA has also completed several reports on its SBIR program. Following the authorizing legislation for the NRC study, NIH launched a major review of the achievements of its SBIR program.

[20]Early examples of evaluations of the SBIR program include S. Myers, R. L. Stern, and M. L. Rorke, *A Study of the Small Business Innovation Research Program*, Lake Forest, IL: Mohawk Research Corporation, 1983, and Price Waterhouse, *Survey of Small High-tech Businesses Shows Federal SBIR Awards Spurring Job Growth, Commercial Sales*, Washington, DC: Small Business High Technology Institute, 1985. A 1998 assessment by Scott Wallsten of a subset of SBIR awardees that were publicly traded (most SBIR awardees are not public) determined that SBIR grants do not contribute additional funding but instead replace firm-financed R&D spending "dollar for dollar." See S. J. Wallsten, "Rethinking the Small Business Innovation Research Program," in *Investing In Innovation*, Lewis M. Branscomb and J. Keller, eds., Cambridge, MA: The MIT Press, 1998. While Wallsten's paper has the virtue of being one of the early attempts to assess the impact of SBIR, Josh Lerner questions whether employing a regression framework to assess the marginal impact of public funding on private research spending is the most appropriate tool in assessing public efforts to assist small high-technology firms. He points out that "it may well be rational for a firm not to increase its rate of spending, but rather to use the funds to prolong the time before it needs to seek additional capital." Lerner suggests that "to interpret such a short run reduction in other research spending as a negative signal is very problematic." See Joshua Lerner, "Public Venture Capital: Rationales and Evaluation" in *The Small Business Innovation Research Program: Challenges and Opportunities*, op. cit., p. 125. See also Joshua Lerner, "Angel Financing and Public Policy: An Overview," *Journal of Banking and Finance*, 22(6-8):773-784, and Joshua Lerner, "The Government as Venture Capitalist: The Long-run Impact of the SBIR Program," *The Journal of Business*, 72(3):285-297, 1999.

[21]See Joshua Lerner, "The Government as Venture Capitalist: The Long-Run Effects of the SBIR Program," op. cit.

[22]See Michael Porter, "Clusters and Competition: New Agendas for Government and Institutions," in *On Competition*, Boston, MA: Harvard Business School Press, 1998.

of existing research, and identification of areas for further research and program improvements.[23]

The Moore Committee reported that:

- SBIR enjoyed strong support in parts of the federal government, as well as in the country at large.
- At the same time, the size and significance of SBIR underscored the need for more research on how well it is working and how its operations might be optimized.
- There should be additional clarification about the primary emphasis on commercialization within SBIR, and about how commercialization is defined.
- There should also be clarification on how to evaluate SBIR as a single program that is applied by different agencies in different ways.[24]

Subsequently, the Department of Defense requested the Moore Committee to review the operation of the SBIR program at Defense with a particular focus on the role played by the Fast Track Initiative. This major review involved substantial original field research, with 55 case studies, as well as a large survey of award recipients. The response rate was relatively high, at 72 percent.[25] It found that the SBIR program at Defense was contributing to the achievement of mission goals—funding valuable innovative projects—and that a significant portion of these projects would not have been undertaken in the absence of the SBIR funding.[26] The Moore Committee's assessment also found that the Fast Track Program increases the efficiency of the Department of Defense SBIR program by encouraging the commercialization of new technologies and the entry of new firms to the program.[27]

More broadly, the Moore Committee found that SBIR facilitates the development and utilization of human capital and technological knowledge.[28] Case studies have shown that the knowledge and human capital generated by the SBIR program has economic value, and can be applied by other firms.[29] And, by acting as a "certifier" of promising new technologies, SBIR awards encourage further private sector investment in an award winning firm's technology.

Based on this and other assessments of public-private partnerships, the Moore Committee's *Summary Report* on U.S. Government-Industry Partnerships recom-

[23]See National Research Council, *The Small Business Innovation Research Program: Challenges and Opportunities*, Charles W. Wessner, ed., Washington, DC: National Academy Press, 1999.

[24]Ibid.

[25]See National Research Council, *The Small Business Innovation Research Program: An Assessment of the Department of Defense Fast Track Initiative*, op. cit., p 24.

[26]Ibid—see Chapter III: Recommendations and Findings, p. 32.

[27]Ibid, p. 33.

[28]Ibid, p. 33.

[29]Ibid, p. 33.

BOX 1-1
The Moore Committee Report on Public-Private Partnerships[a]

In a program-based analysis led by Gordon Moore, Chairman Emeritus of Intel, the National Academies Committee on Government-Industry Partnerships for the Development of New Technologies found that "public-private partnerships, involving cooperative research and development activities among industry, universities, and government laboratories can play an instrumental role in accelerating the development of new technologies from idea to market."

Partnerships Contribute to National Missions

"Experience shows that partnerships work—thereby contributing to national missions in health, energy, the environment, and national defense—while also contributing to the nation's ability to capitalize on its R&D investments. Properly constructed, operated, and evaluated partnerships can provide an effective means for accelerating the progress of technology from the laboratory to the market."

Partnerships Help Transfer New Ideas to the Market

"Bringing the benefits of new products, new processes, and new knowledge into the market is a key challenge for an innovation system. Partnerships facilitate the transfer of scientific knowledge to real products; they represent one means to improve the output of the U.S. innovation system. Partnerships help by bringing innovations to the point where private actors can introduce them to the market. Accelerated progress in obtaining the benefits of new products, new processes, and new knowledge into the market has positive consequences for economic growth and human welfare."

Characteristics of Successful Partnerships

"Successful partnerships tend to be characterized by industry initiation and leadership, public commitments that are limited and defined, clear objectives, cost sharing, and learning through sustained evaluations of measurable outcomes, as well as the application of the lessons to program operations.[b] At the same time, it is important to recognize that although partnerships are a valuable policy instrument, they are not a panacea; their demonstrated utility does not imply that all partnerships will be success-

mended that "regular and rigorous program-based evaluations and feedback is essential for effective partnerships and should be a standard feature," adding that "greater policy attention and resources to the systematic evaluation of U.S. and foreign partnerships should be encouraged."[30]

Drawing on these recommendations, the December 2000 legislation mandated the current comprehensive assessment of the nation's SBIR program. This NRC assessment of SBIR is being conducted in three phases. The first

[30]See National Research Council, *Government-Industry Partnerships for the Development of New Technologies: Summary Report*, Charles. W. Wessner, ed., Washington, DC: National Academy Press, 2002, p. 30.

ful. Indeed, the high-risk—high-payoff nature of innovation research and development assures some disappointment."

Partnerships Are a Complement to Private Finance

"Partnerships focus on earlier stages of the innovation stream than many venture investments, and often concentrate on technologies that pose greater risks and offer broader returns than the private investor normally finds attractive.[c] Moreover, the limited scale of most partnerships—compared to private institutional investments—and their sunset provisions tend to ensure early recourse to private funding or national procurement. In terms of project scale and timing in the innovation process, public-private partnerships do not displace private finance. Properly constructed research and development partnerships can actually elicit 'crowding in' phenomena, with public investments in R&D providing the needed signals to attract private investment."[d]

---

[a]National Research Council, *Government-Industry Partnerships for the Development of New Technologies: Summary Report*, Charles W. Wessner, ed., Washington, DC: The National Academies Press, 2003, p. 30.

[b]Features associated with more successful partnerships are described in the Introduction to this report.

[c]Some programs also support broadly applicable technologies that, while desirable for society as a whole, are difficult for individual firms to undertake because returns are difficult for individual firms to appropriate. A major example is the Advanced Technology Program.

[d]David, Hall, and Toole survey the econometric evidence over the past 35 years. They note that the "findings overall are ambivalent and the existing literature as a whole is subject to the criticism that the nature of the "experiment(s)" that the investigators envisage is not adequately specified." It seems that both crowding out and crowding in can occur. The essential finding is that the evidence is inconclusive and that assumptions about crowding out are unsubstantiated. The outcome appears to depend on the specifics of the circumstance, and these are not adequately captured in available data. See Paul A. David, Bronwyn H. Hall, and Andrew A. Toole, "Is Public R&D a Complement or Substitute for Private R&D? A Review of the Econometric Evidence," NBER Working Paper 7373, October 1999. Relatedly, Feldman and Kelley cite the "halo effect" created by ATP awards in helping firms signal their potential to private investors. See Maryann Feldman and Maryellen Kelley, "Leveraging Research and Development: The Impact of the Advanced Technology Program," in National Research Council, *The Advanced Technology Program*, Charles W. Wessner, ed., Washington, DC: National Academy Press, 2001.

phase developed a research methodology that was reviewed and approved by an independent National Academies panel of experts. Information available about the program was also gathered through interviews with officials at the relevant federal agencies and through two major conferences where these officials were invited to describe program operations, challenges, and accomplishments. These conferences highlighted the important differences in agency goals, practices, and evaluations. They also served to describe the evaluation challenges that arise from the diversity in program objectives and practices.[31]

---

[31]Adapted from National Research Council, *SBIR: Program Diversity and Assessment Challenges*, op. cit.

The second phase of the study implemented the research methodology. The Committee deployed multiple survey instruments and its researchers conducted case studies of a wide variety of SBIR firms. The Committee then evaluated the results, and developed the recommendations and findings found in this report for improving the effectiveness of the SBIR program.

The third phase of the study will provide an update of the survey and related case studies, as well as explore other issues that emerged in the course of this study. It will, in effect, provide a second snapshot of the program and of the agencies' progress and challenges.

1.1.4 The Structure and Diversity of SBIR

Eleven federal agencies are currently required to set aside 2.5 percent of their extramural research and development budget exclusively for SBIR awards and contracts. Each year these agencies identify various R&D topics, representing scientific and technical problems requiring innovative solutions, for pursuit by small businesses under the SBIR program. These topics are bundled together into individual agency "solicitations"—publicly announced requests for SBIR proposals from interested small businesses.

A small business can identify an appropriate topic that it wants to pursue from these solicitations and, in response, propose a project for an SBIR grant. The required format for submitting a proposal is different for each agency. Proposal selection also varies, though peer review of proposals on a competitive basis by experts in the field is typical. Each agency then selects through a competitive process the proposals that are found to best meet program selection criteria, and awards contracts or grants to the proposing small businesses.

In this way, SBIR helps the nation capitalize more fully on its investments in research and development.

1.1.4.1 A Three-Phase Program

As conceived in the 1982 Act, the SBIR grant-making process is structured in three phases:

- Phase I grants essentially fund a feasibility study in which award winners undertake a limited amount of research aimed at establishing an idea's scientific and commercial promise. The 1992 legislation standardized Phase I grants at $100,000. Approximately 15 percent of all small businesses that apply receive a Phase I award.
- Phase II grants are larger—typically about $500,000 to $850,000—and fund more extensive R&D to develop the scientific and technical merit and the feasibility of research ideas. Approximately 40 percent of Phase I award winners go on to this next step.

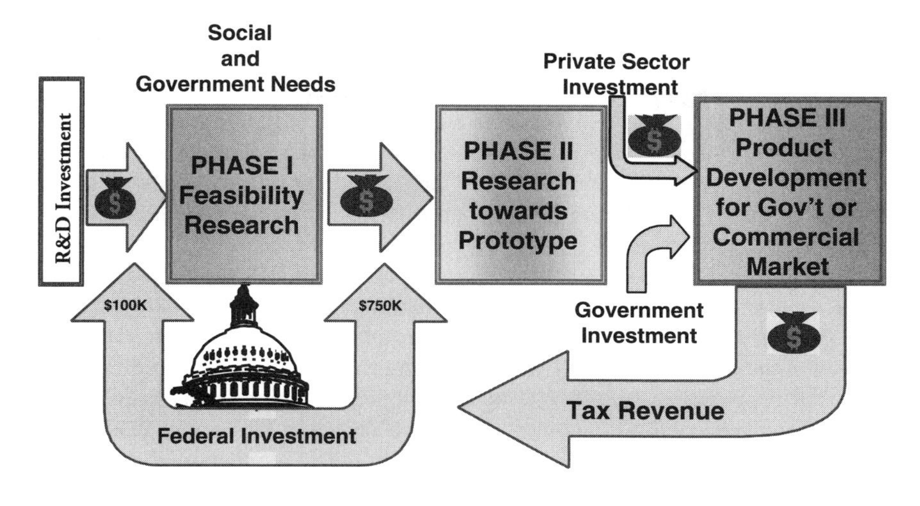

FIGURE 1-2 The structure of the SBIR program.

- Phase III is the period during which Phase II innovation moves from the laboratory into the marketplace. No SBIR funds support this phase. To commercialize their product, small businesses are expected to garner additional funds from private investors, the capital markets, or from the agency that made the initial award. The availability of additional funds and the need to complete rigorous testing and certification requirements at, for example, the Department of Defense or NASA can pose significant challenges for new technologies and products, including those developed using SBIR awards.

Figure 1-2 provides a schematic of the three phases of SBIR, showing how SBIR helps the nation better leverage the federal government's substantial investment in research and development. In describing the program's concept, it also helps to illustrate that the tax system provides a noninvasive means for the federal government to recoup, over time, its investment in small business innovation research. Successful small businesses create employment, with taxes paid on payroll and revenue which, when taxed, defrays some of the costs of the program to the nation's treasury. Of course, innovation spurred by SBIR and its commercialization create value to the nation far beyond tax revenues. Ultimately, the innovation spurred by the SBIR creates products which, in turn, create additions to consumer surplus for the United States and exports around the world.[32]

---

[32]For a discussion of the cumulative effects and the direct and indirect externalities of innovation, see Suzanne Scotchmer, *Innovation and Incentives*, Cambridge MA: The MIT Press, 2004.

1.1.4.2 Significant Program Diversity

Although the SBIR programs at all eleven agencies share the common three-phase structure, they have evolved separately to adapt to the particular mission, scale, and working cultures of the various agencies that administer them. For example, NSF's operation of its SBIR differs considerably from that of the Department of Defense, reflecting in large part differences in size of the agencies as well as the extent to which "research" is coupled with procurement of goods and services. Within the Department of Defense, in turn, the SBIR program is administered separately by ten different defense organizations, including the Navy, the Air Force, the Army, the Missile Defense Agency, and DARPA. Similarly, there are 23 institutes and centers at National Institutes of Health administering their own SBIR program. The number of independent operations has led to a diversity of administrative practices, a point discussed below. (See Table 1-1.) This diversity means that the SBIR program at each agency must be understood in its own context, making the task of assessing the overall program a challenging one.[33]

TABLE 1-1 Variation in Agency Approaches to SBIR

- Number and Timing of SBIR Solicitations.
- Broad vs. Focused Topic Areas.
- Variation in Award Size for Phase I and Phase II.
- Availability and Type of Phase I to Phase II Gap Funding.
- Availability of Post Phase II funding and Commercialization Assistance.
- External vs. Internal Proposal Review Processes.
- Type of Award—Contract or Grant.

1.1.4.3 The Role of the Small Business Administration

The Small Business Administration (SBA) coordinates the SBIR program across the federal government and is charged with directing its implementation at all 11 participating agencies. Recognizing the broad diversity of the program's operations, SBA administers the program with commendable flexibility, allowing the agencies to operate their SBIR programs in ways that best address their unique agency missions and cultures.

SBA is charged with reviewing the progress of the program across the federal government. To do this, SBA solicits program information from all participating agencies and publishes it quarterly in a Pre-Solicitation Announcement.[34] SBA also operates an online reporting system for the SBIR program that utilizes Tech-Net—a Web-based system linking small technology businesses with opportunities

[33]For a review of the diversity within the SBIR program, see National Research Council, *SBIR: Program Diversity and Assessment Challenges*, op. cit.

[34]Access at <*http://www.sba.gov/sbir/mastersch.pdf*>.

BOX 1-2
What Is an Innovation Ecosystem?

An innovation ecosystem describes the complex synergies among a variety of collective efforts involved in bringing innovation to market.[a] These efforts include those organized within, as well as collaboratively across large and small businesses, universities, and research institutes and laboratories, as well as venture capital firms and financial markets. Innovation ecosystems themselves can vary in size, composition, and in their impact on other ecosystems. By linking these different elements, SBIR strengthens the innovation ecosystem in the United States, thereby enhancing the nation's competitiveness.

The idea of an innovation ecosystem builds on the concept of a National Innovation System (NIS) popularized by Richard Nelson of Columbia University. According to Nelson, a NIS is "a set of institutions whose interactions determine the innovative performance . . . of national firms."[b] The idea of an innovation ecosystem highlights the multiple institutional variables that shape how research ideas can find their way to the marketplace. These include, most generally, rules that protect property (including intellectual property) and the regulations and incentives that structure capital, labor, and financial and consumer markets. A given innovation ecosystem is also shaped by shared social norms and value systems—especially those concerning attitudes towards business failure, social mobility, and entrepreneurship.[c]

In addition to highlighting the interdependencies among the various participants, the idea of an innovation ecosystem also draws attention to their ability to change over time, given different incentives. This dynamic element sets apart the idea of an "innovation ecosystem." In this regard, the term "innovation ecosystem" captures an analytical approach that considers how public policies can improve innovation-led growth by strengthening links within the system. Incentives found within intermediating institutions like SBIR can play a key role in this regard by aligning the self-interest of venture capitalists, entrepreneurs and other participants with desired national objectives.[d]

---

[a]Consciously drawing on this ecosystems approach, the Council of Competitiveness' National Innovation Initiative (NII) report and recommendations address the need for new forms of collaboration, governance and measurement that enable U.S. workers to succeed in the global economy. Council on Competitiveness, *Innovate America: Thriving in a World of Challenge and Change*, Washington, DC: Council on Competitiveness, 2005.

[b]See Richard R. Nelson and Nathan Rosenberg, "Technical Innovation and National Systems," in *National Innovation Systems: A Comparative Analysis*, Richard R. Nelson, ed., Oxford, UK: Oxford University Press, 1993.

[c]For a survey of attitudes towards entrepreneurship, see European Commission, "Entrepreneurship—Flash Eurobarometer Survey," January 2004. The survey shows that Europeans have a greater fear of entrepreneurial failure—including loss of property and bankruptcy—than do Americans. Accessed at *<http://europa.eu.int/comm/enterprise/enterprise_policy/survey/eurobarometer83.htm>*.

[d]National Research Council, *Government-Industry Partnerships for the Development of New Technologies: Summary Report*, Charles W. Wessner, ed., Washington, DC: The National Academies Press, 2003.

within federal technology programs.[35] Through Tech-Net, SBA collects and disseminates essential commercialization and other impact data on SBIR.

1.2 ROLE OF SBIR IN THE U.S. INNOVATION ECOSYSTEM

By providing scarce pre-venture capital funding on a competitive basis, SBIR encourages new entrepreneurship needed to bring innovative ideas from the laboratory to the market. Further, by creating new information about the feasibility and commercial potential of technologies held by small innovative firms, SBIR awards aid investors in identifying firms with promising technologies. As noted, SBIR awards appear to have a "certification" function, and often act as a stamp of approval for young firms allowing them to obtain resources from outside investors.[36]

1.2.1 The Importance of Small Business Innovation

Equity-financed small firms are a key feature of the U.S. innovation ecosystem, serving as an effective mechanism for capitalizing on new ideas and bringing them to the market.[37] In the United States, small firms are also a leading source of employment growth, generating 60 to 80 percent of *net* new jobs annually over the past decade.[38] These small businesses also employ nearly 40 percent of the United States' science and engineering workforce.[39] Research commissioned by the Small Business Administration has also found that scientists and engineers working in small businesses produce 14 times more patents than their counterparts in large patenting firms in the United States—and these patents tend to be of higher quality and are twice as likely to be cited.[40]

Small businesses renew the U.S. economy by introducing new products and new lower cost ways of doing things, sometimes with substantial economic benefits. They play a key role in introducing technologies to the market, often re-

---

[35]Access Tech-Net at <*http://tech-net.sba.gov/*>.

[36]Joshua Lerner, "Public Venture Capital," in National Research Council, *The Small Business Innovation Research Program: Challenges and Opportunities*, op. cit.

[37]Zoltan J. Acs and David B. Audretsch, *Innovation and Small Firms*, op. cit.

[38]U.S. Small Business Administration, Office of Advocacy, "Small Business by the Numbers," Washington, DC: U.S. Small Business Administration, 2006. This net gain depends on the interval examined since small businesses churn more than do large ones. For a discussion of the challenges of measuring small business job creation, see John Haltiwanger and C. J. Krizan, "Small Businesses and Job Creation in the United States: The Role of New and Young Businesses" in *Are Small Firms Important? Their Role and Impact*, Zoltan J. Acs, ed., Dordrecht: Kluwer, 1999.

[39]U.S. Small Business Administration, Office of Advocacy, "Small Business by the Numbers," op. cit.

[40]Ibid.

sponding quickly to new market opportunities.[41] By contrast, large firms are less prone to pursue technological opportunities in new and emerging areas. They tend to focus more on improving the performance of existing product lines because they often cannot risk the possibility of large losses on failed breakthrough efforts on their stock price.[42] University research has traditionally focused more on education and publications. The small business entrepreneur often demonstrates the willingness to take on the risks of a new venture, offsetting these risks against the possible rewards of a major (or even moderate) success.

Indeed, many of the nation's large, successful, and innovative firms started out as small entrepreneurial firms. Firms like Microsoft, Intel, AMD, FedEx, Qualcomm, Adobe, all of which grew rapidly in scale from small beginnings, have transformed how people everywhere work, transact, and communicate. The technologies introduced by these firms continue to create new opportunities for investment and sustain the rise in the nation's productivity level.[43]

These economic and social benefits underscore the need to encourage new equity-based high-technology firms in the hope that some may develop into larger, more successful firms that create the technological base for the nation's future competitiveness. This reality is reflected in recent economic theories on the link between increased investments in knowledge creation and entrepreneurship and economic growth. (See Box 1-3.)

1.2.2 Challenges Facing Small Innovative Firms

1.2.2.1 Overcoming Knowledge Asymmetries

Despite their value to the United States economy, small businesses entrepreneurs with new ideas for innovative products often face a variety of challenges in bringing their ideas to market. Because new ideas are by definition unproven, the knowledge that an entrepreneur has about his or her innovation and its com-

[41]For an extended discussion of the empirical evidence supporting the finding of high innovation performance of small firms, see Zoltan J. Acs and David B. Audretsch, "Innovation in Large and Small Firms, An Empirical Analysis," *The American Economic Review*, 78(4):678-690, 1988.

[42]Clayton Christensen, for example, observes that it is highly uncommon for firms that manage established lines of business well to anticipate and respond effectively to a disruptive technology coming from an external agent, much less to commercialize disruptive technologies themselves. Clayton Christensen, *The Innovator's Dilemma*, Boston, MA: Harvard Business School Press, 1997. For a series of papers investigating the various kinds of dynamic capabilities or innovation system elements that contribute towards more successful radical innovation in established companies, see the special issue of the Journal of Engineering and Technology Management, 24(1-2):1-166, March-June 2007—"Research on Corporate Radical Innovation Systems-A Dynamic Capabilities Perspective," edited by Sören Salomo, Richard Leifer, and Hans Georg Gemünden.

[43]National Research Council, *Enhancing Productivity Growth in the Information Age*, Dale W. Jorgenson and Charles W. Wessner, eds., Washington, DC: The National Academies Press, 2007.

BOX 1-3
New Growth Theory and the Knowledge-based Economy

Neoclassical theories of growth long emphasized the role of labor and capital as inputs.[a] Technology was *exogenous*—assumed to be determined by forces external to the economic system. More recent growth theories, by comparison, emphasize the role of technology and assume that technology is *endogenous*—that is, it is actually integral to the performance of the economic system.

The New Growth Theory, in particular, holds that sustaining growth requires continuing investments in new knowledge creation, calling on policy makers to pay careful attention to the multiple factors that contribute to knowledge creation, including research and development, the education system, entrepreneurship, and an openness to trade and investment.[b]

To a considerable extent, knowledge-based economies are distinguished by the changing way that firms do business and how governments respond in terms of policy.[c] Key features of a knowledge-based economy include:

- A capacity to successfully create and exploit scientific knowledge and technology based on a world-class science infrastructure and an entrepreneurial and innovative culture.
- A diffusion and building up of knowledge through effective formal, as well as informal information networks. These networks, facilitated by modern telecommunication technologies and frequently based on public-private partnerships, are designed to encourage cooperation among firms, universities, and government research centers.
- A skilled workforce based on an effective and differentiated educational system and effective job training programs.
- High rates of technological innovation often associated with high-technology industries, underscoring the "virtuous cycle" that these policies can engender.

---

[a]See Robert S. Solow, "Technical Change and the Aggregate Production Function," *Review of Economics and Statistics* 39:312-320, 1957, for a classic expression of the "old" growth theory.

[b]For additional perspective on New Growth Theory, see Richard N. Langlois, "Knowledge, consumption, and endogenous growth," *Journal of Evolutionary Economics*, 11:77-93, 2001.

[c]"Just as the private sector develops innovative institutional arrangements to support and advance research, so should federal policy. In particular, one of the defining features of the knowledge economy is the increased importance of learning and innovation. Partnerships and alliances, among the private sector, universities, and government laboratories, play a key role in facilitating innovation. As a result, federal support, for research in the knowledge economy needs to explicitly encourage research collaboration between industry, government labs and universities." Kenan Patrick Jarboe and Robert D. Atkinson, "The Case for Technology in the Knowledge Economy; R&D, Economic Growth and the Role of Government," Washington, DC: Progressive Policy Institute, June 1, 1998, at *<http://www.ppionline.org/documents/CaseforTech.pdf>*.

mercial potential may not be fully appreciated by prospective investors.[44] For example, few investors in the 1980s understood Bill Gates' vision for Microsoft or, more recently, Bill Page's and Sergey Brin's vision for Google.

1.2.2.2 Overcoming Knowledge Spillovers

Another hurdle for entrepreneurs is the leakage of new knowledge that escapes the boundaries of firms and intellectual property protection. The creator of new knowledge can seldom fully capture the economic value of that knowledge for his or her own firm.[45] The benefits of R&D thus accrue to others who did not make the relevant R&D investment. This public-goods problem can inhibit investment in promising technologies for both large and small firms. Overcoming this problem is especially important for small firms focused on a particular product or process.[46]

1.2.3 The Challenge of Market Commercialization

The challenge of incomplete and insufficient information for investors and the problem for entrepreneurs of moving quickly enough to capture a sufficient return on "leaky" investments can pose substantial obstacles for new firms seeking seed capital. Because the difficulty of attracting investors to support an imperfectly understood, as yet-to-be-developed innovation is especially daunting, the term *"Valley of Death"* has come to describe the period of transition when a developing technology is deemed promising, but too new to validate its commercial potential and thereby attract the capital necessary for its continued development.[47] (See Figure 1-3.)

---

[44]Joshua Lerner, "Public Venture Capital," in National Research Council, *The Small Business Innovation Program: Challenges and Opportunities*, op. cit. For a seminal paper on information asymmetry, see Michael Spence, *Market Signaling: Informational Transfer in Hiring and Related Processes*, Cambridge, MA: Harvard University Press, 1974.

[45]Technological knowledge that can be replicated and distributed at low marginal cost may have a gross social benefit that exceeds private benefit—and in such cases is considered by many as prone to be undersupplied relative to some social optimum. See Richard N. Langlois and Paul L. Robertson, "Stop Crying over Spilt Knowledge: A Critical Look at the Theory of Spillovers and Technical Change," paper prepared for the MERIT Conference on Innovation, Evolution, and Technology, August 25-27, 1996, Maastricht, Netherlands.

[46]Edwin Mansfield, "How Fast Does New Industrial Technology Leak Out?" *Journal of Industrial Economics*, 34(2):217-224.

[47]As the September 24, 1998, Report to Congress by the House Committee on Science notes, "At the same time, the limited resources of the federal government, and thus the need for the government to focus on its irreplaceable role in funding basic research, has led to a widening gap between federally funded basic research and industry-funded applied research and development. This gap, which has always existed but is becoming wider and deeper, has been referred to as the 'Valley of Death.' A number of mechanisms are needed to help to span this Valley and should be considered." See U.S. Congress, House, Committee on Science, *Unlocking Our Future: Toward a New National Sci-*

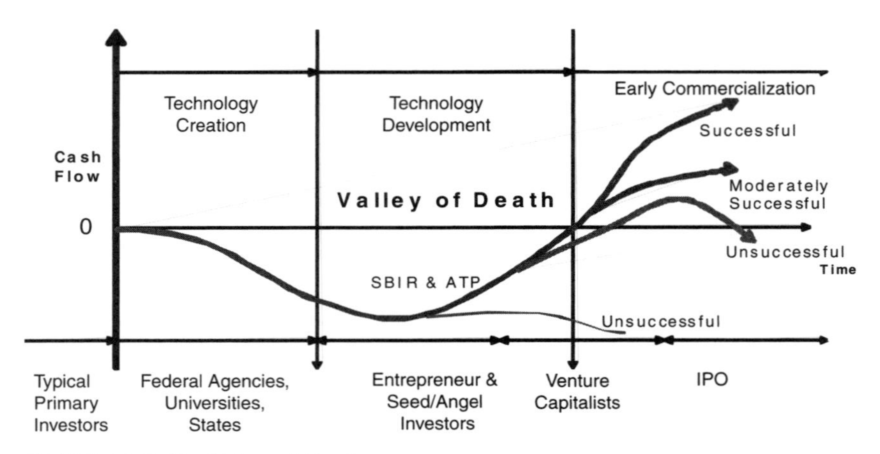

FIGURE 1-3 The Valley of Death.
SOURCE: Adapted from L.M. Murphy and P. L. Edwards, *Bridging the Valley of Death—Transitioning from Public to Private Sector Financing*, Golden, CO: National Renewable Energy Laboratory, May 2003.

This means that inherent technological value does not lead inevitably to commercialization; many good ideas perish on the way to the market. This reality belies a widespread myth that U.S. venture capital markets are so broad and deep that they are invariably able to identify promising entrepreneurial ideas and finance their transition to market. In reality, angel investors and venture capitalists often have quite limited information on new firms. These potential investors are also often focused on a given geographic area. And, as the recent dot-com boom and bust illustrates, venture capital is also prone to herding tendencies, often following market "fads" for particular sectors or technologies.[48]

1.2.3.1 The Limits of Angel Investment

Angel investors are typically affluent individuals who provide capital for a business start-up, usually in exchange for equity. Increasingly, they have been organizing themselves into angel networks or angel groups to share research and

ence Policy: A Report to Congress by the House Committee on Science, Washington, DC: Government Printing Office, 1998. Accessed at <*http://www.access.gpo.gov/congress/house/science/cp105-b/science105b.pdf*>. For an academic analysis of the Valley of Death phenomenon, see Lewis Branscomb and Philip Auerswald, "Valleys of Death and Darwinian Seas: Financing the Invention to Innovation Transition in the United States," *The Journal of Technology Transfer*, 28(3-4), August 2003.

[48]See Tom Jacobs, "Biotech Follows Dot.com Boom and Bust," *Nature*, 20(10):973, October 2002. For an analytical perspective, see Andrea Devenow and Ivo Welch, "Rational Herding in Financial Economics, *European Economic Review*, 40(April):603-615, 1996. Devenow and Welch find that when investment managers are assessed on the basis of their performance relative to their peers (rather than against some absolute benchmark), they may end up making investments similar to each other.

pool their own investment capital. The U.S. angel investment market accounted for over $25.6 billion in the United States in 2006.[49] It is a source of start-up capital for many new firms.

Yet, the angel market is dispersed and relatively unstructured, with wide variation in investor sophistication, few industry standards and tools, and limited data on performance.[50] In addition, most angel investors are highly localized, preferring to invest in new companies that are within driving distance.[51] This geographic concentration, lack of technological focus, and the privacy concerns of many angel investors, make angel capital difficult to obtain for many high-technology start-ups, particularly those seeking to provide goods and services to the federal government.

1.2.3.2 The Limits of Venture Capital

Like angels, venture capital in the United States is concentrated geographically in the country's high-technology regions. This clustering pattern creates large gaps in the availability of venture capital in rural areas and other regions that do not have high technology clusters served by concentrations of venture investors.

Venture capitalists are different from angel investors, however, in that they typically manage the pooled money of others in a professionally managed fund. Given their obligations to their investors, venture capital firms tend not to invest upstream in the higher-risk, early-stages of technology commercialization, and they have been increasingly moving further downstream in recent years. (See Figure 1-4.) In 2005, venture capitalists in the United States invested $21.7 billion over the course of 2,939 deals. However, 82 percent of venture capital in the United States was directed to firms in the later stages of development, with the remaining 18 percent directed to seed and early-stage firms.[52]

Typically, venture capitalists are also interested in larger investments that are easier to manage than is appropriate for many small innovative technology firms.[53] Large venture capital funds are deterred by the costs of meeting due dili-

---

[49]See University of New Hampshire Center for Venture Research, *2006 Angel Market Analysis*, March 19, 2007.

[50]James Geshwiler, John May, and Marianne Hudson, "State of Angel Groups," Kansas City, MO: Kauffman Foundation, April 27, 2006.

[51]See Jeffrey Sohl, John Freear, and William Wetzel, Jr., "Angles on Angels: Financing Technology-based Ventures: A Historical Perspective, *Venture Capital*, 4(4):275-287, 2002. The authors note that angel investors tend to invest close to home, "typically within a day's drive."

[52]Ibid.

[53]As Joshua Lerner notes, "Because each firm in his portfolio must be closely scrutinized, the typical venture capitalist is typically responsible for no more than a dozen investments. Venture organizations are consequently unwilling to invest in very young firms that require only small capital infusions." See Joshua Lerner, "'Public Venture Capital': Rationales and Evaluation" in National Research Council, *The Small Business Innovation Research Program: Challenges and Opportuni-*

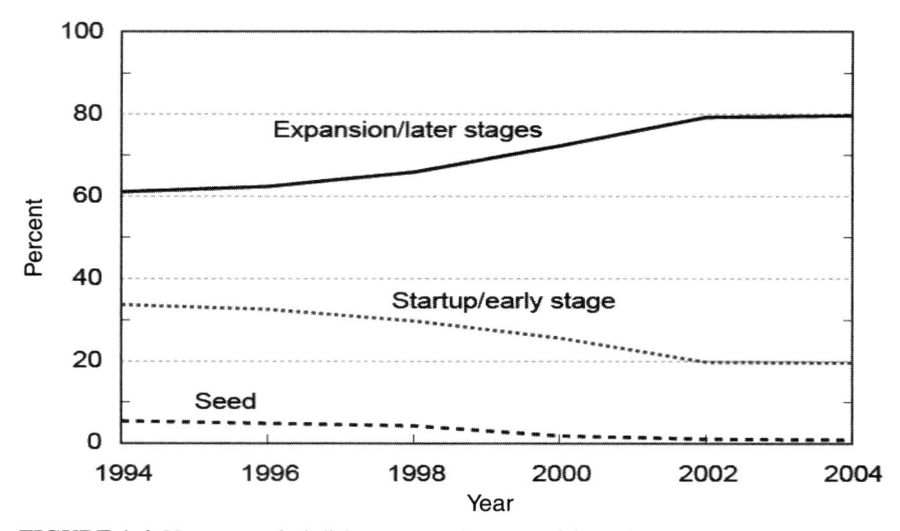

FIGURE 1-4 Venture capital disbursements by state of financing.
SOURCE: National Science Board, *Science and Engineering Indicators 2006*, Arlington, VA: National Science Foundation, 2006.

gence requirements to permit involvement in managing many small investments with remote and uncertain payoffs. The size of the average venture capital deal was $8.3 million in 2006, whereas the average SBIR Phase III project requires $400,000 to $1 million in funds. Venture capital deal size has also been rising over the past decade.[54] This trend is accelerating.

Together, these realities of the angel and venture markets underscore the challenge faced by new firms seeking private capital to develop and market promising innovations within private capital markets.

1.2.3.3 The Challenge of Federal Procurement

Commercializing SBIR-funded technologies though federal procurement is no less challenging for innovative small companies. Finding private sources of funding to further develop successful SBIR Phase II projects for government procurement—those innovations that have demonstrated technical and commercial feasibility—is often difficult because the government's demand for products is unlikely to be large enough to provide a sufficient return for venture investors.

---

ties, op. cit.

[54]PriceWaterhouseCoopers/Venture Economics/National Venture Capital Association Money Tree Survey, 2005.

Venture capitalists also tend to avoid funding firms focused on government contracts, due to the higher costs, regulatory burdens, and limited markets associated with government contracting.[55]

Institutional biases in federal procurement also hinder government funding needed to transition promising SBIR technologies. Procurement rules and practices often impose high costs and administrative overheads that favor established suppliers. In addition, many acquisition officers have traditionally viewed the SBIR program as a "tax" on their R&D budgets, representing a "loss" of resources and control, rather than an opportunity to develop rapid and lower cost solutions to complex procurement challenges.[56] This perception, in turn, can lead to limited managerial attention, less optimal mission alignment, and fewer resources being devoted to the program.

Even when they see the value of a technology, providing "extra" funding to exploit it in a timely manner can be a challenge for government managers that requires time, commitment and, ultimately, the interest of those with budgetary authority for the programs or systems. Attracting such interest and support is not automatic and may often depend on personal relations and advocacy skills rather than on the intrinsic quality of the SBIR project.

1.2.4 The Federal Role in Addressing Early-stage Financing Gap

Although business angels and venture capital firms, along with industry, state governments, and universities provide funding for early stage technology development, the federal role is significant. Research by Harvard University's Lewis Branscomb and Philip Auerswald estimates that the federal government provides between 20-25 percent of all funds for early-stage technology development.[57] (See Figure 1-5.)

This contribution is noteworthy because government awards address segments of the innovation cycle that private investors often do not fund because they find it too risky or too small.

---

[55]See comments by Mark Redding of Impact Technologies in National Research Council, *SBIR and the Phase III Challenge of Commercialization*, Charles W. Wessner, ed., Washington, DC: The National Academies Press, 2007.

[56]Illustrating this point at the NRC's 2005 conference on the SBIR Phase III, Michael McGrath, the Deputy Assistant Secretary of the Navy for Research, Development, Testing and Evaluation, noted that SBIR funds at Navy overwhelmingly came from its advanced development, testing and evaluation functions (often referred to in the DoD idiom as 6.4-6.7 functions) but were spent on basic applied research and technology development (or 6.1-6.3 functions.) This has led to perceptions among managers involved in advanced development, testing and evaluation that SBIR is simply a tax on their programs. See National Research Council, *SBIR and the Phase III Challenge of Commercialization*, op. cit, p. 17.

[57]Lewis Branscomb and Philip Auerswald, *Between Invention and Innovation: An Analysis of Funding for Early-Stage Technology Development*, Gaithersburg, MD: National Institute of Standards and Technology, 2002, p. 23.

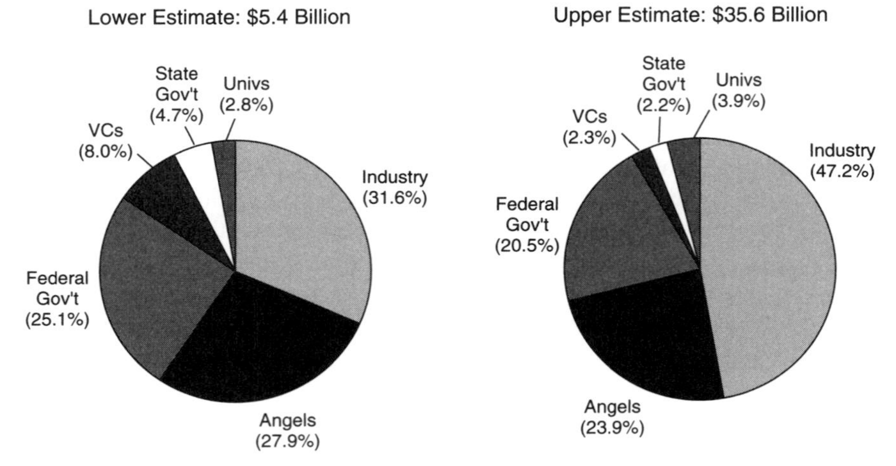

FIGURE 1-5 Estimated distribution of funding sources for early-stage technology development.
SOURCE: Lewis M. Branscomb and Philip E. Auerswald, *Between Invention and Innovation: An Analysis of Funding for Early-Stage Technology Development*, Gaithersburg, MD: National Institute of Standards and Technology, 2002, p. 23.

"It does seem that early-stage help by the government in developing platform technologies and financing scientific discoveries is directed exactly at the areas where institutional venture capitalists cannot and will not go."[a]

David Morgenthaler,
Founding Partner of Morgenthaler Ventures

[a]David Morgenthaler, "Assessing Technical Risk," in L. M. Branscomb, Kenneth P. Morse, and Michael J. Roberts, eds., *Managing Technical Risk: Understanding Private Sector Decision Making on Early Stage Technology-Based Project*, 2000, pp. 107-108.

SBIR is the main source of federal funding for early-stage technology development in the United States. Based on Banscomb and Auerswald's lower estimate of the distribution of funding sources, SBIR provides over 20 percent of funding for early-stage development from all sources and over 85 percent of federal financial support for direct early-stage development. Moreover, SBIR has no public or private substitutes. (See Box 1-4.) Funding opportunities under SBIR can thus provide early stage finance for technologies that are not readily supported by venture capitalists, angel investors, or other sources of early-stage funding in the United States.

BOX 1-4
The Federal Role in Comparative Perspective

"Small technology firms with 500 or less employees now employ 54.8 percent of all scientists and engineers in US industrial R&D. However, these nearly 6 million scientists and engineers are able to obtain only 4.3 percent of extramural government R&D dollars. In contrast, large and medium firms with more than 500 employees combined employ only 45.2 percent but receive 50.3 percent of government R&D funds. Universities receive 35.3 percent, non-profit research institutions 10.0 percent, and states and foreign countries 1.0 percent. Of the 4.3 percent that goes to small firms 2.5 percent is from SBIR and the related Small Business Technology Transfer (STTR) program. Together they receive less than 10 percent of the funding that large firms receive."[a]

Roland Tibbetts, "SBIR Renewal and U.S. Economic Security"[b]

[a]National Science Board, *Science and Engineering Indicators 2006*, Arlington, VA: National Science Foundation, 2006. Figures are for 2005.

[b]*<http://www.nsba.biz/docs/tibbetts_sbir_reauthorization.pdf>*.

1.3 SBIR: STRENGTHS AND LIMITATIONS

SBIR leverages small business innovation to address government and societal needs in such areas as health, security, environment, and energy. The strength of the SBIR concept lies in aligning the interests of each of the participants in the program with the goals of the program. SBIR proposals are industry-initiated based on broad solicitations posted by federal agencies. This bottom-up design promotes a positive interest by small businesses in the outcome of their research. Similarly, the federal agencies can each use the program to advance their own missions; ownership rests with the many agencies, not a single "tech agency."

As described below, both entrepreneurs and agency managers have found SBIR to be a useful tool to help them further their goals in fostering and commercializing new technologies. Although the SBIR concept is a robust one, fundamentally its potential lies in how it is implemented.

1.3.1 SBIR: A Tool for Entrepreneurs to Innovate

A variety of SBIR grant features make the program attractive from the entrepreneur's perspective. Among these is that there is no dilution of ownership and that no repayment is required. SBIR has no recoupment provisions, other than the tax system itself.

BOX 1-5
Advantages and Limitations of the SBIR Concept

Advantages of the SBIR Concept

SBIR plays a catalytic role at an early stage in the technology development cycle. The awards have the virtue that they are not repayable and they do not dilute ownership or control of a firm's management. The awards enable the firm to explore new technological options and in Phase II can often demonstrate a technology's potential. The intellectual property rights remain with the firm, creating an opportunity for downstream contracts. Perhaps most importantly, the awards provide a signaling effect affirming both the quality and a potential market for the technology. These advantages are significant and are not matched by private finance or other public-sector mechanisms.

Few private sector substitutes exist to the SBIR program. The SBIR contributions are also quite distinct from both bank lending and private equity financing of small technology firms. Commercial lending places a financial burden on small businesses with a long product-development cycle. Private equity funding does not require the small business to keep up with interest payments, but it does require the small business to give up a share of ownership and an associated measure of control. Firms that seek equity funding early in their development may be compelled to accept lower valuations that result in a greater loss of control for the same amount of funding than would be the case at a later stage of development.[a]

SBIR does not compete with the private financial markets. On the contrary, often it facilitates the functioning of financial markets by signaling quality—reducing information asymmetries that complicate contracting. Such quality signaling partially reduces impediments (discussed above) typically faced by technology entrepreneurs in seeking financing from both private and public sources. In additional to non-SBIR federal funds, the dominant sources of follow-on funding include self-finance, other domestic private companies, foreign companies, and nonventure capital private equity (angel investors).

Few public-sector substitutes exist to the SBIR program. SBIR continues to offer small firms an opportunity to perform federal R&D in a manner that is minimally burdensome to the firms from a contracting standpoint. While firms in different industries have access to an array of targeted programs to develop technologies for particular government goals, no other government-wide program offers technology-development awards to small business in amounts, or with objectives, comparable to the SBIR program.[b]

Current Limitations of the SBIR Concept

Overhead Costs. SBIR provides many small awards to small businesses to explore technological options to meet federal mission needs. Inherent in this approach is the disadvantage of high overhead costs for the administering agency, as compared with much larger, "bundled" contracts with a single large provider. Agency managers often require more time and energy to solicit, evaluate, and monitor multiple small awards than for larger contracts. The lack of funding in the program to help defray its management costs works against its acceptance within the host agencies and limits the ability of managers to collect data, monitor awardee firm progress over time, and conduct the internal and external assessments needed to effectively manage the program.

Uncertainty. New ideas and new firms involve higher levels of uncertainty than with more established suppliers. Many awards, especially those associated with higher-risk, less proven technologies, will not be successful. Maintaining an environment in which potential benefits and risks are carefully weighed, but where risk taking is encouraged, is a genuine challenge for managers.

Cycle Time. The time required for agencies to solicit, assess, select, and make awards can be a challenge for small firms, especially those newly established. Making the awards in a timely fashion that meets the needs of the small companies, in terms of technology development and potential markets of their product or service, remains a challenge for the program.

Award Size. The amount of the SBIR awards has not kept pace with inflation. In real terms, the resources provided by a Phase I award have declined (by as much as 50 percent depending on the sector). This decline impacts the scope and complexity of work that can be proposed. It may also impact the number and quality of proposals made to the program, although data to support this is not available. What is clear is that the program no longer provides the resources that the Congress initially intended.

Uptake—The Phase III Challenge. Despite significant commercialization, transitioning products and processes developed under the SBIR program into the procurement process or into the private markets remains a challenge for the program. These challenges include appropriate timing for inclusion in procurement activities, sufficient demonstration of project potential and firm capability, and sometimes the inability to communicate a project's potential to acquisition officials. Developing appropriate management and operational incentives for the procurement agencies and encouraging follow-on funding from nonprocurement agencies will require management attention and continued innovation, as reflected in the NSF Phase IIB awards and in the NIH's Phase II Competing Renewal Awards.

---

[a]Paul A. Gompers and Josh Lerner, *The Venture Capital Cycle*, The MIT Press, 1999. Lewis M. Branscomb and Philip E. Auerswald, *Between Invention and Innovation: An Analysis of Funding for Early Stage Technology Development*, NIST GCR 02–841, Gaithersburg, MD: National Institute for Standards and Technology, November 2002; Bronwyn H. Hall, *The Financing of Research and Development*, NBER Working Paper 8773, National Bureau of Economic Research, 2002. Branscomb and Auerswald report that overall, of $266 billion that was spent on national R&D by various sources in the U.S. in 1998, substantially less than 14 percent flowed into early stage technology development activities. The exact figure is elusive, because public financial reporting is not required for these investments.

[b]A program that is comparable along a number of dimensions—although with important differences in objectives, award size, and its emphasis on partnering—is the Advanced Technology Program (ATP), administered by the National Institute of Standards and Technology at the Department of Commerce. While the ATP program similarly stresses the development of novel technologies, that program does not focus exclusively on small firms. Furthermore, award amounts in the ATP program are typically at a factor of ten larger than the standard SBIR phase I award; and awards are not directly related to agency mission.

1.3.1.1 Advantages for Entrepreneurs

Importantly, grant recipients retain rights to intellectual property developed using the SBIR award, with no royalties owed to the government, though the government retains royalty-free use for a period.

As noted previously, being selected to receive SBIR grants also confers a "certification effect" on the small business—a signal to private investors of the technical and commercial promise of the technology held by the small business.[58]

In these ways, SBIR enhances the opportunities for entrepreneurs to turn an innovative idea into a marketable product. Below, we look at some additional reasons why entrepreneurs find SBIR useful.

Some small innovative businesses see SBIR as a strategic asset in their development, but often in different ways. Depending on the firm's size, relationship to capital, and business development strategy, firms can have quite different objectives in applying to the program. Some seek to demonstrate the potential of promising research. Others seek to fulfill agency research requirements on a cost-effective basis. Still others seek a certification of quality (and the investments that can come from such recognition) as they push science-based products towards commercialization.[59]

As they strive to move across the Valley of Death, many small firms see SBIR as one element within a more diversified strategy that includes seeking funding from state programs, angel investors, and other sources of early-stage funding, as well as technology validation through collaboration with universities and other companies.

1.3.1.2 SBIR as a Path to Federal Procurement

Many entrepreneurs are at a disadvantage in traversing the federal acquisition process. New firms are often unfamiliar with government regulations and procurement procedures, especially for defense products, and find themselves at a disadvantage vis-à-vis incumbents. To access the federal procurement process, small companies must learn to deal with a complex and sometimes arcane contracting system characterized by many rules and procedures. A major advantage of SBIR awards is that they enable a successful SBIR firm to obtain a "single source" contract for the subsequent development of the technology and product derived from the SBIR award. SBIR thus assists small firms who lack the re-

[58]This certification effect was initially identified by Josh Lerner, "Public Venture Capital," in National Research Council, *The Small Business Innovation Program: Challenges and Opportunities*, op. cit.

[59]See Reid Cramer, "Patterns of Firm Participation in the Small Business Innovation Research Program in Southwestern and Mountain States," in National Research Council, *The Small Business Innovation Research Program: An Assessment of the Department of Defense Fast Track Initiative*, op. cit.

sources to invest in "contracting overhead," by creating an alternative path for small business to enter the government procurement system.

1.3.2 SBIR's Advantages for Government

SBIR can provide agency officials with technical solutions to help solve operational problems. Faced with an operational puzzle, a program officer can post a solicitation that describes the problem in order to prompt a variety of innovative solutions. These solutions draw on the scientific and engineering expertise found across the nation's universities and innovative small businesses.

Agency challenges successfully addressed through SBIR solicitations range from rapidly deployable high-performance drones for the Department of Defense (see Box 1-6) to needle-free injectors sought by NIH to facilitate mass immunizations. Drawing on SBIR, the government can leverage private sector ingenuity to address public needs. In the process, it helps to convert ideas into potential products, creating new sources of innovation.

1.3.2.1 A Low-cost Technical Probe

A significant virtue of SBIR is that it can act in some cases as a "low-cost technological probe," enabling the government to explore more cheaply ideas that

BOX 1-6
Meeting Agency Challenges—The Case of the Navy Unmanned Air Vehicle

Originally incorporated as a self-financed start-up firm in 1989, Advanced Ceramics Research (ACR) now manufactures products for a diverse set of industries based on its innovative technologies. Over time, it has participated in the SBIR program of several federal agencies, including DoD, NASA, Department of Energy, and the National Science Foundation. ACR is now actively engaged in the development and marketing of the Silver Fox, a small unmanned aerial vehicle (UAV) that was developed with SBIR assistance.

While in Washington, DC, to discuss projects with Office of Naval Research (ONR) program managers in 2000, ACR representatives also had a chance meeting with the SBIR program manager for the Navy, interested in small SWARM unmanned air vehicles (UAVs). At the time, ONR expressed an interest and eventually provided funding for developing a new low cost small UAV as a means to engage in whale watching around Hawaii, with the objective of avoiding damage from the Navy's underwater sonic activities. Once developed however, the UAV's value as a more general-purpose battlefield surveillance technology became apparent, and the Navy provided additional funding to refine the UAV.

BOX 1-7
Case Study Example: Providing Answers to Practical Problems

Aptima, Inc. designed an instructional system for the Navy to improve boat handling safety by teaching strategies that mitigate shock during challenging wave conditions. In addition, the instructional system was to raise skill levels while compressing learning time by creating an innovative learning environment. Phase I of the project developed a training module, and in Phase II, instructional material, including computer animation, videos, images, and interviews were developed. The concept and the supporting materials were adopted as part of the introductory courses for Special Operations helmsmen with the goal of reducing injuries and increasing mission effectiveness.

may hold promise. In some cases, government needs can be met by the "answer" provided through the successful conclusion of the Phase I or Phase II award, with no further research required or a product (e.g., an algorithm or software diagnostic) developed. Here, a small business successfully completes the requirements and objectives of a Phase II contract, meeting the needs of the customer, without gaining additional commercialization revenues. In other cases, awards can provide valuable negative proofs, by identifying dead ends before substantial federal investments are made.

1.3.2.2 Quick Reaction Capability

Because it is flexible and can be organized on an ad hoc basis, SBIR can be an effective means to focus the expertise and innovative technologies dispersed around the country to address new national needs rapidly. In its 2002 report, *Making the Nation Safer*, the National Academies identified SBIR as an existing model of government-industry collaboration that could contribute to current needs, such as the war on terror.[60] SBIR, with its established selection procedures and mechanisms for evaluating and granting awards, offers major benefits in comparison to funding completely new programs; it can "hit the ground running."

Speaking at a National Academies conference on how public-private partnerships can help the government respond to terrorism, Carole Heilman, of the National Institute of Allergy and Infectious Diseases, described how within a month of the September 11, 2001 terrorist attacks, NIAID put out an SBIR solicitation concerning a particular technical problem encountered in preparing the nation's biodefense. This solicitation drew about 300 responses within a month, she noted,

[60] See National Research Council, *Making the Nation Safer: The Role of Science and Technology in Countering Terrorism*, Washington, DC: The National Academies Press, 2002.

adding that "it was a phenomenal expression of interest and capability and good application, with extremely thoughtful approaches."[61]

1.3.2.3 Diversifying the Government's Supplier Base

SBIR awards can help government agencies diversify their supplier base. By providing a bridge between small companies and the federal agencies, SBIR can serve as a catalyst for the development of new ideas and new technologies to meet federal missions in health, transport, the environment, and defense.

This potential role of SBIR is particularly relevant to the Department of Defense as it faces new challenges in an era of constrained budgets, stretched manpower, and new threats. Military capabilities in this new era increasingly depend on the invention of new technologies for new systems and platforms. SBIR can play a valuable role in providing innovative technologies that address evolving defense needs.

1.3.3 SBIR's Role in Knowledge Creation: Publications and Patents

SBIR projects yield a variety of knowledge outputs. These contributions to knowledge are embodied in data, scientific and engineering publications, patents and licenses of patents, presentations, analytical models, algorithms, new research equipment, reference samples, prototypes products and processes, spin-off companies, and new "human capital" through enhanced know-how and expertise. One indication of this knowledge creation can be seen in a NRC survey of NIH projects.[62] Thirty-four percent of firms that have won SBIR awards from NIH reported having generated at least one patent, and just over half of NIH respondents published at least one peer-reviewed article.[63]

Projects funded by SBIR grants often involve high technical risk, implying novel and difficult research rather than incremental change. At NSF, for example, of the 54 percent of Phase I projects that did not get a follow-on Phase II, 32 percent did not apply for a Phase II, and of these, half did not do so for technical reasons.[64] This suggests a robust program that encourages technical risk, and one that recognizes that not all projects will succeed.

---

[61]See Carole Heilman, "Partnering for Vaccines: The NIAID Perspective" in Charles W. Wessner, *Partnering Against Terrorism: Summary of a Workshop*, Washington, DC: The National Academies Press, 2005, pp. 67-75.

[62]National Research Council, *An Assessment of the SBIR Program at the National Institutes of Health*, Charles W. Wessner, ed., Washington, DC: The National Academies Press, 2009, Appendix B.

[63]Because of data limitations, it was not feasible to apply bibliometric and patent analysis techniques to assess the relative importance of these patents and publications.

[64]See National Research Council, *An Assessment of the SBIR Program at the National Science Foundation*, Charles W. Wessner, ed., Washington, DC: The National Academies Press, 2008.

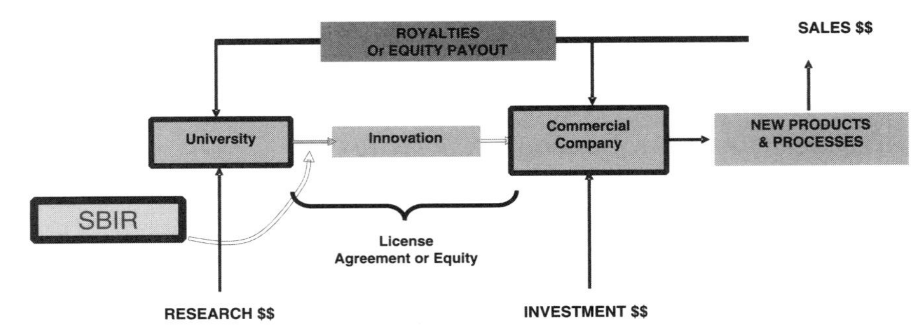

FIGURE 1-6 SBIR and the commercialization of university technology.
SOURCE: Adapted from C. Gabriel, Carnegie Mellon University.

There is also anecdotal evidence concerning "indirect path" effects of SBIR. Investigators and research staff often gain new knowledge and experience from projects funded with SBIR awards that may become relevant in a different context later on—perhaps for another project or another company. These effects are not directly measurable, but interviews and case studies affirm their existence and importance.[65]

1.3.4 SBIR and the University Connection

SBIR is increasingly recognized as providing a bridge between universities and the marketplace. In the NRC Firm Survey, conducted as a part of this study, over half of respondents reported some university involvement in SBIR projects.[66] Of those companies, more than 80 percent reported that at least one founder was previously an academic.

SBIR encourages university researchers to found companies based on their research. Importantly, the availability of the awards and the fact that a professor can apply for an SBIR award without founding a company, encourages applications from academics who might not otherwise undertake the commercialization of their own discoveries. In this regard, previous research by the NRC has shown that SBIR awards directly cause the creation of new firms, with positive benefits in employment and growth for the local economy.[67] Of course, not all universities

[65]See, for example, the NRC-commissioned case study of National Recovery Technologies, Inc. Early SBIR support for NRT technology that uses optoelectronics to sort metals and alloys at high speeds is now being evaluated for its use in airport security by the Transportation Security Administration. See National Research Council, *An Assessment of the SBIR Program at the National Science Foundation*, op. cit.

[66]National Research Council, NRC Firm Survey. See Appendix A.

[67]National Research Council, *The Small Business Innovation Research Program: An Assessment of the Department of Defense Fast Track Initiative*, op. cit., p. 35.

in the United States have a strong commercialization culture, and there is great variation in the level of success among those universities that do.[68]

1.4 ASSESSING SBIR

Regular program and project analysis of SBIR awards are essential to understand the impact of the program. A focus on analysis is also a means to clarify program goals and objectives and requires the development of meaningful metrics and measurable definitions of success. More broadly, regular evaluations contribute to a better appreciation of the role of partnerships among government, university, and industry. Assessments also help inform public and policy makers of the opportunities, benefits, and challenges involved in the SBIR innovation award program.

As we have noted before, despite its large size and 25-year history, SBIR has not done especially well with regard to evaluation. As a whole, the program has been the object of relatively limited analysis. This assessment of the SBIR program is the first comprehensive assessment ever conducted among the departments and agencies charged with managing the bulk of the program's resources.

A major challenge has been the lack of data collection and assessment within the agencies and the limited number and nature of external assessments. Despite the challenges of assessing a diverse and complex program, the NRC assessment has sought to document the program's achievements, clarify common misconceptions about the program, and suggest practical operational improvements to enhance the nation's return on the program.

1.4.1 The Challenges of Assessment

At its outset, the NRC's SBIR study identified a series of assessment challenges that must be addressed.[69]

1.4.1.1 Recognizing Program Diversity

One major challenge is that the same administrative flexibility that allows each agency to adapt the SBIR program to its particular mission, scale, and working culture makes it difficult, and often inappropriate, to compare programs across agencies. NSF's operation of SBIR differs considerably from that of the Department of Defense, for example, reflecting, in large part, differences in the

---

[68]Donald Siegel, David Waldman, and Albert Link, "Toward a Model of the Effective Transfer of Scientific Knowledge from Academicians to Practitioners: Qualitative Evidence from the Commercialization of University Technologies," *Journal of Engineering and Technology Management*, 21(1-2): March-June 2004, pp. 115-142.

[69]See National Research Council, *SBIR: Program Diversity and Assessment Challenges*, op. cit.

extent to which "research" is coupled with procurement of goods and services. Although SBIR at each agency shares the common three-phase structure, the SBIR concept is interpreted uniquely at each agency and each agency's program is best understood in its own context.

1.4.1.2 Different Agencies, Divergent Programs

The SBIR programs operated by the five study agencies (DoD, NIH, NSF, NASA, and DoE) are perhaps as divergent in their objectives, mechanisms, operations, and outcomes as they are similar. Commonalities include:

- The three-phase structure, with an exploratory Phase I focused on feasibility, a more extended and better funded Phase II usually over two years, and for some agencies a Phase III in which SBIR projects have some significant advantages in the procurement process but no dedicated funding;
- Program boundaries largely determined by SBA guidelines with regard to funding levels for each phase, eligibility, and
- Shared objectives and authorization from Congress, including adherence to the fundamental congressional objectives for the program, i.e., compliance with the 2.5 percent allocation for the program.

However important this shared framework, there is also a profusion of differences among the agencies. In fact, the agencies differ on the objectives assigned to the program, program management structure, award size and duration, selection process, degree of adherence to standard SBA guidelines, and evaluation and assessment activities. No program shares an electronic application process with any other agency; there are no shared selection processes, though there are "shared" awards companies (but not projects). There are shared outreach activities, but no systematic sharing (or adoption) of best practices.

The following section summarizes some of the most important differences, drawing directly on a more detailed discussion in Chapter 5, focused on program management.

1.4.1.3 Award Size and Duration

Most agencies follow the SBA guidelines for award size ($100,000 for Phase I and $750,000 for Phase II) and duration (6 months/2 years) most of the time. However, some agencies have reduced the size of awards (e.g., several DoD components for Phase I, NSF for Phase II), partly in order to create funding either to bridge the gap between Phase I and Phase II, or to create incentives for companies to find matching funding for an extended Phase II award (e.g., NSF Phase IIB).

NIH, however, has in many cases extended both the size and duration of Phase I and Phase II awards. The 2006 GAO study indicated that more than 60 percent of NIH awards from 2002 to 2005 were above the SBA guidelines; discussions with NIH staff indicate that no-cost extensions have become a standard feature of the NIH SBIR program.

These operational differences reflect the differences in agency objectives, means (or lack thereof) for follow-on funding, and circumstances (e.g., time required for clinical trials at NIH).

1.4.1.4 Balancing Multiple Objectives

Congress, as indicated earlier in this chapter, has mandated four core objectives for the agencies, but has not, understandably, set priorities among them.

Recognizing the importance the Congress has attached to commercialization, all of the agencies have made efforts to increase commercialization from their SBIR programs. They also all make considerable efforts to ensure that SBIR projects are in line with agency research agendas. Nonetheless, there are still important differences among them.

The most significant difference is between acquisition agencies and nonacquisition agencies. The former are focused primarily on developing technologies *for the agency's own use*. Thus at DoD, a primary objective is developing technologies that will eventually be integrated into weapons systems purchased or developed by DoD.

In contrast, the nonacquisition agencies do not, in the main, purchase outputs from their own SBIR programs. These agencies—NIH, NSF, and parts of DoE—are focused on developing technologies *for use in the private sector*.

This core distinction largely colors the way programs are managed. For example, acquisition programs operate almost exclusively through contracts—award winners contract to perform certain research and to deliver certain specified outputs; nonacquisition programs operate primarily through grants—which are usually less tightly defined, have less closely specified deliverables, and are viewed quite differently by the agencies—more like the basic research conducted by university faculty in other agency-funded programs.

Thus, contract-type research focuses on developing technology to solve specific agency problems and/or provide products; grant-type research funds activities by researchers.

1.4.1.5 Topics, Solicitations, and Deadlines

Technical topics are used to define acceptable areas of research at all agencies except NIH, where they are viewed as guidelines, not boundaries. There are three kinds of topic-usage structures among the agencies:

- Procurement-oriented approaches, where topics are developed carefully to meet the specific needs of procurement agencies;
- Management-oriented approach, where topics are used at least partly to limit the number of applications;
- Investigator-oriented approaches, where topics are used to publicize areas of interest to the agency, but are not used as a boundary condition.

Acquisition agencies (NASA and DoD) use the procurement-oriented approach; NIH uses the investigator-oriented approach, and NSF and DoE use the management-oriented approach.

Agencies publish these topics of interest in their solicitations—the formal notice that awards will be allocated. Solicitations can be published annually or more often, depending on the agency, and agencies can also offer multiple annual deadlines for applications (as do NIH and some DoD components), or just one (DoE and in effect NSF).

1.4.1.6 Award Selection

Agencies differ in how they select awards. Peer review is widely used, sometimes by staff entirely from outside the agency (e.g., NIH), sometimes entirely from internal staff (e.g., DoD), sometimes a mix of internal and external staff, e.g., NSF. Some agencies use quantitative scoring (e.g., NIH); others do not (e.g., NASA). Some agencies have multiple levels of review, each with specific and significant powers to affect selection; others do not.

Companies unhappy with selection outcomes also have different options. At NIH, they can resubmit their application with modifications at a subsequent deadline. At most other agencies, resubmission is not feasible (where topics are tightly defined, the same topic may not come up again for several years, or at all). Most agencies do not appear to have widely used appeal processes, although this is not well documented.

The agencies also differ in how they handle the specific issue of commercial review—assessing the extent to which the company is likely to be successful in developing a commercial product, in line with the congressional mandate to support such activities.

DoD, for example, has developed a quantitative scorecard to use in assessing the track record of companies that are applying for new SBIR awards. Other agencies do not have such formal mechanisms, and some, such as NIH, do not provide any mechanism for bringing previous commercialization records formally into the selection process.

This brief review of agency differences underscores the need to evaluate the different agency programs separately. At the same time, opportunities to apply best practices concerning selected aspects of the program do exist.

1.4.1.7 Assessing SBIR: Compared to What?

The high-risk nature of investing in early-stage technology means that the SBIR program must be held to an appropriate standard when it is evaluated. An assessment of SBIR should be based on an understanding of the realities of the distribution of successes and failures in early-stage finance. As a point of comparison, Gail Cassell, Vice President for Scientific Affairs at Eli Lilly, has noted that only one in ten innovative products in the biotechnology industry will turn out to be a commercial success. Similarly, venture capital firms generally anticipate that only two or three out of twenty or more investments will produce significant returns.[70] In setting metrics for SBIR projects, it is therefore important to have realistic expectations of success rates for new firms, for firms with unproven but promising technologies, and for firms (e.g., at DoD and NASA) which are subject to the uncertainties of the procurement process. Systems and missions can be cancelled or promising technologies can not be taken up, due to a perception of risk and readiness that understandably condition the acquisition process. It may even be a troubling sign if an SBIR program has too high a success rate, because that might suggest that program managers are not investing in a sufficiently ambitious portfolio of projects.

[70]While venture capitalists are a referent group, they are not directly comparable insofar as the bulk of venture capital investments occur in the later stages of firm development. SBIR awards often occur earlier in the technology development cycle than where venture funds normally invest. Nonetheless, returns on venture funding tend to show the same high skew that characterizes commercial returns on the SBIR awards. See John H. Cochrane, "The Risk and Return of Venture Capital," *Journal of Financial Economics*, 75(1):3-52, 2005. Drawing on the VentureOne database, Cochrane plots a histogram of net venture capital returns on investments that "shows an extraordinary skewness of returns. Most returns are modest, but there is a long right tail of extraordinary good returns. 15 percent of the firms that go public or are acquired give a return greater than 1,000 percent! It is also interesting how many modest returns there are. About 15 percent of returns are less than 0, and 35 percent are less than 100 percent. An IPO or acquisition is not a guarantee of a huge return. In fact, the modal or 'most probable' outcome is about a 25 percent return." See also Paul A. Gompers and Josh Lerner, "Risk and Reward in Private Equity Investments: The Challenge of Performance Assessment," *Journal of Private Equity*, 1(Winter 1977):5-12. Steven D. Carden and Olive Darragh, "A Halo for Angel Investors," *The McKinsey Quarterly*, 1, 2004, also show a similar skew in the distribution of returns for venture capital portfolios.

1.4.2 Addressing SBIR Challenges

Realizing the potential of the SBIR program depends on how well the program is managed, adapted to the various agency missions, and evaluated for impact and improvement. As a part of this evaluation, the National Research Council Committee assessing SBIR has highlighted several structural and operational challenges faced by the program.[71] These include:

1.4.2.1 Improving the Phase III Transition

SBIR is designed with two funded Phases (I and II) and a third, nonfunded, Phase III. The transition from Phase II to the nonfunded Phase III is often uncertain. Projects successfully completing Phase II are expected to attract Phase III funding from non-SBIR sources within federal agencies, or to deploy products directly into the marketplace. There are, however, widely recognized challenges in transitioning to Phase III. As alluded to above, these challenges include appropriate timing for inclusion in procurement activities, sufficient demonstration of project potential and firm capability, and sometimes the inability to communicate a project's potential to acquisition officials. Many of these obstacles can impede take-up in the agencies that can procure SBIR funded technologies.

Some agencies have sought, with the approval of SBA, to experiment with SBIR funding beyond Phase II in order to improve the commercialization potential of SBIR funded technologies. NSF, notably, has pioneered use of the Phase IIB grant, which allows a firm to obtain a supplemental follow-on grant ranging from $50,000 to $500,000 provided that the applicant is backed by two dollars of third-party funding for every one dollar of NSF funding provided.[72] This private-sector funding validates the technology's potential for the Phase IIB award, while providing a positive incentive for potential private investors.

Ongoing program experimentation to improve program outcomes, as seen at NSF, is a sign of proactive management and is essential to the future health of the SBIR program. However, such experimentation has to be coupled with more regular evaluation to assess if these experiments are yielding positive results that should be reinforced, or if the program needs to be fine-tuned or substantially revised.

1.4.2.2 Increased Risks for Agency Procurement Officials

As with private markets, Phase III transitions can be difficult to achieve in the case of agency procurement. This can occur when agency acquisition offi-

[71] For a full list of the Committee's proposals, see Chapter 2: Findings and Recommendations.

[72] See National Research Council, *An Assessment of the SBIR Program at the National Science Foundation*, op. cit.

BOX 1-8
DoD Risks Associated with SBIR Procurement from Small, Untested Firms

Technical Risks. This includes the possibility that the technology would not, in the end, prove to be sufficiently robust for use in weapons systems and space missions.

Company Risks. SBIR companies are by definition smaller and have fewer resources to draw on than prime contractors have. In addition, many SBIR companies have only a very limited track record, which limits program manager confidence that they will be able to deliver their product on time and within budget.

Funding Limitations. The $750,000 maximum for Phase II might not be enough to fund a prototype sufficiently ready for acquisition, necessitating other funds and more time.

Testing Challenges. SBIR companies are often unfamiliar with the very high level of testing and engineering specifications (mil specs) necessary to meet DoD acquisition requirements.

Scale and Scope Issues. Small companies may not have the experience and resources necessary to scale production effectively to amounts needed by DoD. In addition, the procurement practices of an agency like DoD may not be adapted to include small firms in its large awards.[a]

Timing Risks. DoD planning, programming, and budgets work in a two-year cycle, and it is difficult for Program Executive Officers to determine whether a small firm will be able to create a product to meet program needs in a timely manner, even if the initial research has proven successful.

[a]This is a correction of the text in the prepublication version released on July 27, 2007.

cials hesitate to employ the results of SBIR awards because of the added risks, both technical and personal, associated with contracting through small firms. As a result of these disincentives, program managers in charge of acquisitions have traditionally not seen SBIR as part of their mainstream activities, often preferring to take the cautious route of procuring technology through the established prime contractors.

Risk aversion is by no means peculiar to the Department of Defense. Program officers at NASA usually have only one opportunity to get their projects right, given limited opportunities for in-flight adjustments. Recognizing this constraint, some NASA program managers are uncertain that SBIR can deliver reliable technology on time and at a manageable level of risk.

It is important to recognize those risks associated with the program and to develop measures to reduce negative incentives that cause procurement officers to avoid contracting with SBIR companies. Enhanced use of the program can help introduce innovative and often low-cost solutions to the mission needs of the Department of Defense and NASA.

1.4.2.3 Growing Alignment in Incentives

As we noted earlier, there are advantages for entrepreneurs as well as agency officials in the SBIR concept. It is interesting that some services (e.g., Navy) and many prime contractors are finding the SBIR program to be directly relevant to their interests and objectives. With the right positive incentives and management attention, the performance and contributions of the SBIR program can be improved.

The experience of the Navy's SBIR program demonstrates that the SBIR program works well when each of these participants recognizes program benefits and is willing to take part in facilitating the program's operations.

BOX 1-9
Attributes of the Navy's Submarine SBIR Program

The Submarine Program Executive Office is widely considered to be one of the more successful Phase III program at DoD. The program takes a number of steps to use and support SBIR as an integral part of the technology development process.

Acquisition Involvement. SBIR opportunities are advertised through a program of "active advocacy." Program managers compete to write topics to solve their problems.

Topic Vetting. Program Executive Officers keep track of all topics. Program managers compete in a rigorous process of topic selection. SBIR contracts are considered a reward, not a burden.

Treating SBIR as a Program, including follow-up and monitoring of small businesses to keep them alive till a customer appears. Program managers are encouraged to demonstrate commitment to a technology by paying half the cost of a Phase II option.

Providing Acquisition Coverage, which links all SBIR awards to the agency's acquisition program.

Awarding Phase III Contracts within the $75 million ceiling that avoids triggering complex Pentagon acquisition rules.

Brokering Connections between SBIR and the prime contractors.

Recycling unused Phase I awards, a rich source for problem solutions.

Most notably, the Program Executive Office (PEO) for the Navy's submarine program has shown how SBIR can be successfully leveraged to advance mission needs. Its experience demonstrates that senior operational support and additional funding for program management provides legitimacy and the means needed for the program to work more effectively. In addition to funding program operations, this additional support allows for the outreach and networking initiatives (such as the Navy Forum), as well as other management innovations, that contribute to enhanced matchmaking, commercialization, and to the higher insertion rates of SBIR technologies into the Navy's ships, submarines, and aircraft.

1.4.2.4 Creating a Culture of Evaluation

Continuous improvement comes through institutionalized learning. Regular internal assessments that are transparent and based on objective criteria are necessary for the agencies to continue to experiment with and adapt the SBIR program to meet their changing mission needs. In addition, external assessments are important to improve the public's understanding of the nature of high-risk, high-payoff investments and the increasingly important role that SBIR plays in the nation's innovation system. The failure of individual projects does not indicate program failure, yet a failure to evaluate expenditures against outcomes is a program failure.

Prior to the start of the Academies' assessment, only very limited assessment and evaluation had been done at any of the SBIR agencies. Insufficient data collection, analytic capability, and reporting requirements, together with the decentralized character of the program, mean that there is very little data to use in evaluating the connection between outcomes and program management and practices.

One encouraging development in the course of the NRC study is that agencies are showing increasing interest in regular assessment and evaluation.

1.4.2.5 Monitoring and Assessment

The agencies differ in their degree of interest in monitoring and assessment. DoD has led the way in external, arms-length assessment. All agencies have now completed at least one effort to assess outcomes from their programs.

At NIH, a survey of Phase II recipients has been followed by a second survey tracking the same recipients, which has generated important results. At DoD, the Company Commercialization Report is being used by some components to track outcomes; efforts are also underway to track Phase III contracts directly through the DD350 reporting mechanism. At NASA, NSF, and DoE less formal surveys have been used for similar purposes. The extent and rigor of these efforts vary widely, partly because the available resources for such activities vary widely as well.

1.5 CHANGING PERCEPTIONS ABOUT SBIR

As the SBIR program matures and more is known about its accomplishments and potential, it is increasingly viewed by the agencies to be part of their wider portfolio of agency R&D investments. This is a welcome change. SBIR was for many years viewed as the stepchild by the research community at each agency. Created by Congress in 1982, SBIR imposed a specific mandate on agencies funding significant amounts of extramural R&D.

The perception that SBIR was a tax on "real" research meant that a number of agencies made limited efforts to integrate SBIR into agency R&D strategy. Often, agencies tried to simply piggyback SBIR activities on existing approaches. Selection procedures, for example, were often copied from elsewhere in the agency, rather than being designed to meet the specific needs of the small business community. Furthermore, in the early years agencies generally made no effort to determine outcomes from the program, or to design and implement management reforms that would support improved outcomes.

Beginning in the late 1990s, this limited view of SBIR began to change. Initiatives taken by senior management at the Department of Defense and by SBIR program managers at the operational level, which demonstrated that SBIR projects can make a real difference to agency missions, began to alter wider agency perspectives about the program's potential. As a result, more acquisition and technical officials have become interested in cooperating with the program. As noted above, in some parts of some agencies, SBIR is now viewed as a valuable vehicle within the wider portfolio of agency R&D investments.

Indeed, the agencies are increasingly coming to see that SBIR offers some unique benefits in terms of accessing pools of technology and capabilities otherwise not easily integrated into research programs dominated by prime contractors and agency research labs. At the NRC's conference on the challenge of the SBIR Phase III transition, for example, senior representatives of the Department of Defense affirmed the program's role in developing innovative solutions for mission needs. Further underscoring the program's relevance, several prime contractors' representatives at the conference stated that they have focused management attention, shifted resources, and assigned responsibilities within their own management structures to capitalize on the creativity of SBIR firms.

1.5.1 SBIR Around the World

Increasingly, governments around the world view the development and transformation of their innovation systems as an important way to promote the competitiveness of national industries and services. They have adopted a variety of policies and programs to make their innovation systems more robust, normally developing programs grounded in their own national needs and experiences.

Nevertheless, governments around the world are increasingly adopting SBIR-type programs to encourage the creation and growth of innovative firms in their economies. Sweden and Russia have adopted SBIR-type programs. The United Kingdom's SIRI program is similar in concept. The Netherlands has a pilot SBIR program underway and is looking to expand its scope. Asia, Japan, Korea, and Taiwan have also adopted the SBIR concept with varying degrees of success, as a part of their national innovation strategies. This level of emulation across national innovation systems is striking and speaks to the common challenges addressed by SBIR awards and contracts.

1.6 CONCLUSION

This report provides a summary of the study that Congress requested when reauthorizing the SBIR program in 2000. Drawing on the results of newly commissioned surveys, case studies, data and document analyses, and interviews of program staff and agency officials, the NRC assessment of SBIR has examined how the program is meeting the legislative objectives of the program. As with any analysis, this assessment has its limitations and methodological challenges. These are described more fully below. Nonetheless, this cross-agency report, along with `individual reports on the SBIR programs at the Department of Defense, the National Institutes of Health, the Department of Energy, the National Aeronautics and Space Administration, and the National Science Foundation, provide the most comprehensive assessment to date of the SBIR program. In addition to identifying the challenges facing SBIR today, the NRC Committee responsible for this study has also recommended operational improvements to the program. By strengthening the SBIR program, the Committee believes that the capacity of the United States to develop innovative solutions to government needs and promising products for the commercial market will be enhanced.

2

Findings and Recommendations

I. NATIONAL RESEARCH COUNCIL (NRC) STUDY FINDINGS

A. **The Small Business Innovation Research (SBIR) Program Is Making Significant Progress in Achieving the Congressional Goals for the Program.** The SBIR program is sound in concept and effective in practice. With the programmatic changes recommended here, the SBIR program should be even more effective in achieving its legislative goals.

1. **Overall, the Program Has Made Significant Progress in Achieving its Congressional Objectives by:**

   a. **Stimulating Technical Innovation.** By a variety of metrics, the program is contributing to the nation's stock of new scientific and technical knowledge. (See Finding F.)

   b. **Using Small Businesses to Meet Federal Research and Development Needs.** SBIR program objectives are aligned with, and contribute significantly to fulfilling the mission of each studied agency.[1] In some cases, closer alignment and greater integration should be possible. (See Finding C).

   c. **Fostering and Encouraging Participation by Minority and Disadvantaged Persons in Technological Innovation.** The SBIR program supports the growth of a diverse array of small businesses, including minority- and woman-owned business, by providing market access, funding, and recognition. To better assess this support, enhanced efforts to collect better data and to monitor outcomes more closely are required at some agencies. Also, more analysis is needed to improve understanding of the obstacles faced by minority- and woman-owned

[1]The Department of Defense (DoD), National Institutes of Health (NIH), Department of Energy (DoE), National Aeronautics and Space Administration (NASA), and National Science Foundation (NSF).

businesses and the nature of the measures needed to address these obstacles. (See Findings D and E.)

d. **Increasing Private Sector Commercialization of Innovation Derived from Federal Research and Development.** The program enables small businesses to contribute to the commercialization of the nation's R&D investments, both through private commercial sales, as well as through government acquisition, thereby enhancing American health, welfare, and security through the introduction of new products and processes. (See Finding B.)

2. **SBIR Is Meeting Federal R&D Needs.** SBIR plays an important role in introducing innovative, science-based solutions that address the diverse mission needs of the federal agencies.

   a. The program has the potential to make systematic investments in high-potential, high-risk technologies that can meet specific mission needs and/or offer the potential of significant commercial development. Balancing the potential gains from high risk, but potentially high payoff, against technologies with more promising immediate commercial benefits, represents a challenge for management and appropriate evaluation.

3. **Improving SBIR.** As structured, the management of SBIR exhibits considerable flexibility, allowing program experimentation needed to address varied agency missions and technologies. This flexibility is commendable and has enabled the program to contribute to multiple agency missions. At the same time, more regular assessment is required, both internal and external, to inform agency management of program outcomes and to improve performance through analysis of program outcomes and regular adoption of agency "best practices." (See Section II on Recommendations.)

B. **Commercialization.**

1. **Commercialization Despite High Risk.** Small technology companies use SBIR awards to advance projects, develop firm-specific capabilities, and ultimately create and market new commercial products and services.

   a. Given the very early stage of SBIR investments, and the high degree of technical risk often involved (reflected in risk assessment scores developed during agency selection procedures), the fact that a high proportion of projects reach the market place in some form is significant, even impressive.

   b. Although the data vary by agency, responses to the NRC Phase II Survey indicate that just under half of the projects do reach the marketplace.[2]

[2]See Figure 4-1. In addition, the response to the survey may reflect a degree of self-selection bias,

2. **Multiple Indicators of Commercial Activity.** The full extent of commercialization for both government and private use is insufficiently captured by simply recording sales of a product or service. The Committee's analysis uses five measures of commercial activity, although sales in the marketplace are obviously of particular importance. These indicators are:

   a. Sales (using a range of different benchmarks to indicate different degrees of commercial activity)

   b. Additional non-SBIR research funding and contracts

   c. Licensing revenues

   d. Third-party investment (including both venture funding and other sources of investment)

   e. Additional SBIR awards for related work

3. **A Skewed Distribution of Sales.** SBIR awards result in sales numbers that are highly skewed, with a small number of awards accounting for a very large share of the overall sales generated by the program.[3] This is to be expected in funding early-stage technological innovation and is broadly consistent with the general experience of other sources of early technology financing by angel investors.[4] Most projects, however, do not achieve significant commercial success; a few companies do.[5]

---

e.g., the successful companies (those still in existence) are more likely to respond. Survey bias can manifest itself in several ways, however, and in some cases understate returns. For a fuller discussion of this important caveat, see Box 4-1 in Chapter 4: SBIR Program Outputs.

[3]See Figure 4-2.

[4]Even venture capital—a proximate referent group not appropriate for direct comparison because venture investments do not provide the same kind of project financing—shows a high returns skew. Nonetheless, returns on venture funding tend to show a similar high skew as characterize commercial returns on the SBIR awards. See John H. Cochrane, "The Risk and Return of Venture Capital," *Journal of Financial Economics*, 75(1):3-52, 2005. Drawing on the VentureOne database, Cochrane plots a histogram of net venture capital returns on investments that "shows an extraordinary skewness of returns. Most returns are modest, but there is a long right tail of extraordinary good returns. 15 percent of the firms that go public or are acquired give a return greater than 1,000 percent! It is also interesting how many modest returns there are. About 15 percent of returns are less than 0, and 35 percent are less than 100 percent. An IPO or acquisition is not a guarantee of a huge return. In fact, the modal or 'most probable' outcome is about a 25 percent return." See also Paul A. Gompers and Josh Lerner, "Risk and Reward in Private Equity Investments: The Challenge of Performance Assessment," *Journal of Private Equity*, 1(Winter 1977):5-12. Steven D. Carden and Olive Darragh, "A Halo for Angel Investors," *The McKinsey Quarterly*, 1, 2004, also show a similar skew in the distribution of returns for venture capital portfolios.

[5]This is expected with research that pushes the state-of-the-art where technical failures are expected and when companies face the vagaries of the procurement process. A small proportion (3-4 percent) of projects generate more than $5 million in cumulative revenues. At NIH for example, only 6 of 450 projects in the recent NIH survey were identified as generating more than $5 million in revenues.

a. As an example, just eight of the NSF-supported projects—each of which had $2.3 million or more in sales—accounted for over half of the total sales dollars reported by respondents.

b. Similarly, one firm selected for the NRC Phase II Survey accounted for over half of all licensing income that the 162 respondents reported they earned from the NSF SBIR-supported projects.[6]

4. **SBIR Is an Input, Not a Panacea.** SBIR can be a key input to encourage small business commercialization, but most major commercialization successes require substantial post-SBIR research and funding from a variety of sources. SBIR awards will have been in many cases a major, even critical input—but only one of many inputs.

5. **SBIR Projects Attract Significant Additional Funding.** SBIR funded research projects enable small businesses to develop the technical know-how needed to attract third-party interest from a variety of public and private sources, including other federal R&D funds, angel investors, and venture funds. The NRC survey revealed that 56 percent of surveyed projects were successful in attracting additional funding from a variety of sources.[7]

a. **Federal Funding.** Responses to the NRC Phase II Survey indicate that about 10 percent of Phase II projects were eventually supported by other federal research funding; over half received at least one ad-

---

Not all valuable technology innovations lead directly to high revenues, however, and SBIR has led to valuable technology innovations across the spectrum of work funded by NIH. Technologies such as the bar-coding developed by Savi using SBIR funds have been of great importance to DoD, enabling massive savings in logistics. This degree of skew underscores the limitations of random surveys that can easily miss important outcomes in terms of mission and/or sales. SBIR companies that achieve significant sales and show promise of more are often acquired. Once acquired, they normally do not respond to surveys. For example, one firm, Digital Systems Resources, Inc., was acquired by General Dynamics in September of 2003. At the time of acquisition, DSR had received 40 Phase II SBIR awards and had reported $368 million in resultant sales and investment. See Findings and Recommendations chapter of National Research Council, *An Assessment of the SBIR Program at the Department of Defense*, Charles W. Wessner, ed., Washington, DC: The National Academies Press, 2009.

[6]See National Research Council, *An Assessment of the SBIR Program at the National Science Foundation*, Charles W. Wessner, ed., Washington, DC: The National Academies Press, 2008, p. 214.

[7]See NRC Phase II Survey, Questions 22 and 23, in Appendix A. SBIR awards are related to a range of other funding events for recipients, and interviews strongly suggest that SBIR does provide a positive validation effect for third party funders. This perspective is confirmed by some angel investors who see SBIR awards as a valuable step in providing funds to develop proof of concept and encourage subsequent angel investment. Steve Weiss, Personal Communication, December 12, 2006. Weiss, an angel investor, cites SBIR awards as an example of how the public and private sectors can collaborate in bringing new technology to markets. Further, see National Research Council, *An Assessment of the SBIR Program at the National Institutes of Health*, Charles W. Wessner, ed., Washington, DC: The National Academies Press, 2009.

ditional related Phase I SBIR award, and more than half received at least one additional related Phase II award;[8]

b. **Licensing.** Sixteen percent of projects report licensing agreements in place for their technologies, with an additional 16 percent engaged in licensing negotiations at the time of the NRC Phase II Survey.[9] Some companies are making substantial use of the technology.[10]

c. **Venture Funding.** Venture capital (VC) funding has, in some cases, also played a significant role for SBIR award winners. At NIH, for example, some 25 percent of the 200 companies that received the highest number of SBIR awards had also obtained funding from venture capital, totaling some $1.59 billion.[11]

d. **Acquisition.** In some cases, the SBIR funded technology developed sufficient commercial potential that investors bought the grantee company outright. For example, in 2000, Philips bought out SBIR recipient Optiva, reportedly for the sum of $1 billion.[12]

6. **Better Documentation of Commercialization Is Needed.** While the recorded successes of the program are significant, commercialization outcomes are not yet adequately documented in most of the agencies reviewed. Indeed, some agencies have made very limited efforts to track the evolution of firms having received SBIR awards. Understanding outcomes and the impact of program modifications on those outcomes is essential for effective program management.

---

[8]See response to NRC Phase II Survey, Question 23.

[9]See response to NRC Phase II Survey, Question 12.

[10]While 5.6 percent of respondents to the NRC Phase II Survey reported licensee sales greater than $0, three licenses reported more than $70 million in sales each and accounted for more than half of all reported sales, with one project alone reporting more than $200 million in licensee sales. See Section 4.2.3.4.

[11]See Figure 4-5. The impact of the 2004 SBA ruling that excluded firms with majority ownership by venture firms is to be empirically assessed in a follow-on study by the National Research Council. On December 3, 2004, the Small Business Administration issued a final rule saying that to be eligible for an SBIR award "an entity must be a for-profit business at least 51 percent owned and controlled by one or more U.S. individuals." U.S. Congress, House, Committee on Science, Subcommittee on Environment, Technology, and Standards, Hearing on "Small Business Innovation Research: What is the Optimal Role of Venture Capital," Hearing Charter, June 28, 2005.

[12]With SBIR funding, David Giuliani, together with University of Washington professors David Engel and Roy Martin, formed a company in 1987, eventually known as Optiva Corporation Inc., to promote a new, innovative dental hygiene device based on sonic technology. In October 2000, Royal Philips Electronics, acquired Optiva Corporation, Inc., now known as Philips Oral Healthcare. The final terms were not disclosed. Royal Philips Electronics, accessed at <*http://www.homeandbody.philips.com/sonicare/gb_en/03d-story.asp*>.

C. **Effective and Flexible Alignment of SBIR with Agency Mission.**

1. **Flexible Program Management.** The effective alignment of the program with widely varying mission objectives, needs, and modes of operation is a central challenge for an award program that involves a large number of departments and agencies. The SBIR program has been adapted effectively by the management of the individual departments, services, and agencies, albeit with significant operational differences, reflecting distinct missions and operational cultures. This flexibility in program management is one of the great strengths of the program.

2. **Mission Support.** Each of the agency SBIR programs is effectively supporting the mission of the respective funding agency:[13]

   a. **Defense.**[14] At DoD, considerable progress has been made in aligning SBIR-funded research with the strategic objectives of agency research and acquisition. In some parts of DoD, e.g., some Program Executive Officers within the Navy, significant success has been achieved in improving the insertion of SBIR-funded technologies into the acquisition process. The commitment of upper management to the effective operation of the program and the provisioning of additional funding appear to be a key element of success. Teaming among the agency, the SBIR awardees, and the prime contractors is important in the transition of technologies to products to integration in systems. The growing importance of the SBIR program within the defense acquisition system is reflected in the increasing interest of primes, who are seeking opportunities to be involved with SBIR projects—a key step toward acquisition.[15]

   b. **NIH.**[16] NIH and other public health agencies increasingly see SBIR as an important element in the agency's translational strategy—designed to move technologies from the lab into the marketplace.[17] The NIH

[13]This assessment is based both on procedures for selecting topics and awardees aligned with agency mission, as well as outcomes analysis. Each of the five separate volumes prepared by the Committee on the SBIR program at the five individual agencies under review explores this issue in considerable depth. The Outcomes chapter of this volume summarizes these findings in more detail.

[14]National Research Council, *An Assessment of the SBIR Program at the Department of Defense*, op. cit.

[15]The growing interest of Defense prime contractors is recorded in National Research Council, *SBIR and the Phase III Challenge of Commercialization*, Charles W. Wessner, ed., Washington, DC: The National Academies Press, 2007.

[16]National Research Council, *An Assessment of the SBIR Program at the National Institutes of Health*, op. cit.

[17]The importance of this strategy to NIH is reflected in the NIH RoadMap, announced in September 2003. Recent interest at NIH, led by the National Cancer Institute in conjunction with the SBIR program offices, has underscored the growth in interest in the role and potential of the program. See *<http://nihroadmap.nih.gov/>*.

awards cases examined focused on public health and biomedical science and technology. The NRC's review of the SBIR program at NIH documents that SBIR program funds projects have a significant impact on public health: NIH considers a project's impact on public health during the selection process, and recipients and NIH staff note that examining potential impacts is an important component in every application review.[18] The FDA's Critical Path Initiative is focused on similar concerns.[19]

c. **NASA.**[20] At NASA, the primary metric for project success is the deployment of SBIR-funded technologies on space missions, where the agency can point to a number of significant impacts.[21]

d. **Energy.**[22] At DoE, as at NASA, SBIR topics and project selection are heavily influenced by the agency research staff's wider R&D responsibilities. At DoE, all technical topics in the SBIR solicitation are constructed to support the overall mission of each of the agency's technical program areas.

e. **NSF.**[23] At NSF, SBIR-funded research meets the agency's primary mission of expanding scientific and technical knowledge.[24] However, the SBIR program does so by tapping the scientific and engineering capabilities of an important sector—small businesses—that is almost entirely unserved by the remainder of NSF's programs.[25]

---

[18]The NIH Mission is "science in pursuit of fundamental knowledge about the nature and behavior of living systems and the application of that knowledge to extend healthy life and reduce the burdens of illness and disability."

[19]The FDA describes its Critical Path Initiative as "an effort to stimulate and facilitate a national effort to modernize the scientific process through which a potential human drug, biological product, or medical device is transformed from a discovery or "proof of concept" into a medical product." Accessed at *<http://www.fda.gov/oc/initiatives/criticalpath/>*.

[20]National Research Council, *An Assessment of the SBIR Program at the National Aeronautics and Space Administration*, Charles W. Wessner, ed., Washington, DC: The National Academies Press, 2009.

[21]NASA SBIR managers report that several SBIR projects relate to the Mars Rover. While these are considered to be valuable contributions to that mission, by definition they do not represent opportunities for high-volume sales. Interview with NASA Management, December 21, 2006.

[22]National Research Council, *An Assessment of the SBIR Program at the Department of Energy*, Charles W. Wessner, ed., Washington, DC: The National Academies Press, 2008.

[23]National Research Council, *An Assessment of the SBIR Program at the National Science Foundation*, op. cit.

[24]NSF's mission is "To promote the progress of science; to advance the national health, prosperity, and welfare; to secure the national defense." Accessed at *<http://www.nsf.gov/nsf/nsfpubs/straplan/mission.htm>*.

[25]Nearly 40 percent of U.S. scientific and technical workers are employed by small businesses. U.S. Small Business Administration, Office of Advocacy (2005) data drawn from U.S. Bureau of the Census; Advocacy-funded research by Joel Popkin and Company (Research Summary #211); Federal Procurement Data System; Advocacy-funded research by CHI Research, Inc. (Research Summary

D. **The SBIR Program Provides Significant Support for Small Business.**

1. **Substantial Benefits.** The NRC Phase II Survey and NRC Firm Survey show that the SBIR program has provided substantial benefits for participating small businesses at all agencies, in a number of different ways. These benefits include:

   a. **Encouraging Company Foundation.** According to the NRC Firm Survey, just over 20 percent of companies indicated that they were founded entirely or partly because of an SBIR award;[26]

   b. **Supporting the Project Initiation Decision.** Over two-thirds of SBIR projects reportedly would not have taken place without SBIR funding.[27] This finding reflects the known difficulties in funding commercial applications for early-stage technologies. The NRC Phase II Survey shows that SBIR seems to provide funding necessary to initiate early stage projects.

   c. **Providing an Alternative Development Path.** Companies often use SBIR to fund alternative development strategies, exploring technological options in parallel with other activities.

   d. **Partnering and Networking.** SBIR funding helps small businesses access outside resources, especially academic consultants and partners, helping to create networks and facilitating the transfer of university knowledge to the private sector.

   e. **Commercializing Academic Research.** In addition to fostering partnerships between academic institutions and private firms, SBIR plays an important role in encouraging academics to found new firms that can commercialize their research.[28]

2. **SBIR Support Is Widely Distributed.** SBIR provides support for innovation activity that is widely distributed across the small business research community.

   a. **Large Number of Firms.** During the fourteen years between 1992 and

#225); Bureau of Labor Statistics, Current Population Survey; U.S. Department of Commerce, International Trade Administration.

[26]For the breakout, see Table 4-7. Chapter 4 also includes a case study of Sociometrics, Inc., which illustrate the role of SBIR in firm formation.

[27]See Figure 4-19 for a breakdown of the survey findings. Thirty-eight percent reported that they definitely would not have proceeded with the project without SBIR funding and an additional 33 percent reported that they probably would not have proceeded with the project without SBIR funding.

[28]A significant number of firms involved in the program report that the founder was previously an academic. See Finding F-3.

2005, inclusive, more than 14,800 different firms received at least one Phase II award, according to the SBA Tech-Net database.[29]

b. **Many New Participants.** The agencies tracking new winners indicate that at least a third of awards at all agencies go to companies that had not previously won awards at that agency.[30] This steady infusion of new firms is a major strength of the program and suggests that SBIR is encouraging innovation across a broad spectrum of firms, creating additional competition among suppliers for the agencies with procurement responsibilities, and providing agencies with new mission-oriented research and solutions.

E. **Benefits for Minority- and Woman-owned Small Businesses.**

1. **A Key Program Objective.** One of the four congressional objectives for the SBIR program is to enhance opportunities for woman- and minority-owned businesses. SBIR's competitive awards provide a source of capital for small innovative firms with pre-prototype technologies—the phase where funding is most difficult to obtain. The certification effect that SBIR awards generate can be especially valuable as a signal to the early-stage capital markets of the potential of the project and hence of the firm.

2. **A Mixed Record.** Woman- and minority-owned firms face substantial challenges in obtaining early-stage finance.[31] Recognizing these challenges, the legislation calls for fostering and encouraging the participation of women and minorities in SBIR. Given this objective, some current trends are troubling. Agencies do not have a uniformly positive record in funding research by woman- and minority-owned firms. Efforts to document and monitor their participation have been inadequate at, for example, NIH. It is also true, however, that developing appropriate measures in this area is complex given, *inter alia*, the variations in the demographics of the applicant pool.[32]

a. **Trends in Support for Woman-owned Businesses Vary Across Agencies.**

---

[29]See U.S. Small Business Administration, Tech-Net Database, accessed at <*http://www.sba.gov/sbir/indextechnet.html*>.

[30]See Section 4.5.3 for a discussion of the incidence of new entrants. See also Figures 4-21, 4-22, 4-23, and 4-24 for data on new entrants at NSF, NIH, and DoD.

[31]Academics represent an important future pool of applicants, firm founders, principal investigators, and consultants. Recent research shows that owing to the low number of women in senior research positions in many leading academic science departments, few women have the chance to lead a spinout. "Under-representation of female academic staff in science research is the dominant (but not the only) factor to explain low entrepreneurial rates amongst female scientists." See Peter Rosa and Alison Dawson, "Gender and the commercialization of university science: academic founders of spinout companies," *Entrepreneurship & Regional Development*, 18(4):341-366, July 2006.

[32]See Chapters 3, 4, and 5 for additional discussion.

i. At **DoD**, which accounts for over half of the SBIR program funding, the share of Phase II awards going to woman-owned businesses increased from 8 percent at the time of the 1992 reauthorization (1992-1994) to 9.5 percent in the most recent years covered by the NRC Phase II Survey (1999-2001).[33] This percentage increase occurred in a program that expanded significantly over the same time period, both in terms of the R&D base and the SBIR allocation.

ii. At **NSF**, in recent years, woman-owned businesses have submitted an average of 213 Phase I proposals annually, and they have received an average of 26.6 Phase I awards annually. With the exception of the bump-up in 2002 and 2003, there is no upward trend. In other words, while woman-owned businesses do participate in the NSF SBIR program, that participation does not seem to be increasing over time.

iii. At **NIH**, awards to woman-owned businesses have increased, and their share of all awards is trending upward.[34] However, the percentage of female life scientists has been growing much faster.[35]

b. **Support for Minority-owned Firms Has Not Increased Proportionately.**

i. **DoD.** The share of Phase I awards to minority-owned firms at DoD has declined quite substantially since the mid 1990s and fell below 10 percent for the first time in 2004 and 2005.[36] Data on Phase II awards suggest that the decline in Phase I award shares for minority-owned firms is reflected in Phase II.

ii. **NSF** minority firm participation is higher than that of woman-owned firms.[37] However, application rates have not been rising in line with changes in the demographics of the scientific and technical workforce, and success rates for woman- and minority-owned companies continue to lag those of other small businesses.[38]

iii. **NIH.** Data only recently provided by NIH raises significant con-

[33]See Figure 2-4 and Figure 3-12 in National Research Council, *An Assessment of the SBIR Program at the Department of Defense*, op. cit.

[34]See Figure 2-1 in National Research Council, *An Assessment of the SBIR Program at the National Institutes of Health*, op. cit.

[35]See Section 3.2.6.4 in National Research Council, *An Assessment of the SBIR Program at the National Institutes of Health*, op. cit.

[36]This statistic is drawn from the DoD Awards Database. See Figure 4-13.

[37]See Figure 4.2-15 and Figure 4.2-16 in National Research Council, *An Assessment of the SBIR Program at the National Science Foundation*, op. cit.

[38]See Figure 4.2-14 in National Research Council, *An Assessment of the SBIR Program at the National Science Foundation*, op. cit.

cerns about the shares of awards being made to minority-owned firms. The numbers of awards and applications, as well as success rates, have all declined for minorities, for both Phase I and Phase II. The lack of accurate and complete data suggests insufficient management attention to this component of the program's objectives.[39]

F. **Stimulating Scientific and Technological Knowledge Advances.** SBIR companies have generated many patents and publications, the traditional measures of activity in this area.

1. **Multiple Knowledge Outputs.** SBIR projects yield a variety of knowledge outputs. These contributions to knowledge are embodied in data, scientific and engineering publications, patents and licenses of patents, presentations, analytical models, algorithms, new research equipment, reference samples, prototypes products and processes, spin-off companies, and new "human capital" (enhanced know-how, expertise, and sharing of knowledge).
2. The NRC Phase II Survey found that 34 percent of NIH projects surveyed generated at least one patent, and just over half of NIH respondents published at least one peer-reviewed article.[40]
3. **Linking Universities to the Public and Private Markets.** The SBIR program supports the transfer of research into the marketplace, as well as the general expansion of scientific and technical knowledge, through a wide variety of mechanisms. With regard to SBIR's role in linking universities to the market, about a third of all NRC Phase II and Firm Survey respondents indicated that there had been involvement by university faculty, graduate students, and/or a university itself in developed technologies. This involvement took a number of forms.[41] Among the responding companies—

[39]See Figures 2-1, 2-2, 3-10, and 3-11 in National Research Council, *An Assessment of the SBIR Program at the National Institutes of Health*, op. cit.

[40]See Table 4-23 in National Research Council, *An Assessment of the SBIR Program at the National Institutes of Health*, op. cit. The patenting activity of SBIR firms has also been highlighted in recent Congressional testimony. Representatives of the Small Business Technology Council claimed that SBIR-related companies produced some 45,000 patents, compared to just over 2,900 patents produced by universities in 2006. The same testimony, citing the National Science Foundation, reported that some 32 percent of U.S. scientists and engineers are employed by small businesses. (Testimony of Michael Squillante before the Senate Committee on Small Business and Entrepreneurship on July 12, 2006) The Small Business Administration reports that small businesses employ 41 percent of high-tech workers, including scientists, engineers, and computer workers. U.S. Small Business Administration, *Frequently Asked Questions*, June 2006, accessed at *<http://www.sba.gov/advo/stats/sbfaq.pdf>*.

[41]See Section 4.6.3.1 on university faculty and company formation. Also see Table 4-13 on university involvement in SBIR projects.

a. More than two-thirds had at least one academic founder, and more than a quarter had more than one;

b. About one-third of founders were most recently employed in an academic environment before founding the new company;

c. In some 27 percent of projects, university faculty were involved as principal investigators or consultants on the project;

d. 17 percent of Phase II projects involved universities as subcontractors; and

e. 15 percent of Phase II projects employed graduate students.

These data underscore the significant level of involvement by universities in the program and highlight the program's contribution to the transition of university research to the marketplace.

4. **High-risk Focus.** Projects funded by SBIR often involve high technical risk, implying novel and difficult research rather than incremental change. At NSF, for example, just more than half of the Phase I projects did not get a follow-on Phase II. One third of the Phase I awardees did not apply for Phase II awards, and of those that did, many did not get a follow-on Phase II for technical reasons. The chief reason for discontinuing Phase II projects was technical failure or difficulties.[42] This effort to push the technological frontier is a strength of the program and necessarily involves project, but not program, failure.

5. **Indirect Path Effects.** There is strong anecdotal evidence concerning "indirect path" effects—that investigators and research staff gain knowledge from projects that may become relevant in a different technical context later on, and often in another project or even another company. These effects are not directly measurable, but interviews and case studies affirm their existence and importance.[43]

G. **The Measurement Challenge.** The diversity of missions among agencies involved with SBIR means that metrics and indicators that capture common goals and others that reflect the diversity in mission and mechanisms (e.g., contracts or grants) must be employed to measure outcomes and impacts.

1. **Differing Goals.** A fundamental difference exists between those agencies that procure SBIR-funded products for their missions and those that

[42]See also National Research Council, *An Assessment of the SBIR Program at the National Science Foundation*, op. cit.

[43]For a discussion of indirect path effects, see National Research Council, *The Advanced Technology Program: Assessing Outcomes*, Charles W. Wessner, ed., Washington, DC: National Academy Press, 2001.

normally do not. At DoD, the program is driven by the need to put the results of research into products and systems to support war-fighters; at NIH, the focus of the SBIR program is to conduct research that contributes to improvements in public health, almost always through eventual take-up by entities outside NIH.[44]

2. **Differing Mechanisms.** These distinctly different goals are reflected in the mechanisms used by the agency to implement the SBIR program.[45] For example, the DoD employs contracts whereas the NIH generally employs grants for its SBIR awards.

3. **Different Metrics.** Developing metrics that can capture impacts in both of these areas is a challenge. Measuring the extent to which the program supports the agency mission in any nonprocurement agency like NSF is substantially different from measures applicable to a procurement agency like DoD or NASA. In some cases, the products of NIH-funded research need to be taken up by the private sector in order to be further developed before their contribution can be realized. In other cases, such as diagnostics, software, and educational materials to encourage better health, the SBIR awards can be sufficient in themselves.

H. **SBIR Program Flexibility Is a Positive Feature and Should Be Preserved.** As structured, the management of SBIR exhibits commendable flexibility. SBA and agency management have allowed the program managers the room to adapt the program to the needs of specific technologies and unique mission needs.

1. **Multiple Missions.** The varied missions of the agencies that fund SBIR have encouraged SBA to take a flexible approach to implementing program guidelines. SBA has provided waivers on funding size and other programmatic matters, and agencies have used considerable ingenuity in exploring the boundaries of the guidelines, particularly in terms of support for commercialization.[46]

2. **Commendable Flexibility.** This flexibility is highly commendable and should be expanded where possible. It is precisely the relatively wide

[44]See the discussion of agencies differing goals, modes of operation, and their implications for the assessment in the Introduction chapter. See also the section on assessment challenges, and for a more complete review, see the study's methodology report, National Research Council, *An Assessment of the Small Business Innovation Research Program—Project Methodology*, Washington, DC: The National Academies Press, 2004.

[45]The difference between grants and contracts is discussed in Finding H.

[46]For a discussion of expanded award size and flexibility at NIH, see the Awards chapter in National Research Council, *An Assessment of the SBIR Program at the National Institutes of Health*, op. cit. For a discussion of the Phase IIB awards at NSF, see the Program Management chapter in National Research Council, *An Assessment of the SBIR Program at the National Science Foundation*, op. cit.

range of options available to the agencies and their SBIR program managers that helps to account for the effectiveness of the program. The need for flexibility is grounded in the very wide differences in objectives, structure, and culture across the agency SBIR programs as well as in the uncertainties of early stage project finance.[47]

3. **Program Diversity.**

   a. **Mission Agencies.** As noted above, agencies with substantial hardware and systems procurement responsibilities rely primarily on contracts as the vehicle for commissioning research—including SBIR. Contracts are oriented toward the production of results for which the contracting agency has a specific need, often via a contract with strict terms and conditions, and clear deliverables. In a contracts environment, Phase I is seen as a means of testing the feasibility of solutions for a problem, and a single Phase II may be selected to implement the most promising of these options.

   b. **Research Agencies.** Agencies without significant hardware or systems procurement generally use grants. Grants awarded to support promising academic proposals are often initiated by researchers with interests in a specific area. At NIH, for example, topics in the SBIR solicitation serve only as guidance to areas of interest, underscoring the NIH commitment to investigator-driven research. For research agencies, almost all sales are to the private sector, and the sequential funding approach of DoD is replaced by a wider competition among all Phase I and Phase II proposals in a given technical area.

   c. **Consequences of Different Approaches.** Contracts and grants carry with them important differences in approach and agency mindset, and consequently in outcomes, metrics, and evaluation benchmarks. Such differences permeate all aspects of the program, and underscore the need for a very flexible approach to program guidelines. The grants-oriented approach at NIH has included substantial flexibility in the application of deadlines for completion, award size, and supplemental funding among other program attributes. The contracts-driven approach at DoD has resulted in tight deadlines, more limited funding flexibility within the program, and program goals based on clear deliverables.

4. **Commendable Innovation in Program Operation.** Since the late 1990s, the SBIR program has seen considerable innovation in its operations, a trend that has accelerated during this study. As the list below shows, some SBIR program managers have increasingly utilized the flexibility inherent

[47]See the discussion of the challenges of early-stage finance in the Introduction to this volume.

in the structure of such a widely dispersed program to introduce a range of initiatives covering almost all aspects of SBIR program management.

a. These include innovations and initiatives related to:
    - Award size and duration.
    - Topic development.
    - Gap funding.
    - Phase III commercialization support.
    - Matching funds for commercialization.
    - Commercialization training for SBIR awardees.
    - Outreach into the small business community, especially to underrepresented states.
    - Increased evaluation and assessment activity.

b. While not all of these changes have been made at each agency, these changes taken as a whole do indicate that the program is increasingly benefiting from an innovative management culture and is becoming better adapted to serve its goals. These changes also suggest that there are likely to be further innovations and adaptations of the program to the changing needs of agencies' missions.

5. **Procedural Issues.** A review of topic development and selection procedures, which vary among agencies, indicated that in general, agencies work hard to ensure that applicants are treated fairly, and that topics are developed in line with agency needs. Specific recommendations for improving program management are provided in subsequent sections of this report.

6. **The Need for Regular Internal and External Evaluation.** While the initiatives taken by several agencies to adapt their program to serve their mission needs more effectively are welcome and their flexibility to do so is commendable, these efforts require periodic assessment to determine if these 'operational experiments' are successful and to identify the appropriate lessons learned. Moreover, some important innovations do not appear to be based on any detailed evaluation of problems with the existing mechanisms.

7. **SBA and SBIR.** The SBA has oversight responsibility for the eleven SBIR programs underway across the federal government. The agency is to be commended for its flexibility in exercising its oversight responsibilities. As noted above, the agencies adaptation of the program to fit their needs and methods of operation has proven fundamental to the program's success.

    a. **Modest Resources.** In general, SBA has been allocated modest resources to oversee SBIR. For example, efforts to build a unified database continue to encounter difficulties and resources are apparently a constraint.

b. **More Flexibility.** Despite the generally flexible approach, there is evidence from the agencies that in some cases, SBA has believed itself obliged to limit flexibility in the application of its guidelines to the program. For example, projects initially funded as a Phase I STTR cannot be shifted to a Phase II SBIR, even if circumstances make this the most effective way to proceed.[48]

I. **Venture Funding and SBIR.**

1. **Synergies.** There can often be useful synergies between angel and venture capital investments and SBIR funding; each of these funding sources tends to select highly promising companies and technologies. Moreover, an SBIR award provides information to potential investors by validating the technology concept and commercial potential of an innovation.

   a. **Angel Investment.** Angel investors often find SBIR awards to be an effective mechanism to bring a company forward in its development to the point where risk is sufficiently diminished to justify investment.[49]

   b. **Venture Investment.** Reflecting this synergy, initial NRC review indicates about 25 percent of the top 200 NIH Phase II award winners (1992-2005) have acquired some venture funding in addition to the SBIR awards.[50]

---

[48]The Small Business Technology Transfer (STTR) was established by Congress in 1992 with a similar statutory purpose as SBIR. A major difference between SBIR and STTR is that the STTR requires the small business to have a research partner consisting of a university, a Federally Funded Research and Development Center (FFRDC), or a qualified nonprofit research institution. In STTR, the small business must be the prime contractor and perform at least 40 percent of the work, with the research partner performing at least 30 percent of the work. The balance can be done by either party and/or a third party. Currently, the STTR set-aside is 0.3 percent of an agency's extramural R&D budgets.

[49]See the presentation, "The Private Equity Continuum," by Steve Weiss, Executive Committee Chair of Coachella Valley Angel Network, at the Executive Seminar on Angel Funding, University of California at Riverside, December 8-9, 2006, Palm Springs, CA. In a personal communication, Weiss points out the critical contributions of SBIR to the development of companies such as Cardio-Pulmonics. The initial Phase I and II SBIR grants allowed the company to demonstrate the potential of its products in animal models of an intravascular oxygenator to treat acute lung infections and thus attract angel investment and subsequently venture funding. Weiss cites this case as an example of how the public and private sectors can collaborate in bringing new technology to markets. Steve Weiss, Personal Communication, December 12, 2006.

[50]There may be substantially more venture funding than the NRC research has, as yet, revealed. The GAO report on venture funding within the NIH and DoD SBIR programs used a somewhat different methodology to identify firms with VC funding. As a result of the approach adopted, no conclusions can be drawn from the study as to whether firms identified as VC funded are in fact excluded from the SBIR program on ownership grounds. In addition, the number of VC-funded firms—reportedly

2. **Program Change.** During the first two decades of the program, some venture-backed companies participated in the program, receiving SBIR awards in conjunction with outside equity investments. During this lengthy period, no negative impact on the program's operation or effectiveness was apparent.

   In a 2002 directive, the Small Business Administration said that to be eligible for SBIR the small business concern should be "at least 51 percent owned and controlled by one or more individuals who are citizens of, or permanent resident aliens in, the United States, except in the case of a joint venture, where each entity to the venture must be 51 percent owned and controlled by one or more individuals who are citizens of, or permanent resident aliens in, the United States."[51] The effect of this directive has been to exclude companies in which VC firms have a controlling interest.[52,53]

   a. It is important to keep in mind that the innovation process often does not follow a crisp, linear path. Venture capital funds normally (but not always) seek to invest when a firm is sufficiently developed in terms of products to offer an attractive risk-reward ratio.[54] Yet even firms benefiting from venture funding may well seek SBIR awards as a

---

18 percent of all NIH firms receiving Phase II awards from 2001-2004—is considerably higher than suggested by preliminary NRC analysis. U.S. Government Accountability Office, *Small Business Innovation Research: Information on Awards made by NIH and DoD in Fiscal years 2001-2004*, GAO-06-565, Washington, DC: U.S. Government Accountability Office, 2006.

[51]Access the SBA's 2002 SBIR Policy Directive, Section 3(y)(3) at <*http://www.zyn.com/sbir/sbres/sba-pd/pd02-S3.htm*>.

[52]This new interpretation of "individuals" resulted in the denial by the SBA Office of Hearings and Appeals of an SBIR grant in 2003 to Cognetix, a Utah biotech company, because the company was backed by private investment firms in excess of 50 percent in the aggregate. Access this decision at <*http://www.sba.gov/aboutsba/sbaprograms/oha/allcases/sizecases/siz4560.txt*>. The ruling by the Administrative Law Judge stated that VC firms were not "individuals," i.e., "natural persons," and therefore SBIR agencies could not give SBIR grants to companies in which VC firms had a controlling interest. The biotechnology and VC industries have been dismayed by this ruling, seeing it as a new interpretation of the VC-small business relationship by SBA. See for example, testimony by Thomas Bigger of Paratek Pharmaceuticals before the U.S. Senate Committee on Small Business and Entrepreneurship, July 12, 2006.

[53]This paragraph is a correction of the text in the prepublication version released on July 27, 2007.

[54]The 1994-2004 period saw a decline in venture investments in seed and early stage and a concomitant shift away from higher-risk early-stage funding. The last few years have seen some recovery in early-stage finance. See the discussion of this point in the Introduction chapter. See also National Science Board, *Science and Technology Indicators 2006*, Arlington, VA: National Science Foundation, 2006. This decline is reportedly particularly acute in early-stage technology phases of biotechnology where the investment community has moved toward later-stage projects, with the consequence that early-stage projects have greater difficulty raising funds. See the testimony by Jonathan Cohen, founder and CEO of 20/20 GeneSystems, at the House Science Committee Hearing on "Small Business Innovation Research: What is the Optimal Role of Venture Capital," July 28, 2005.

means of exploring a new concept, or simply as a means of capitalizing on existing research expertise and facilities to address a health-related need or, as one participant firm explained, to explore product-oriented processes not "amenable to review" by academics who review the NIH RO1 grants.[55]

b. Some of the most successful NIH SBIR award winning firms—such as Martek—have, according to senior management, been successful only because they were able to attract substantial amounts of venture funding as well as SBIR awards.[56]

c. Other participants in the program believe that companies benefiting from venture capital ownership are essentially not small businesses and should therefore not be entitled to access the small percentage of funds set aside for small businesses, i.e., the SBIR Program. They believe further that including venture-backed firms would decrease support for high-risk innovative research in favor of low-risk product development often favored by venture funds.[57]

3. **Limits on Venture Funding.** The ultimate impact of the 2004 SBA ruling remains uncertain. What is certain is that no empirical assessment of its impact was made before the ruling was implemented. At the same time, the claims made by proponents and opponents of the change appear overstated.

   a. Preliminary research indicates that approximately 25 percent of the most prolific NIH SBIR Phase II winners have received VC funding; that some of these are now graduates of the program (having grown too large or left for other reasons), and some are also not excluded by the

---

[55]See the statements by Ron Cohen, CEO of Acorda Technologies, and Carol Nacy, CEO of Sequella Inc, at the House Science Committee Hearing on "Small Business Innovation Research: What is the Optimal Role of Venture Capital," July 28, 2005. Squella's Dr. Nacy's testimony captures the multiple sources of finance for the 17-person company (June 2005). They included—founder equity investments; angel investments; and multiple, competitive scientific research grants, including SBIR funding for diagnostics devices, vaccines, and drugs. SBIR funding was some $6.5 million out of a total of $18 million in company funding. Dr. Nacy argues that SBIR funding focuses on research to identify new products while venture funding is employed for product development.

[56]The biomedical context of this finding on venture capital funding reflects the recent attention to this issue in the NIH SBIR program.

[57]See the testimony by Jonathan Cohen, founder and CEO of 20/20 GeneSystems, at the House Science Committee Hearing on "Small Business Innovation Research: What is the Optimal Role of Venture Capital," July 28, 2005. In the same hearing Mr. Fredric Abramson, President and CEO of AlphaGenics, Inc., argues that "any change that permits venture owned small business to compete for SBIR will jeopardize biotechnology innovation as we know it today." For additional arguments against including VC-backed firms in the SBIR program, see testimony by Robert Schmidt at the House Science Committee Subcommittee on Technology and Innovation Hearing on "Small Business Innovation Research Authorization on the 25th Program Anniversary," April 26, 2007.

ruling because they are still less than 50 percent VC owned. Yet it is important to recognize that these companies may be disproportionately among the companies most likely to succeed—such as previous highly successful SBIR companies that were simultaneously recipients of VC funding.[58] What is not known is how many companies are excluding themselves from the SBIR program as a result of the ruling.

b. For firms seeking to capitalize on the progress made with SBIR awards, venture funding may be the only plausible source of funding at the levels required to take a product into the commercial marketplace. Neither SBIR nor other programs at NIH are available to provide the average of $8 million per deal currently characterizing venture funding agreements.[59]

c. For firms with venture funding, SBIR may allow the pursuit of high-risk research or alternative path development that is not in the primary commercialization path, and hence is not budgeted for within the primary development path of the company.[60]

4. **An Empirical Assessment.** As noted above, the SBA ruling concerning eligibility alters the way the program has operated during the period of this review (1992-2002) and, presumably, from the program's origin. Anecdotal evidence and initial analysis indicate that a limited number of venture-backed companies have been participating in the program. To better understand the impact of the SBA exclusion of firms receiving venture funding (resulting in majority ownership), the NIH recently commissioned an empirical analysis by the National Academies. This is a further positive step towards an assessment culture and should provide data necessary to illuminate the ramifications of this ruling.[61]

---

[58] Included in this group of firms are highly successful SBIR companies like Invitrogen, MedImmune, and Martek. For discussion of the factors affecting the returns to venture capital organizations, including incentive and information problems and the role venture funds have played in supporting a limited number of highly successful firms, see Paul Gompers and Josh Lerner, *The Venture Capital Cycle*, Cambridge, MA: The MIT Press, 2000, Ch. 1.

[59] See National Venture Capital Association, Money Tree Report, November, 2006. The mean venture capital deal size for the first three quarters of 2006 was $8.03 million. This trend has been accelerated by the growth of larger venture firms. See Paul Gompers and Josh Lerner, *The Venture Capital Cycle*, op. cit.

[60] Firms that have used SBIR in this manner include Neurocrine and Illumina. The latter indicated in interviews that these alternative paths later become critical products that underpinned the success of the company. See also Dr. Nacy's testimony cited above.

[61] This research will address questions such as—which NIH SBIR participating companies have been or are likely to be excluded from the program as a result of the 2002 rule change on Venture Capital Company ownership; and what is the likely impact of the 2002 ruling had it been applied during the 1992-2006 timeframe and what is its probable current impact? Key variables will include the presence and amount of SBIR support, the receipt of venture capital funding or other outside funding, and output measures including those related to commercialization and knowledge generation.

II. NRC STUDY RECOMMENDATIONS

The recommendations in this section are designed to improve the operation of the nation's SBIR program. They complement the core findings that the program is largely addressing its legislative goals—that significant commercialization is occurring, that the awards are making valuable additions to nation's stock of scientific and technical knowledge, and that SBIR is developing products that apply this knowledge to agency missions, and hence to the nation's needs.

A. **Preserve Program Flexibility.** Agencies, SBA, and the Congress should seek to ensure that any program adjustments made should not reduce the program's flexibility.

1. The SBIR program is effective across the agencies partly because a "one-size-fits-all" approach has not been imposed.
2. This flexible approach should be continued, subject to appropriate monitoring, across the departments and agencies in order to adapt the program to their evolving needs and to improve its operation and output.

B. **Regular Evaluations Are Needed.**

1. **The SBIR Program Is Currently Not Sufficiently Evidence-based.** Some aspects of the SBIR program have been subject to internal and external evaluation.[62] Nonetheless, insufficient data collection, analytic capability, and reporting requirements, together with the decentralized character of the program, mean that there is limited ability to make connections between program outcomes and program management and practices. There are several issues. These include:

   a. **Limited Collection of Data and Tracking of Outcomes.** There is insufficient tracking of awards, as well as a widespread lack of systemic data collection and analysis, needed to support an evidence-based program culture and to enable program managers to be more effective in their stewardship of the program.

   b. **Limited Analyses and Use of Metrics.** Changes to the program need to be better understood and the impact tracked. Because of the factors listed above, decisions that affect the operation of the program are necessarily taken with limited data and analysis, and without clear

[62] DoD has been particularly diligent in seeking outside evaluations of its program's operations and innovations, e.g., the Fast Track initiative, evaluated in 2000 and again in 2006 in a further study by the NSF. Similarly, NIH recently commissioned outside evaluation by the NRC to assess the impact of the SBA ruling on venture-funded firms.

benchmarks or metrics for assessing the success or failure of a given initiative.[63]

c. **Inadequate Management Funding.** Agencies administering SBIR can improve the management of the program by providing greater support for data collection and analysis. The SBIR program involves a significant amount of public funding, some $1.851 billion in FY2005. The necessary resources to manage the program and evaluate outcomes should be provided.[64]

2. **Single Benchmarks of Achievement Are Problematic.** The current study represents a major research effort in an area with limited previous research. Differences in both program structures and agency objectives mean that no simple benchmark for success is applicable across the entire program. Existing studies, such as those by GAO, have generally been limited in scope, focusing in most cases only on a limited aspect of the program.[65] This is not surprising, given the size and diversity of the program and the difficulties in developing the comprehensive data required for accurate assessments of program outcomes (described above). To address this assessment challenge, regular, multidimensional, multiagency evaluation is required.

---

[63]The recent increase in the average and median size of Phase I and Phase II awards provides an example. NIH staff has offered a number of different justifications, but no systematic analysis or review appears to have preceded such an important change to the program.

[64]According to recent OECD analysis, the International Benchmark for program evaluation of large SME and Entrepreneurship Programs is between 3 percent (for small programs) and 1 percent (for large-scale programs). See Organization for Economic Cooperation and Development, "Evaluation of SME Policies and Programs: Draft OECD Handbook," OECD Handbook CFE/SME(2006)17. See the recommendation for supplementary management funding.

[65]See U.S. General Accounting Office, *Federal Research: Small Business Innovation Research Participants Give Program High Marks*, Washington, DC: U.S. General Accounting Office, 1987; U.S. General Accounting Office, *Federal Research: Assessment of Small Business Innovation Research Program*, Washington, DC: U.S. General Accounting Office, 1989; U.S. General Accounting Office, *Small Business Innovation Research Program Shows Success But Can Be Strengthened*, RCED–92–32, Washington, DC: U.S. General Accounting Office, 1992; U.S. General Accounting Office, *Federal Research: DoD's Small Business Innovation Research Program*, RCED–97–122, Washington, DC: U.S. General Accounting Office, 1997; U.S. General Accounting Office, *Federal Research: Evaluations of Small Business Innovation Research Can Be Strengthened*, RCED–99–198, Washington, DC: U.S. General Accounting Office, 1999. U.S. General Accountability Office, *Small Business Innovation Research: Information on Awards made by NIH and DoD in Fiscal Years 2001 through 2004*, GAO-06-565, Washington, DC: U.S. General Accountability Office, 2006. U.S. General Accountability Office, *Small Business Innovation Research: Agencies Need to Strengthen Efforts to Improve the Completeness, Consistency, and Accuracy of Awards Data*, GAO-07-38, Washington, DC: U.S. General Accountability Office, 2006. See also Bruce Held, Thomas Edison, Shari Lawrence Pfleeger, Philip Anton, and John Clancy, *Evaluation and Recommendations for Improvement of the Department of Defense Small Business Innovation Research (SBIR) Program*, Arlington, VA: RAND National Defense Research Institute, 2006.

3. **Developing a Culture of Evaluation.** Recent agency initiatives toward more and better evaluation, e.g., at DoD, NIH, and NSF, represent positive steps. However, considerable further assessment is needed. The lack of an adequate assessment effort is, in some cases, the result of shortfalls in the resources needed to carry out assessments, and in others because agencies do not have the support of agency leadership for a sustained effort of evaluation and assessment.

4. **Recommendations.** More evaluations of the SBIR program's outcomes are needed. Such assessment can provide the data needed to inform program management decisions and to enhance understanding of the program's contributions.

   a. **Annual Reports.** Top agency management should make a direct annual report to Congress on the state of the SBIR program at their agency. The report should describe management goals and commitment to the program, and measures taken to enhance the implementation of the program. This report should include a statistical appendix, which would provide data on awards, processes, outcomes, and survey information.

   b. **Internal Evaluation.** Agencies should be encouraged—and funded—to develop improved data collection technologies and evaluation procedures. Where possible, agencies should be encouraged to develop interoperable standards for data collection and dissemination.[66]

   c. **External Evaluation.** Agencies should be directed to commission an external evaluation of their SBIR programs on a regular basis.

   d. **Special Topic Studies.** A number of further studies on special topics might help to improve program outcomes. These include:

      i. The role of venture capital funding in the SBIR program.

      ii. The linkages and synergies between state programs and SBIR.

      iii. Improving links between SBIR and the acquisition community at DoD: Effects of matching fund programs.

      iv. Tools for reducing cycle time.

      v. Gap funding mechanisms and results.

      vi. Phase III transition incentives.

      vii. Impact of increased award size.

---

[66]The agencies should consider providing data from a range of sources—including agency databases, agency surveys, the patent office, and bibliographic databases, along with data from award recipients themselves.

viii. Assessment of selection mechanisms.

ix. Evaluation of professional commercialization consultancy services.

C. **Additional Management Resources Are Needed.**

1. To enhance program utilization, management, and evaluation, the program should be provided with additional funding for management and evaluation.

2. Effective management and evaluation requires adequate funding.[67] An evidence-based program requires high quality data and systematic assessment. As noted above, sufficient resources are not currently available for these functions.

3. Increased funding is needed to provide effective oversight, including site visits, program review, systematic third-party assessments, and other necessary management activities.

4. In considering how to provide additional funds for management and evaluation, there are three ways that this might be done:

   a. Additional funds might be allocated internally, within the existing budgets of the services and agencies, as the Navy has done.

   b. Funds might be drawn from the existing set-aside for the program to carry out these activities.

   c. The set-aside for the program, currently at 2.5 percent of external research budgets, might be marginally increased, with the goal of providing management resources necessary to maximize the program's return to the nation.[68] At the same time, increased resources should

[67]As noted above, a recent OECD report, the International Benchmark for program evaluation of large SME and Entrepreneurship Programs is between 3 percent for small programs and 1 percent for large-scale programs. See "Evaluation of SME Policies and Programs: Draft OECD Handbook," op. cit.

[68]Each of these options has its advantages and disadvantages. For the most part, over the past 25 years, the departments, institutes, and agencies responsible for the SBIR program have not proved willing or able to make additional management funds available. Without direction from Congress, they are unlikely to do so. With regard to drawing funds from the program for evaluation and management, current legislation does not permit this and would have to be modified. This would also limit funds for awards to small companies, the program's core objective. The third option, involving a modest increase to the program, would also require legislative action and would perhaps be more easily achievable in the event of an overall increase in the program. In any case, the Committee envisages an increase of the "set-aside" of perhaps 0.03 percent to 0.05 percent on the order of $35 million-40 million per year, or roughly double what the Navy currently makes available to manage and augment its program. In the latter case (0.05 percent), this would bring the program "set aside" to 2.55 percent, providing modest resources to assess and manage a program that is approach-

not be used to create new or separate management. Increased resources should go to enable existing management to enhance their efforts with respect to the program. The Committee recommends this third option.

D. **Improve Program Processes.**

1. The processing periods for awards vary substantially by agency, and appear to have significant effects on recipient companies.[69]

   a. Cycle time is a concern for small high-technology businesses with limited resources. Such small companies often have a substantial "burn rate" (outgoing monthly cash flow), and the speed with which projects are approved and funded can have a material impact on small business ventures.

   b. Numerous entrepreneurs interviewed for this study identified long cycle time—that is between initial application and Phase I awards, and between Phases—as a significant problem. Some case study interviewees noted that these gaps discourage otherwise excellent companies from applying to the SBIR program.[70]

   c. Processing schedules that are adequate for managing larger-scale research projects or academic research may not be appropriate for the small, high-technology firms that apply for SBIR awards.

2. **Agencies Should Be Encouraged to Employ a Full Range of Tools to Reduce the Time Between Applications and Awards.**

   a. **Monitoring and Reporting.** As part of a new and expanded annual report, agencies should closely monitor and report on cycle times for each element of the SBIR program: topic development and publication, solicitation, application review, contracting, Phase II application and selection, and Phase III contracting. Agencies should also specifically report on initiatives to shorten decision cycles.

   b. **Best Practice Adoption.** Agencies should seek to ensure that they adopt best practice in all areas of the program. Specifically, agencies should consider agency procedures that the Committee has identified that are expected to shorten the time from idea to implementation.

---

ing an annual spend of some $2 billion. Whatever modality adopted by Congress, the Committee's call for improved management, data collection, experimentation, and evaluation may prove moot without the benefit of additional resources.

[69]Drawn from case study interviews and gap data from the NRC Phase II Survey.

[70]See case studies at all agencies. In addition, the NRC Phase II Survey indicated that most companies had stopped work in between phases, and many had experienced other funding gaps. See Chapter 5: Program Management.

Below we have provided a list of potential initiatives. Not all of these options would be appropriate for every agency, and agencies should be encouraged to experiment and assess the impact of such changes. Identified practices include:

i. Multiple annual solicitations.

ii. Broader topic definitions.

iii. Solicitations where no topic is necessary, or where some funding is available to projects outside the topic framework.

iv. Rapid review and decisions on applications.

v. Clear and tight deadlines for completing reviews.

vi. Permitting resubmission, and providing rejected applicants with reviews fast enough to ensure that they can apply again during the next cycle.

vii. Use of electronic tools to augment or replace traditional selection meetings.

viii. Rapid electronic delivery of results to applicants.

ix. Clear and tight deadlines for contracts and grants negotiations to be completed. This may require more resources and better training for awards management staff.

x. Alignment of Phase I/Phase II applications, so that completion of Phase I comes at an appropriate time for Phase II application.

xi. Fast Track mechanisms such as that operating at NIH, where Phase I and Phase II are considered as a single application.

xii. Phase IIB awards which add additional agency resources on the condition that equal, or larger, amounts of private funding are also obtained.

xiii. Further training of technical points of contact (TPOCs) and contracting officer's technical representatives (COTARs) to help ensure that they understand the importance of shortening cycle times.

E. **Agencies Should Take Steps to Increase the Participation and Success Rates of Woman- and Minority-owned Firms in the SBIR Program.**

1. **Encourage Participation.** Develop targeted outreach to improve the participation rates of woman- and minority-owned firms, and strategies to improve their success rates. These outreach efforts and other strategies

should be based on causal factors determined by analysis of past proposals and feedback from the affected groups.[71]

2. **Encourage Emerging Talent.** The number of women and, to a lesser extent, minorities graduating with advanced scientific and engineering degrees has been increasing significantly over the past decade, especially in the biomedical sciences. This means that many of the woman and minority scientists and engineers with the advanced degrees usually necessary to compete effectively in the SBIR program are relatively young and may not yet have arrived at the point in their careers where they own their own companies. However, they may well be ready to serve as principal investigators (PIs) and/or senior co-investigators (Co-Is) on SBIR projects. Over time, this talent pool could become a promising source of SBIR participants.

3. **Improve Data Collection and Analysis.** The Committee also strongly encourages the agencies to gather the data that would track woman and minority firms as well as principal investigators (PIs), and to ensure that SBIR is an effective road to opportunity. The success rates of woman and minority PIs and Co-Is is the traditional measure of their participation in the non-SBIR research grants funded by agencies like NIH and NSF. It can also be a measure of woman and minority participation in the SBIR program.

F. **Agencies that Do Not Now Have an Independent Advisory Board that Draws Together Senior Agency Management, SBIR Managers, and Other Stakeholders as Well as Outside Experts Should Consider Creating a Board to Review Current Operations and Achievements and Recommend Changes to the SBIR Program.**

1. The purpose of such an advisory board is to provide a regular monitoring and feedback mechanism that would address the need for upper management attention, and encourage internal evaluation and regular assessment of progress towards definable metrics.

2. Each agency's annual program report could be presented to the board. The board would review the report that would include updates on program progress, management practices, and make recommendations to senior agency officials in charge.

3. The board could be assembled on the model of the Defense Science Board

[71]This recommendation should not be interpreted as lowering the bar for the acceptance of proposals from woman- and minority-owned companies, but rather as assisting them to become able to meet published criteria for grants at rates similar to other companies on the basis of merit, and to ensure that there are no negative evaluation factors in the review process that are biased against these groups.

BOX 2-1
Some Examples of Agency Best Practices

A major strength of the SBIR program is its flexible adaptation to the diverse objectives, operations, and management practices at the different agencies. In some cases, however, there are examples of best practice that should be examined for possible adoption by other agencies. Examples of these best practices include:

DoD: The Pre-Release Period. DoD announces the contents of its upcoming solicitations some time before the official start date of the solicitation. By attaching detailed contact information, prospective applicants can talk directly to the technical officers in charge of specific topics. This helps companies determine whether they should apply and gives the prospective applicant a better understanding of the agency needs and objectives. This informal approach provides an efficient mechanism for information exchange. Federal Acquisition Regulations prevent such discussion after formal release.

DoD: Help Desk and Web Support. DoD maintains an extensive and effective Web presence for the SBIR program, which can be used by companies to resolve questions about their proposals. In addition, DoD staffs a help desk aimed at addressing nontechnical questions. This is appreciated by companies, and is strongly supported by program staff because it reduces the burden of process oriented calls on technical staff.

DoD: Commercialization Tracking. DoD's approach requires companies with previous Phase II awards to enter data into a commercialization tracking database each time these companies apply for SBIR awards at DoD. The database captures outcomes (both financial, such as sales and additional funding by source, and other benefits resulting from SBIR; e.g., public health, cost savings, improved weapon system capability, etc.) from these companies for all their previous SBIR awards, including those at other agencies. It also captures information on firm size and growth since entering the SBIR program, as well as the percent of annual revenue derived from SBIR awards. These historical results of prior awards are then used in proposal evaluation.

Non-DoD agencies should consider adapting both this approach and the DoD technology and contributing to the DoD database. This would provide a unified tracking system. Adaptations could be made to track additional data for specific agencies, but this would provide a cost-effective approach to enhance data collection on award outcomes.

NIH: Resubmission. In many cases, changes in topics between successive solicitations may not allow for resubmission of Phase I proposals. However, agencies should

(DSB) or perhaps the National Science Foundation's Advisory Board.[72] In any case, it should include senior agency staff and the Director's Office on an *ex officio* basis, and bring together, *inter alia*, representatives

[72]The intent here is to use the DSB or the NSF Board as a model, not something necessarily to be copied exactly.

consider selective use of resubmission of Phase II as a useful way of improving application quality.

NIH: Investigator-driven Topics. While DoD and NASA have focused primarily on topics designed to address specific agency needs, NIH accepts applications that are not directly linked to an explicit topic. Other agencies might consider putting aside some portion of SBIR funds to encourage investigator driven research that does not fall within the "standard" solicitations.

Multiple Agencies: Gap-reduction Strategies. The agencies have, to different degrees, recognized the importance of reducing funding gaps. While details vary, best practice would involve the development of a formal gap-reduction strategy with multiple components covering application, selection, contract negotiation, the Phase I-Phase II gap, and support after Phase II.

NSF: Phase IIB, DoD Phase II Enhancement. The matching fund approach adopted by NSF for Phase IIB and by DoD for Phase II Enhancement might be explored at other agencies. The NSF matching requirement represents an important tool for helping companies to enter Phase III at non-procurement agencies. The DoD funding match by acquisition programs provides a transition link into Phase III contracts with the agency.

DoD-Navy: Technology Assistance Program. The Navy has developed the most comprehensive suite of support mechanisms for companies entering Phase III, and has also developed new tools for tracking Phase III outcomes. These are important initiatives, and other components and agencies should consider them carefully.

NIH: Program Flexibility. NIH is unique in the range of award size, the use of supplementary funding, the competing continuation awards (now called Phase II Competing Renewals Awards) that provide funding for post-Phase II activities, and the use of no cost extensions for awards. While this flexibility may require additional review at NIH, as recommended in the NIH Report, it is also an effective way for agencies to meet the varying needs of small research companies.

DoE: Commercialization Assistance. DoE has the longest running commercial assistance program, and a broad menu of possible services. These have now been widened to include Phase I recipients. This extended range of offerings seems well adapted to improve fits between the varied Departmental needs of small business capabilities.

NASA: Innovative Electronic Interface. NASA has pioneered the development of an innovative electronic interface designed to help applicants navigate the complex process of applying for awards and managers track and assess program activities.

from industry (including award recipients), academics, and other experts in program management.

G. **Preserve the Basic Program Structure.**

1. **The Phase I Bypass Proposal.** Some agency staff and recipient companies have suggested that promising research has been excluded from

Phase II funding because all Phase II recipients must first receive a Phase I award.[73] To address this concern, some program participants and agency staff have suggested that consideration be given to changing the requirement that SBIR recipients apply for and receive a Phase I award before applying for Phase II. They suggest that the application of this requirement excludes promising research that could help agencies meet their congressionally mandated goals.

2. **Recommendation.** The Committee recommends that fundamental changes to the Phase I, Phase II program structure should not be made.

   a. Permitting companies to apply directly to Phase II could significantly change the program. In particular, it could shift the balance of both awards and funding significantly away from Phase I toward Phase II.

   b. Every additional Phase II award represents funding approximately equivalent to 7.5 Phase I awards. If "direct to Phase II" were as attractive to applicants as proponents suggest, it might become a significant component of the program. This, in turn, could make a very substantial difference to funding patterns in SBIR, to the detriment of Phase I.[74]

   c. This change could be detrimental to the program and should not be undertaken. Phase I is an important component in achieving congressional objectives and deemphasizing it should—at a minimum—require prior review and assessment.

H. **Encourage Program Experimentation.**

1. **The Fast Track Program**, whose principal objective is to reduce the funding gap between Phase I and Phase II of the SBIR award process, began as an experiment at the Department of Defense. It was subsequently evaluated by the NRC, which found that the Fast Track Program increases the effectiveness of the SBIR program at DoD by encouraging the commercialization of new technologies. That report, published in 2000, recommended that DoD consider expanding the Fast Track Program within appropriate services, organizations, and agencies within DoD. That report also called for cross-agency comparisons of the impact of Fast Track, noting that it could prove useful for the continued refinement of the program.[75] DoD has since commissioned the NRC to conduct a follow up study of the Fast Track Program.

---

[73]Discussions with NIH SBIR program managers, June 13, 2006.

[74]Phase I awards may have particular importance in meeting noncommercial objectives of the program, for example, helping academics to transition technologies out of the lab into startup companies.

[75]National Research Council, *The Small Business Innovation Research Program: An Assessment of the Department of Defense Fast Track Initiative*, op. cit., pp. 33-39.

2. **Funding Beyond Phase II.** SBIR is designed with two funded Phases (I and II) and a third nonfunded Phase III. Projects successfully completing Phase II are expected to be well placed to obtain Phase III funding from non-SBIR sources, e.g., through procurement contracts, to attract private investment, or to deploy products directly into the marketplace.

    a. While this three-phase approach has often proved successful, there are widely recognized difficulties in obtaining funding for Phase III, and consequently in achieving external commercial success or agency take-up in the agencies with major procurement responsibilities.[76]

    b. Some agencies have sought, with the approval of SBA, to experiment with SBIR funding beyond Phase II in order to improve the commercialization potential of SBIR-funded technologies. NIH is now experimenting with tools for supporting projects through the early stages of regulatory review.

    c. The NSF Phase IIB initiative and the NIH Competing Continuation Awards are positive examples that might well be adapted elsewhere.

3. **Other Improvements**. The agencies should be encouraged to develop program modifications to address possible improvements to the SBIR program, including but not limited to the pilot projects suggested above. (See Box 2-1.) SBA should make every effort to accommodate agency initiatives.

4. **Evaluation of Change.** Agencies should equally ensure that program modifications are designed, monitored, and evaluated, so that positive and negative results can be effectively evaluated. Where feasible and appropriate, the agencies should conduct scientifically rigorous experiments to evaluate their most promising SBIR approaches—experiments in which SBIR program applicants, awardees, and/or research topics are randomly assigned to the new approach or to a control group that participates in the agency's usual SBIR process.[77]

---

[76]For a discussion of Phase III commercialization issues, see National Research Council, *SBIR and the Phase III Challenge of Commercialization*, op. cit.

[77]Randomized evaluations are recognized as a highly rigorous means for evaluating the effectiveness of a strategy or approach across many diverse fields. When properly conceived and executed, they enable one to determine to a high degree of confidence whether the new approach itself, as opposed to other factors, causes the observed outcomes. Such randomized evaluations are recognized as an effective means for evaluating an intervention's effectiveness across many diverse fields. When properly conceived and executed, they enable one to determine with some confidence whether the intervention itself, as opposed to other factors, causes the observed outcomes.

See, for example, U.S. Department of Education, "Scientifically-Based Evaluation Methods: Notice of Final Priority," *Federal Register*, 70(15):3586-3589, January 25, 2005; the Food and Drug Administration's standard for assessing the effectiveness of pharmaceutical drugs and medical devices, at 21 C.F.R. §314.12; "The Urgent Need to Improve Health Care Quality," Consensus statement of the

I. **Readjust Award Sizes.**

1. **Erosion of Award Value.**

a. The real value of SBIR awards, last increased in 1995, has eroded due to inflation. It is now 14 years since the Congress increased the standard limits on the size of Phase I and Phase II awards. Many agency staff and award recipients have noted that in the face of continuing low but steady inflation, the amount of research funded by these awards has declined. Calculated using the NIH BioMedical Research and Development Price Index, the real value of the awards has declined by just over 35 percent.[78]

b. Given that Congress did not indicate that the real value of awards should be allowed to decline, this erosion in the value of awards needs to be addressed. However, many recipients contacted over the course of this study expressed concern that increases in the size of award would lead to a corresponding decline in the number of awards, if the SBIR budget is held constant.[79]

c. Steady increases in the number of applications for SBIR awards at most agencies (NIH is a notable recent exception) may be evidence that declining real award size has not led to a corresponding decline in interest among potential applicants.

2. **Recommendations.**

a. **Phase I Increase.** In order to restore the program to the approximate initial levels, adjusted for inflation, the Congress should consider making a one-time adjustment that would give the agencies latitude to increase the guidance on standard Phase I awards to $150,000.

b. **Phase II Increase.** In order to restore the program to the approximate initial levels, adjusted for inflation and able to sustain quality applications, the Congress should consider making a one-time adjustment that

---

Institute of Medicine National Roundtable on Health Care Quality, *Journal of the American Medical Association*, 280(11):1003, September 16, 1998; American Psychological Association, "Criteria for Evaluating Treatment Guidelines," *American Psychologist*, 57(12):1052-1059, December 2002; Society for Prevention Research, *Standards of Evidence: Criteria for Efficacy, Effectiveness and Dissemination*, April 12, 2004, at <*http://www.preventionresearch.org/softext.php*>; Office of Management and Budget, *What Constitutes Strong Evidence of Program Effectiveness*, pp. 4-8, accessed at <*http://www. whitehouse.gov/omb/part/2004_program_eval.pdf, 2004*>.

[78]Accessed at <*http://officeofbudget.od.nih.gov/PDF/BRDPI_2_5_07.xls*>.

[79]For example, a 25 percent increase in the size of Phase II awards implies a 25 percent reduction in the number of Phase II awards, all other things being equal—or a much larger reduction in the number of Phase I awards.

would give the agencies latitude to increase the guidance on standard Phase II awards to approximately $1,000,000.[80]

c. **Flexibility in Size of Awards.** It should be stressed that recommendations are intended as *guidance* for standard award size. The SBA should continue to provide the maximum flexibility possible with regard to award size and the agencies should continue to exercise their judgment in applying the program standard. Recognizing agencies' need for flexibility to meet new technical or mission challenges expeditiously—such as countermeasures for biological threats or Improvised Explosive Devices—strict limits on the minimum or maximum amount for awards should be avoided. The agencies, as well, should consider whether pilot programs offering larger (or indeed smaller) awards might be useful in some cases, and whether close evaluation of large awards made in the past could help guide future practice.[81]

d. **Flexibility—Duration.** Contracting agencies might wish to experiment with a limited increase in flexibility, especially for Phase I where radical changes in technical direction are more likely to be required. Innovation implies uncertainty, so a more flexible timeline might help to attract approaches that are more technically innovative.

e. **Flexibility in Additional Funding**. Agencies might consider providing supplementary awards to small businesses with promising technologies at the discretion of the program manager. Supplementary awards are made at NIH, and the NSF has, for some time, implemented a Phase IIB and Phase IIB+ program. The results of these initiatives should be assessed, not least because the results may be of significant value to other agencies.[82]

[80]Recognizing that these values are not identical, the erosion by inflation of the amounts available for Phase I appear more constraining than for Phase II. One may argue that, for parity, the Phase II should be raised to approximately $1,125,000; the trade-offs involve attracting adequate numbers of quality proposals, and providing sufficient resources to enable firms to actually carry out the necessary work, while also recognizing the impact on the number of awards. While recognizing that there is necessarily an arbitrary element in fixing these amounts, the Committee is confident that a $1,000,000 Phase II award will maintain the program's attraction to innovative small businesses.

[81]Program innovations that offer flexibility in award size introduced at NIH appear promising, though there are currently no data to determine whether this approach has resulted in more successful projects. Several agencies, NIH, NSF, and NASA, have successfully made their own judgments as to the most appropriate award sizes to meet their agency needs and continue to draw adequate "deal flows." Agencies with smaller SBIR programs, not reviewed here, have adopted much smaller award sizes.

[82]See National Research Council, *An Assessment of the SBIR Program at the National Institutes of Health*, op. cit., Awards chapter. See also the discussion of the Phase IIB program in National Research Council, *An Assessment of the SBIR Program at the National Science Foundation*, op. cit.

This *de facto* focus on the Phase III transition is especially important to those agencies that generally do not acquire the products of the firms receiving their awards.

f. **Transition.** Larger award sizes should be phased in over two to three years in order to avoid a sudden reduction in the number of awards. The impact of the larger awards may be mitigated by the larger R&D budgets currently under consideration in the Congress. As suggested above, the transition in award sizes should be left to the individual agencies to implement in light of their current needs, capabilities, and priorities.

J. **Understanding and Managing Firms Winning Multiple Awards.**

1. **Unproductive Multiple Award Winners Do Not Appear to Be a Major Problem, Although Continued Monitoring Is Necessary.**

a. The common perception about the prevalence of mills in the SBIR program—i.e., that they have captured a large percentage of the awards, that they rarely commercialize, and that they do not meet agency research needs—is not substantiated by the evidence. Nonetheless, belief that the phenomenon of unproductive "mills," that win repeated awards, is widespread. This perception has led to calls for a limit on the number of awards that a single firm can win.

b. The NRC data suggest that firms that repeatedly win SBIR awards but fail to commercialize are not a significant problem at any of the granting agencies.[83] As the list of the top five Phase II winners (those with projects with $10 million or more in sales or investment) found in Table 2-1 further illustrates, most companies with multiple awards (but not all) are actually producing significant commercial products, often with substantial sales.

2. **Characteristics of Multiple Award Winners.** Companies that do receive large numbers of awards at the contracting agencies (notably DoD) share two characteristics:

a. If the award process is relatively objective, then one can argue that multiple award winners are high performers, able to address and meet agency needs, just as leading universities and major defense companies repeatedly win contracts for work based on a proven track record.

b. As they grow in size, the dependence of frequent award winners on SBIR declines. They typically cease to fit the SBIR-dependent model of the mills outlined above.

[83] See Section 3.8 and Section 4.4.4.2 for related data and discussion.

3. **Recommendations.** Because there is little evidence to suggest that multiple award winners are, in themselves, a problem, efforts to limit their participation seem misplaced. Limiting participation by unproductive winners is, of course, appropriate and already occurs. Setting a necessarily arbitrary limit on the number of awards seems neither necessary nor desirable in light of the contributions made by these firms, nor is it likely to be effective as firms could easily change names or spin out subsidiaries.

   a. **Improved Monitoring.** This phenomenon, however, should be monitored. As agencies improve their evaluation capacity, one area of focus could be multiple award winners. DoD, for example, should continue its efforts to track both awards to multiple winners and outcomes to ensure that they generate sufficient value to the defense mission, recognizing the varied contributions made by program participants. Better information on numbers of awards and especially on outcomes would, of course, enable management to make judgments that are more informed.

   b. **The Use of Quotas to Reduce Applications by Multiple Winners Should Be Discontinued.**

      i. In the case of multiple award winners who qualify in terms of the selection criteria, the acceptance/rejection decision should be based on both the strength of their applications and their past performance rather than on the number of grants received or applications made. Firms able to provide quality solutions to solicitations should not be excluded, *a priori*, from the program except on clear and transparent criteria (e.g., quality of research and/or past performance).

      ii. Agencies should avoid imposing quotas unless there are compelling reasons to do so. Arbitrary limitations run the risk of limiting innovative ideas and of unnecessarily restricting opportunities among prospective principal investigators in the larger, eligible small companies to provide high quality solutions to the government.[84]

III. SUMMARY: AN EFFECTIVE PROGRAM

In summary, the program is proving effective in meeting congressional objectives. It is increasing innovation, encouraging participation by small companies in federal R&D, providing support for small firms owned by minorities and women, and resolving research questions for mission agencies. Should the Congress wish to provide additional funds for the program in support of these objectives, with the programmatic changes recommended above, those funds could be employed effectively by the nation's SBIR program.

[84]Creare, Inc., has received a significant number of SBIR awards and now has 21 patents resulting from SBIR-funded work. Staff members have published dozens of papers. The firm has licensed technologies including high-torque threaded fasteners, a breast cancer surgery aid, corrosion preventative coverings, an electronic regulator for firefighters, and mass vaccination devices (pending). Products and services developed at Creare include thermal-fluid modeling and testing, miniature vacuum pumps, fluid dynamics simulation software, network software for data exchange. A significant technical accomplishment was achieved with the NCS Cryocooler used on the Hubble Space Telescope to restore the operation of the telescope's near-infrared imaging device.

In some cases, the company has developed technical capabilities that have remained latent for years until a problem arose for which those capabilities were required. The cryogenic cooler for the Hubble telescope is an example. The technologies that were required to build that cryogenic refrigerator started being developed in the early 1980s as one of Creare's first SBIR projects. Over 20 years, Creare received over a dozen SBIR projects to develop the technologies that ultimately were used in the cryogenic cooler.

TABLE 2-1 Top Five Phase II Winners: Projects with $10 Million or More in Sales or Investment

Firm Name	Number of Phase II Awards	Commercialization Achievement Index (CAI)	Agency	Award Year	Project Title	Total Sales ($)	Total Investment ($)
Foster-Miller, Inc.	260	80	DARPA	1994	Lemmings—A Swarming Approach to Shallow Water Mine Field Clearance	63,190,000	1,150,000
			ARMY	1996	Hard Faced Lightweight Composite Armor Materials for Helmets	45,110,015	1,500,000
			Air Force	1992	Novel Aircraft Door/Panel Fastening System	30,379,559	0
			NASA	1992	Liquid Crystal Polymers for CTE Matched PWBs	11,203,111	2,295,421
			Air Force	1996	In Situ Biological Remediation of Hydrazine Spills Using Gel Encapsulated Enzymes	3,000,000	17,466,000
			Air Force	1996	Integrated In-Line Wear and Lubricant Condition Sensor	1,868,000	12,126,105
			Air Force	1994	Novel Infrared F/O Sensor for In Field Identification of Hazardous Waste Solutions	267,505	49,545,842
			NAVY	1997	Electric Energy Absorption System for Aircraft Recovery (EARS)	0	25,388,571
Physical Optics Corporation	234	70	NASA	1994	Embedded Distributed Moisture Sensor for Nondestructive Inspection of Aircraft Lap Joints	$11,040,000	$1,732,000
			AF	1996	Visually Transparent Films	$10,120,000	$962,000
			NSF	1997	The Photochemical Generation of Planar Waveguides in Sol-Gel Glasses	$1,077,000	$10,500,000

continued

TABLE 2-1 Continued

Firm Name	Number of Phase II Awards	Commercialization Achievement Index (CAI)	Agency	Award Year	Project Title	Total Sales ($)	Total Investment ($)
Physical Sciences, Inc.	205	30	Air Force	1985	Space Shuttle Plasma Interactions	20,711,803	1,409,755
			Air Force	1997	Airborne Hyperspectral Imager	14,570,673	999,525
			Air Force	1997	Micromachined Sensor Array for Optical Meas. of Surface Pressure	44,872	18,614,672
Creare, Inc.	190	100	NASA	1986	Reliable Long-Lifetime Closed-Cycle Cryocooler for Space	1,212,529	39,417,294
			NASA	1989	Three-Phase Inverter for High Speed Motor Drive	37,403	15,662,926
Radiation Monitoring Devices, Inc.	140	95	HHS/NIH	1991	Intraoperative Imaging Probe for Delineation of Tumors	132,500,000	1,600,000
			HHS/NIH	1993	A New Portable XRF System for Lead Paint Analysis	91,343,990	4,339,000
			NASA	1985	Portable Nuclear Cardiac Ejection Fraction Monitor	26,600,000	4,500,000
			HHS/NIH	1992	Emboli Monitor to Reduce Surgical Neurological Deficits	25,500,000	470,000
			HHS/NIH	1986	Sensitive Probe for Intra-Operative Bone Scanning	19,700,000	1,000,000
			DoE	1984	Advanced Avalanche Photodiode for Positron Emission Tomography	588,000	21,870,000

SOURCE: U.S. Small Business Administration, Tech-Net Database.

3

Statistics of SBIR Awards

3.1 INTRODUCTION

This chapter outlines the information available on Small Business Innovation Research (SBIR) program applications and awards, with a focus on the five study agencies: Department of Defense (DoD), National Institutes of Health (NIH), National Aeronautics and Space Administration (NASA), National Science Foundation (NSF), and Department of Energy (DoE). The objective is to provide a quantitative overview of award patterns, while highlighting both some data issues and some areas of possible concern.

There is no authoritative source for SBIR data. While the U.S. Small Business Administration (SBA) maintains the Tech-Net Database, which includes data submitted by the agencies, agency databases do not exactly match SBA databases (as we shall see in Section 3.7). And a more detailed review of agency data indicates that there are some areas where current data collection efforts are insufficient (see below).

We begin by providing an overview of the SBIR program, which today totals about $2 billion annually, divided between Phase I and Phase II. We consider each separately.

3.2 SBIR PHASE I AWARDS

In FY2005, all SBIR awarding agencies made a total of 4,208 Phase I awards (see Table 3-1).

Fifty-six percent of FY2005 Phase I awards were made by DoD. NIH accounted for a further 20 percent. Overall, the number of Phase I awards has fallen

TABLE 3-1 Phase I Awards

Year	Number of Phase I Awards												Total
	DHS	DoC	DoD	DoE	DoT	ED	EPA	HHS	NASA	NRC	NSF	USDA	
1992		19	1,078	196	29	23	41	636	346	17	169	45	2,599
1993		19	1,292	167	49	29	34	659	384	8	255	55	2,951
1994		39	1,394	209	33	20	35	596	413	12	307	62	3,120
1995		72	1,245	196	20	27	47	666	309	8	295	72	2,957
1996		38	1,367	167	30	18	27	560	349		249	62	2,867
1997		63	1,519	194	29	50	35	764	339		252	72	3,317
1998		45	1,276	204	21	41	37	725	344		211	77	2,981
1999		33	1,398	185	17	40	47	931	290		243	84	3,268
2000		40	1,380	292	25	0	58	1,233	287		320	125	3,760
2001		37	1,606	310	38	55	54	1,293	307		291	127	4,118
2002		54	2,308	328	19	72	63	1,339	267		342	125	4,917
2003		53	2,319	323	15	31	51	1,393	312		510	122	5,129
2004	97	65	2,078	257	12	54	43	1,135	291		236	101	4,369
2005	80	34	2,364	259	8	22	38	862	297		152	92	4,208
Total	177	611	22,624	3,287	345	482	610	12,792	4,535	45	3,832	1,221	50,561

NOTE: This is a correction of the prepublication version released on July 27, 2007.
SOURCE: U.S. Small Business Administration, Tech-Net Database; and National Aeronatuics and Space Administration.

from a peak of 5,129 in FY2003, to 4,208 in FY2005, a decline of 18 percent (see Figure 3-1).

This overall decline primarily reflects a reduction in the number of Phase I awards at NIH (down 38 percent), NASA (down 4.8 percent), and DoE (down 19.8 percent). Since the number of Phase II awards has increased by 16.1 percent over the past few years (2002-2005), agencies appear to be making a shift in emphasis away from Phase I and toward Phase II.

The growth and then decline in the number of Phase I awards are not directly reflected in the amounts spent by the agencies on Phase I. Funding levels for Phase I peaked in 2004, and funding committed to Phase I has fallen by only 8.6 percent (see Figure 3-2). This suggests that while the number of awards has been falling, the size of awards has been growing. And this is in fact the case (see Figure 3-3).

The decline in award size from 1999 to 2001 may be related to the introduction of Phase I options at DoD, which reduced the size of "standard" awards to $70,000 in most cases. The sharp increase in award size from 2001-2004 (up 56 percent in three years) has been driven by changes at the two largest programs.

At DoD, average award size increased from $71,056 to $90,508 (up 27 percent). At HHS (the home agency of NIH), average award size increased from

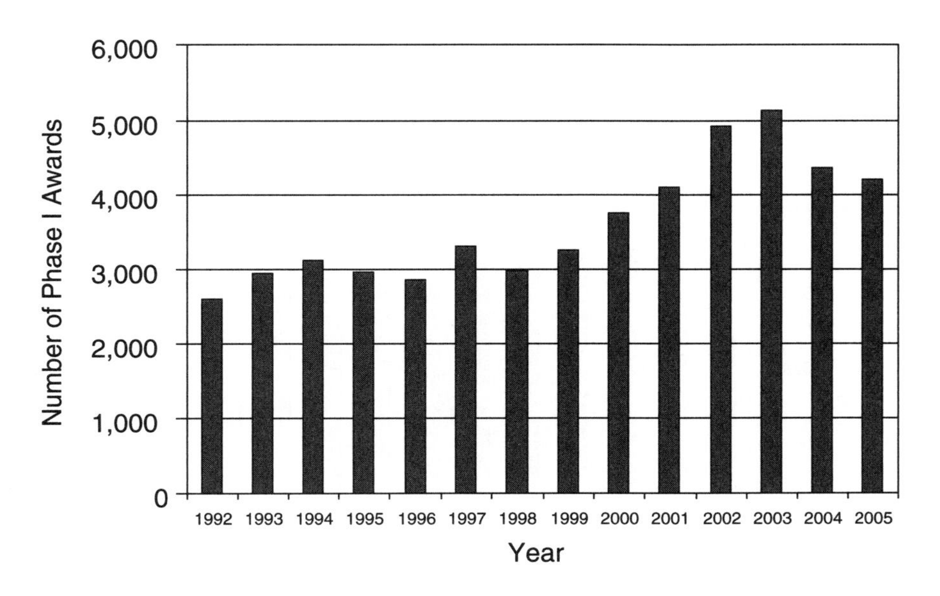

FIGURE 3-1 Total Phase I awards, 1992-2005.
NOTE: This is a correction of the prepublication version released on July 27, 2007.
SOURCE: U.S. Small Business Administration, Tech-Net Database.

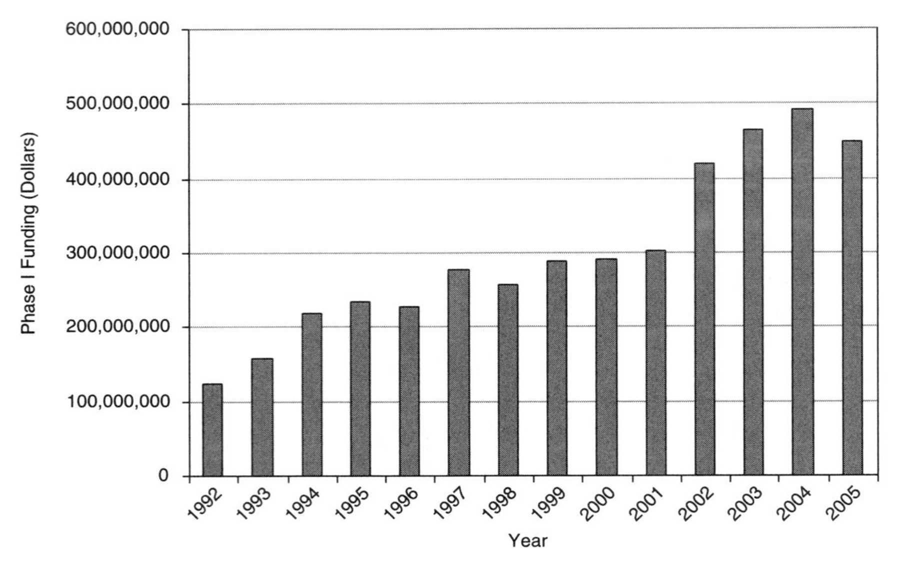

FIGURE 3-2 Funding for Phase I, 1992-2005.
SOURCE: NIH Awards Database.

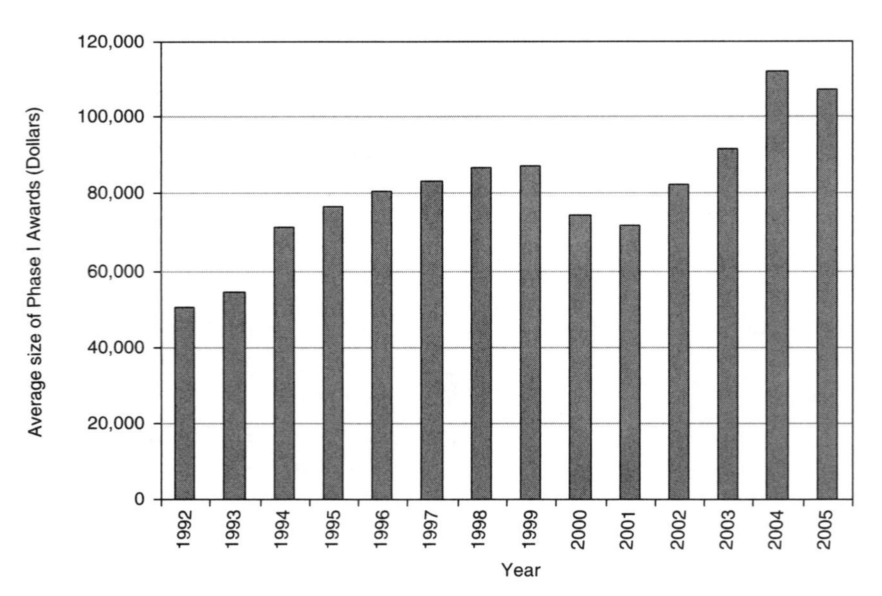

FIGURE 3-3 Average size of Phase I awards, 1992-2005.
SOURCE: NIH Awards Database.

$85,937 to $179,108 (up more than 108 percent). Doubling the average size of awards in three years, while reducing the number of awards by 37 percent, represents a fundamental shift within the NIH program.

3.3 SBIR PHASE II AWARDS

The number of Phase II awards made since 1992 does not fluctuate as dramatically, as shown in Figure 3-4.

Here the increase in numbers is more gradual, and the decline both more recent and less pronounced, with awards falling in only one year since 1999, and by only 6.6 percent.

Overall funding for Phase II grew considerably faster than the number of awards. It is up by 97.5 percent between 2000 and 2004 (see Figure 3-5).

As with Phase I, this reflects growth in the average size of Phase II awards. The award data appear to show an increase of about 40 percent between 2001 and 2005 (see Figure 3-6).

3.4 OVERSIZED AWARDS—NIH

The formal size limits for awards were last changed in the 1992 reauthorization, to $100,000 for Phase I and $750,000 for Phase II. More recently, NIH

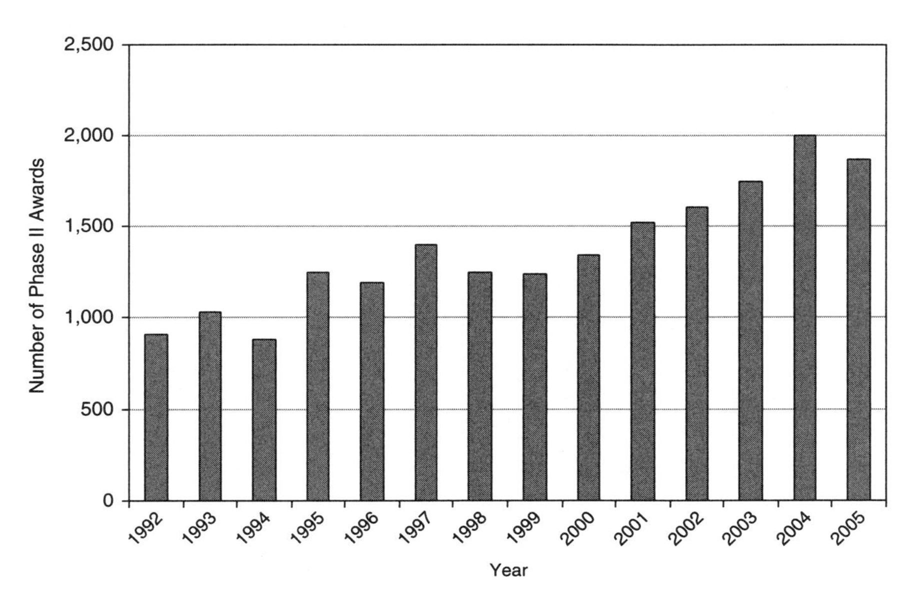

FIGURE 3-4 Number of Phase II awards, 1992-2005.
SOURCE: NIH Awards Database.

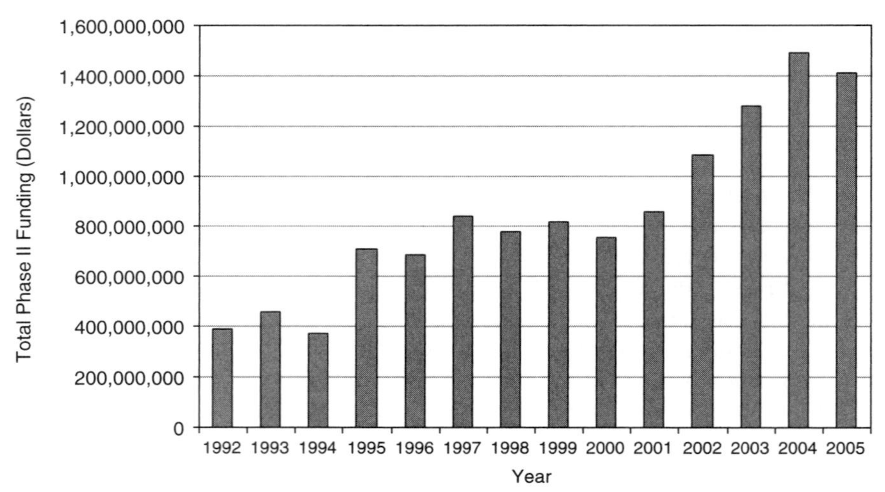

FIGURE 3-5 Funding for Phase II, 1992-2005.
SOURCE: NIH Awards Database.

expanded the definition of what is possible in terms of award size for SBIR. The agency has awarded at least one $3 million Phase I award, and SBA data indicates that 18 Phase I awards have been for more than $1 million (all since 2000).

While DoD has always had a few Phase II awards of more than $2 million (28 over 13 years), NIH has more recently made such awards in greater numbers:

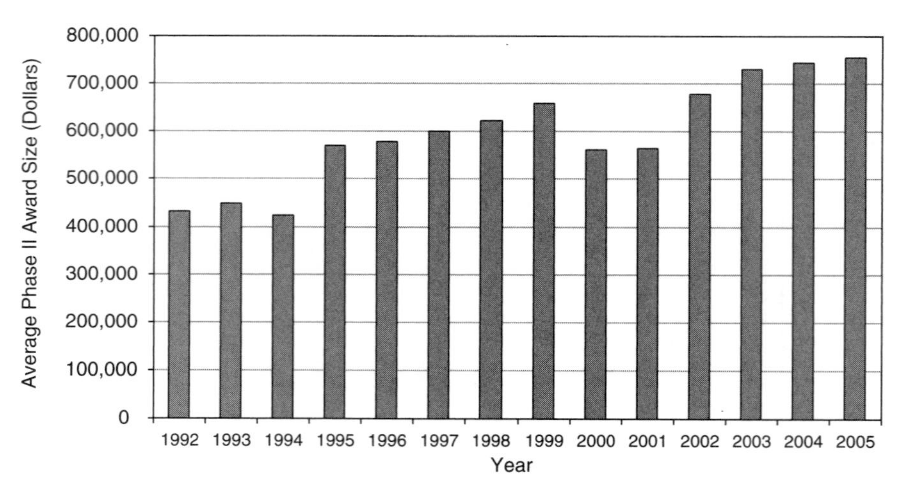

FIGURE 3-6 Average size of Phase II awards, 1992-2005.
SOURCE: NIH Awards Database.

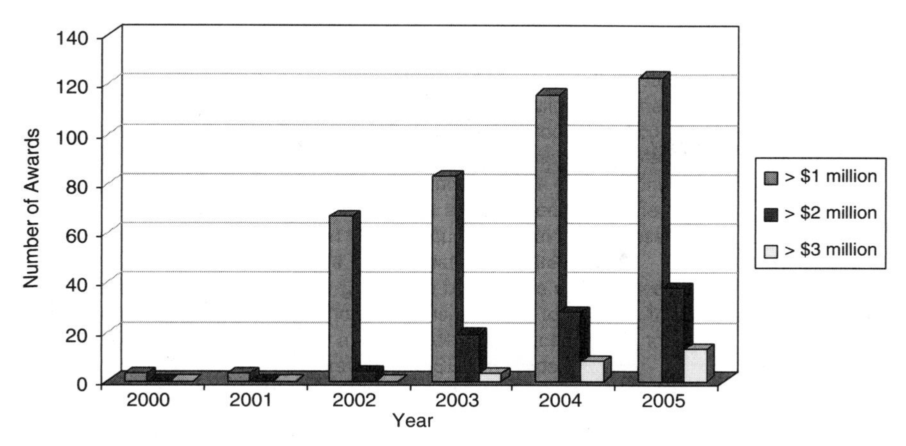

FIGURE 3-7 Over-sized Phase II awards at NIH.
SOURCE: NIH Awards Database.

91 overall, with all except 6 in the last three years. In FY2005, NIH made 38 awards (out of 373) of over $2 million. In FY2005, more than 10 percent of NIH Phase II awards were for more than $2 million, and a further 85 (22 percent) were for more than $1 million. Thirteen awards were for more than $3 million.

Sharp growth in award size in recent years has led to an average Phase II award size of $1,089,294 in FY2005—almost 40 percent above the mandated maximum of $750,000. The agency has been given a blanket waiver from SBA for this departure, but the change does raise some questions.[1]

These data and other aspects of award size at NIH are explored in more detail in the NIH volume.

3.5 APPLICATIONS AND SUCCESS RATES

As SBA does not provide information on the number of applications and success rates, we have used data from the agencies for this purpose. Figure 3-8 shows success rates (awards as a percentage of applications) by agency.

The data show that in general, success rates have fluctuated around 15 percent, with the exception of NIH, where Phase I success rates have averaged 23.9 percent since 1994. Neither at NIH nor at NSF has the recent increase in the amount of overall funding for SBIR been reflected in growing success rates. On the contrary, success rates at NIH have fallen since 2002, reflecting sharp growth in the number of applications, shown in Figure 3-9. Although agency-specific

[1]This is addressed in more detail in the report focused on NIH. See National Research Council, *An Assessment of the SBIR Program at the National Institutes of Health*, Charles W. Wessner, ed., Washington, DC: The National Academies Press, 2009.

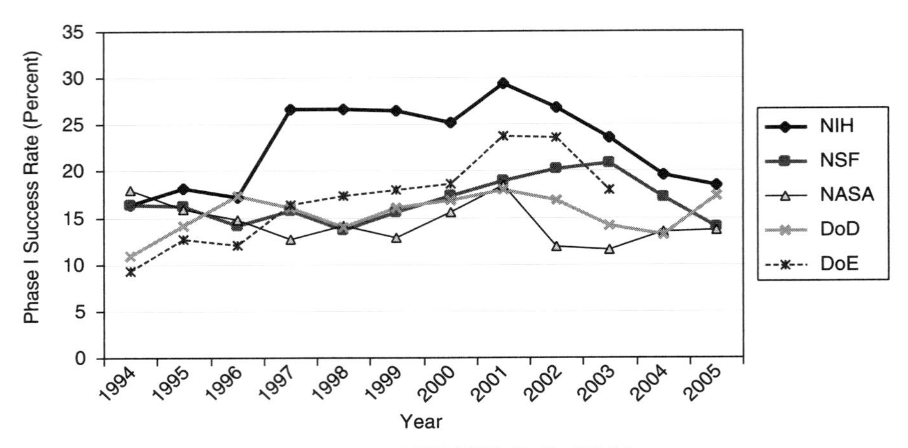

FIGURE 3-8 Phase I success rates at NSF, NIH, DoD, NASA.
SOURCE: Agency data.

features are also likely to have an impact, the number of applications appears, in general, to be more closely related to changes in the high-tech economy than to any other broad factor. Notably, the rise in Phase I applications in 2002 appears correlated to the collapse of the venture capital bubble that reached its peak in 2000. The correlation suggests that as private venture capital became scarce following the collapse of the venture capital bubble (Figure 3-10), small businesses looked to SBIR as a means to continue funding for their innovative activities.

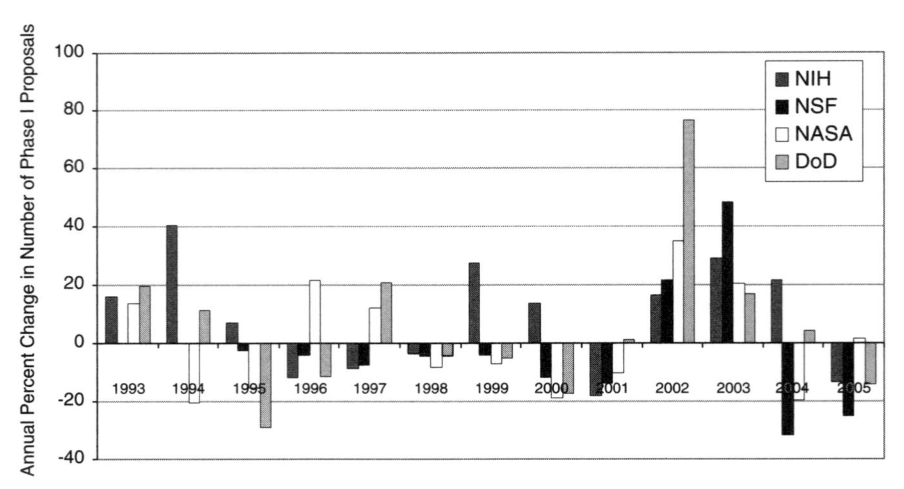

FIGURE 3-9 Annual change in the number of Phase I applications.
SOURCE: Agency data.

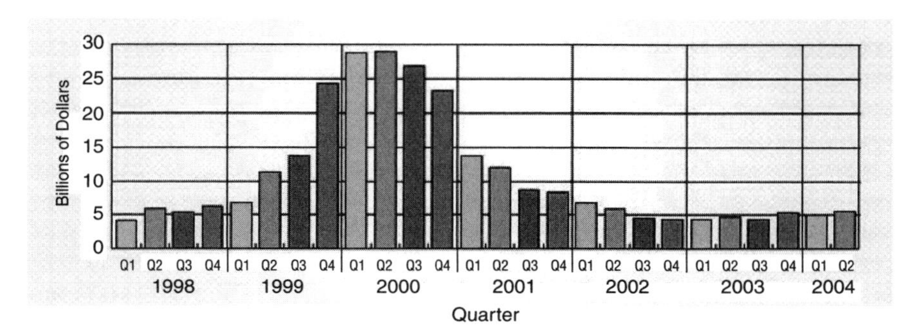

FIGURE 3-10 The venture capital bubble.
SOURCE: PriceWaterhouseCoopers, Thomson Venture Economics, National Venture Captial Association.

3.6 GEOGRAPHICAL DISTRIBUTION

The geographical distribution of awards reflects but does not mirror the distribution of science and engineering talent.

For Phase I, six states received more than 2,000 Phase I awards between 1992 and 2005. Together, these six states accounted for 54 percent of all awards during the study period (see Table 3-2).

Conversely, eight states received fewer than 100 Phase I awards, and together these states accounted for 1.1 percent of all awards.

A similarly skewed distribution is found when examining Phase II awards. Again, the top six states—winning at least 700 Phase II awards each—accounted for 54 percent of all Phase II awards (although it should be noted that this is not the same group of top winning states as Phase I). (See Table 3-3.)

The lowest-winning eight states accounted for 1.03 percent of all awards.

Many variables affect the geographical distribution of applications and awards. These include the populations of the state, the number of eligible high-

TABLE 3-2 States with 2,000 or More Phase I Awards

State	Percent of All Phase I Awards
California	20.8
Massachusetts	14.5
Virginia	5.5
Maryland	5.1
New York	4.2
Texas	4.0
Total	54.0

SOURCE: U.S. Small Business Administration.

TABLE 3-3 Top Six Phase II-Winning States

State	Percent of All Phase II Awards
California	21.0
Massachusetts	14.4
Virginia	5.8
Maryland	4.8
Colorado	4.6
Ohio	3.7
Total	54.4

SOURCE: U.S. Small Business Administration.

tech companies, the science and engineering talent in the state, state expenditures related to R&D, total R&D conducted, and the number of research universities.[2]

Analysis from the NIH Report[3] shows that if Phase I awards were expressed per 1,000 life scientists employed, the top "winning" states would change dramatically (see Table 3-4).

Massachusetts, Maryland, Virginia, and California remain in the top ten winners for both gross number of awards and awards per 1,000 life scientists, but New York and Texas—two other "top six" states for gross number of awards, fall into the lower half of the state rankings per 1,000 life scientists, while Oregon and Connecticut move into the top five states.

It is important to look at state award levels, not by assessing gross data, but by placing that data in the context of the scientific and engineering talent available: It does not make sense to criticize that there are insufficient Phase II awards in a given state if there are few scientists and engineers qualified to apply for one.

[2]From Fiscal Year 1993 through Fiscal Year 1996, companies in one-third of the states received 85 percent of the program's awards, largely because companies in these states submitted the most proposals. Companies from California and Massachusetts won the highest number of awards. To broaden the geographic distribution of awards, agencies have made efforts to encourage the submission of proposals from companies in states with fewer awards. U.S. General Accounting Office, *Federal Research: Evaluations of Small Business Innovation Research Can Be Strengthened*, RCED–99–198, Washington, DC: U.S. General Accounting Office, 1999. See also U.S. Small Business Administration, "An Analysis of the Distribution of SBIR Awards by States, 1983-1996." Washington, DC: Small Business Administration, 1998. The SBA report found that one-third of the states received 85 percent of all SBIR awards and SBIR funds. It also found that, with some exceptions, SBIR awards were related closest to the R&D resources in each state and the total R&D conducted in dollars.

[3]National Research Council, *An Assessment of the SBIR Program at the National Institutes of Health*, op. cit.

TABLE 3-4 Phase I Awards by State at NIH, per 1,000 Life Scientists Employed, 2003

	Life & Physical Sciences	Phase I Awards	Phase I Awards per 1,000 L&PS
New Hampshire	1,480	14	9.5
Vermont	850	6	7.1
Massachusetts	20,380	140	6.9
Maryland	17,910	90	5.0
Oregon	5,870	23	3.9
Connecticut	5,670	22	3.9
California	64,390	248	3.9
Virginia	13,030	40	3.1
Ohio	15,100	45	3.0
Iowa	3,130	9	2.9
Colorado	11,710	33	2.8
Indiana	4,070	11	2.7
Rhode Island	1,580	4	2.5
Michigan	9,390	23	2.4
Arizona	5,580	13	2.3
Nevada	2,510	5	2.0
Wyoming	1,510	3	2.0
Delaware	2,020	4	2.0
Minnesota	11,200	22	2.0
Washington	16,940	33	1.9
New Mexico	3,200	6	1.9
Wisconsin	11,220	21	1.9
New Jersey	17,530	32	1.8
Utah	5,060	9	1.8
South Carolina	4,610	8	1.7
Maine	1,830	3	1.6
Pennsylvania	25,080	41	1.6
District of Columbia	5,210	8	1.5
Oklahoma	3,350	5	1.5
North Dakota	1,420	2	1.4
New York	30,330	41	1.4
North Carolina	17,770	24	1.4
Missouri	9,240	12	1.3
Kansas	3,910	5	1.3
Florida	19,440	24	1.2
Alabama	5,170	6	1.2
Texas	42,440	49	1.2
Illinois	18,300	21	1.1
Arkansas	2,700	3	1.1
Louisiana	5,540	5	0.9
Nebraska	3,920	3	0.8
Kentucky	2,660	2	0.8
Alaska	2,800	2	0.7
South Dakota	1,420	1	0.7
Georgia	11,410	8	0.7
Tennessee	7,130	4	0.6
Hawaii	1,790	1	0.6
Montana	2,790	1	0.4
Idaho	3,100	1	0.3
Mississippi	3,650	1	0.3
West Virginia	2,510	0	0.0
Average			3.2

SOURCE: NIH data; and National Science Board, *Science and Engineering Indicators 2005*, Arlington, VA: National Science Foundation, 2005.

3.7 GAUGING PARTICIPATION BY WOMEN AND MINORITIES

One congressional objective for the program concerns support for woman and minority firms. Understanding current trends is more complex than it might appear.[4]

A major problem concerns data collection and monitoring. NRC analysis revealed a systemic failure by NIH to capture this data effectively from applications forms during several recent years. At the request of the NRC, NIH has now re-entered that data for 2002-2005; and it is believed that the data prior to 2000 are accurate.[5] At DoD, the problem lies with the companies: Spot checks have indicated that companies with many applications often label them inconsistently: a firm can label some applications as woman-owned, some as minority-owned, some as neither—sometimes all in the same year. While DoD makes an effort to correct these errors once the application has become an award, this means that the data for woman and minority success rates at DoD are not reliable.

Finally, there are inconsistencies described earlier with discrepancies between SBA and agency data.

Still, bearing these points in mind, we use the SBA data in this section to provide comparability.

Figure 3-11 shows the award data for woman-owned and minority-owned businesses reported by SBA. The data show that the number of Phase I awards to woman- and minority-owned businesses have increased, especially since 1998, and are up by 172 percent since then. The apparent surge in 2000 is not understood.

The data on minority-owned firms is different. It shows a much flatter trend—again marked by the surge in 2000, with no increase at all over the last five years, at a time when the number of awards made overall has increased sharply.

The number of awards provides only part of the story. Context is provided by the overall number of awards made, and hence by the shares of Phase I awards to woman- and minority-owned firms.

Overall, the percentage of Phase I awards going to minority-owned firms have been declining, while the share going to woman-owned firms has been growing (see Figure 3-13). The trends are broadly consistent, with woman-owned

[4]The traditional benchmark for this has been the inclusion of woman- and minority-owned businesses, addressed below. However, it is worth noting that support for woman and minority principal investigators may be another important way to view "inclusion," although currently data are not collected in a way that would address this possibility.

[5]Following discussions with the NRC staff, the NIH made an effort to recalculate the data for women and minority owners' participation in the SBIR program. In September, 2007, the NIH provided corrected data. However, apparent anomalies in the NIH data on the participation of women and minorities in 2001-2002 could not be resolved by the time of publication of this report. This qualification applies to all charts in this section of the report. (This is a correction of the text in the prepublication version released on July 27, 2007.)

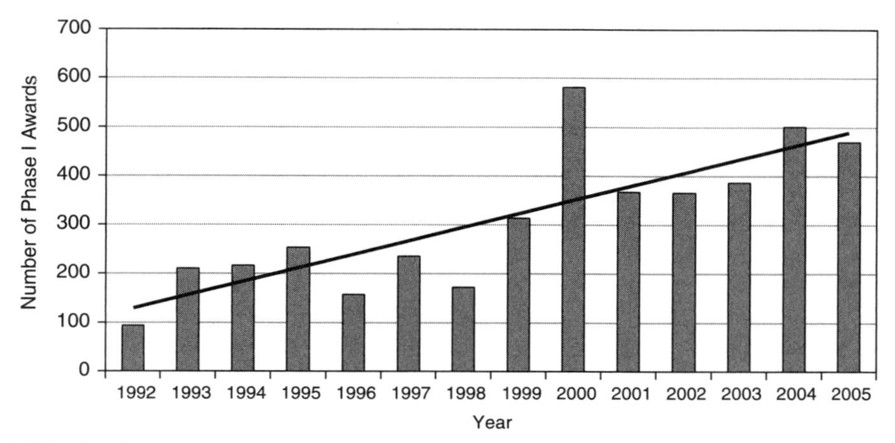

FIGURE 3-11 Phase I awards to woman- and minority-owned businesses.
SOURCE: U.S. Small Business Administration, Tech-Net Database.

firms growing from about 6.5 percent of all awards in 1998, to over 11 percent in 2004-2005. In the same period, award shares for minority-owned firms fell from just over 10 percent in 1992-1998 to just over 8 percent in 2001-2005, with a slight uptick in 2005.

In total, the share of awards to woman- and minority-owned firms has remained almost flat over the period, at 17 percent to 20 percent, while the absolute number of awards has increased substantially. Data for Phase II are comparable (see Figure 3-15).

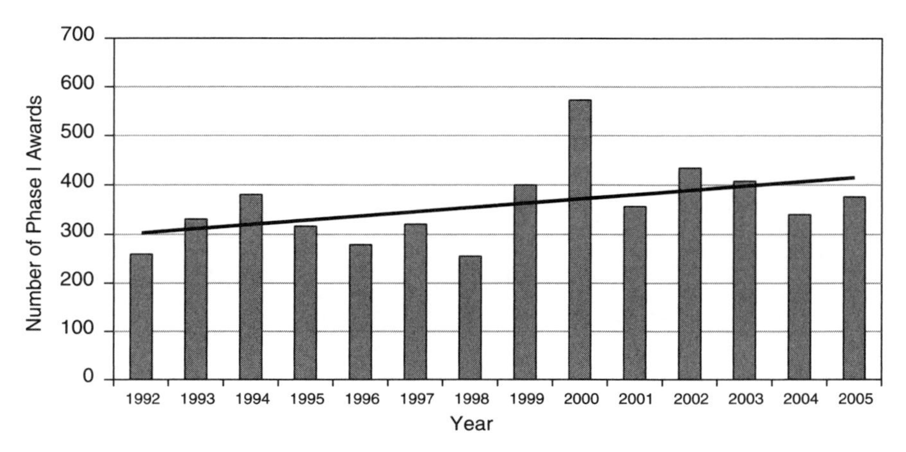

FIGURE 3-12 Phase I awards to minority-owned firms, 1992-2005.
SOURCE: U.S. Small Business Administration, Tech-Net Database.

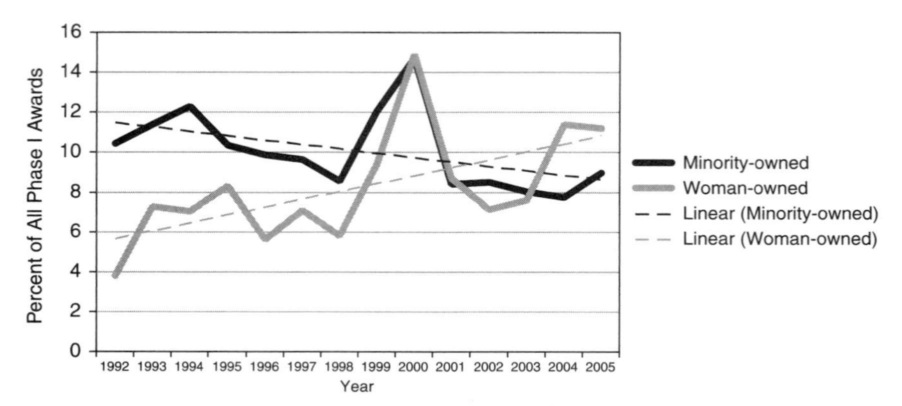

FIGURE 3-13 Woman- and minority-owned firms: Shares of Phase I awards, 1992-2005 (with trendlines).
SOURCE: U.S. Small Business Administration.

It is worth noting that while the share of awards going to woman-owned firms has increased, from 7.7 percent in 1999 to 11.7 percent in 2005, the percentage of female doctorates and scientists and engineers in the workplace has grown considerably faster, according to NSF data.

Finally, while the different patterns emerging from this analysis are interesting, detailed assessment requires that accurate data be collected on applications and on success rates (the number of applications per award for woman- and minority owned firms, and for all other firms). This will help to indicate whether

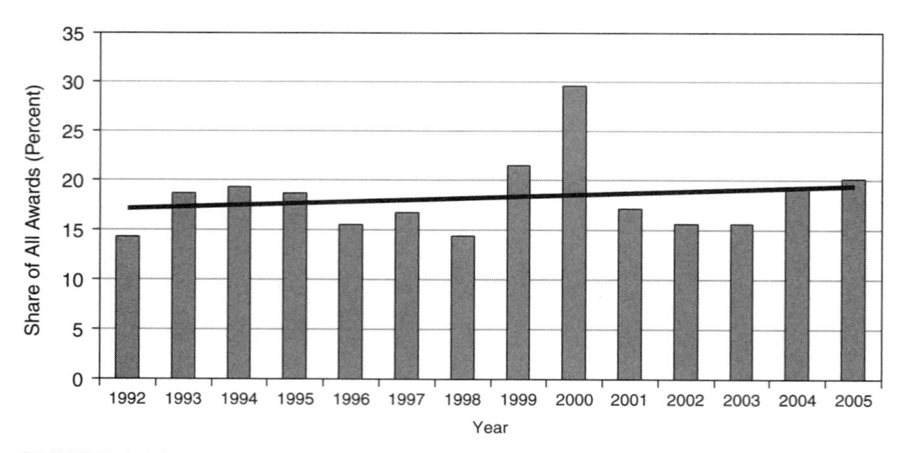

FIGURE 3-14 Share of woman- and minority-owned firms in Phase I awards (percent of total).
SOURCE: U.S. Small Business Administration.

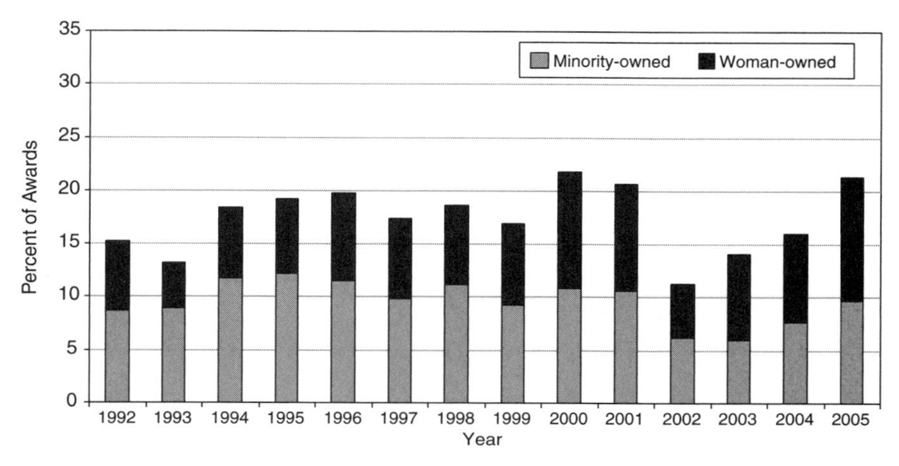

FIGURE 3-15 Phase II—award shares to minority- and woman-owned businesses, 1992-2005.
SOURCE: U.S. Small Business Administration, Tech-Net Database.

there are issues of outreach—insufficient applications from these groups—or whether there is something in the selection process that might need to be addressed (plenty of applications, lower success rates).

3.8 MULTIPLE-AWARD WINNERS AND NEW ENTRANTS

Data on multiple winners and on the percentage of awardees who are new to the program can be used as an indicator of the openness of the program—the extent to which good ideas and projects from outside the circle of previous winners can find a hearing and funding among the SBIR agencies and departments.

At the same time, the existence of multiple-award winners is sometimes used as an argument to justify limits on participation in the program. These proposed limits appear misconceived. They are discussed in detail in Chapter 5 of this volume, on Program Management.

Unfortunately, the absence of an effective centralized data management system for SBIR means that data on multiple-award winners are difficult to locate and verify and may not be especially accurate. Critics of multiple-award winners have tended to focus on DoD, partly because the numbers at DoD are higher than at other agencies. Analysis of NIH awards from 1992-2003 indicates that only two companies had as many as thirty Phase II awards during these 12 years.

By contrast, the DoD data show that ten firms had at least 75 Phase II awards.

TABLE 3-5 Multiple Award Winners, from the DoD Commercialization Database

Number of Phase II Awards per Firm	Number of Firms
≥125 projects	5
≥75 and <110 (no firms had between 111 and 124 projects)	5
≥50 and <75	17
≥25 and <50	77
≥15 and <25	101
>0 and <15	2,715

NOTES: The DoD commercialization database covers awards from all agencies; however, records are preponderantly for DoD projects. In the data collected from the agencies on Phase II awards made from 1992 to 2002, there were 2,257 firms who had at least 1 Phase II, but were not in the DoD database These firms were not included in this table. Of these 2,257 firms, only 6 had received 15 or more Phase II awards during the 10 years for which we received award data. Although inclusion of pre-1992 and post-2001 awards would have increased this number, the firms in the DoD data represent by far the majority of multiple winners.
SOURCE: Department of Defense, commercialization database.

3.8.1 New Program Participants

Much less attention is normally devoted to the extent to which the SBIR program is open to new entrants. Once again, this is not a metric tracked by SBA, but agencies provide their own figures on new entrants into the program.

At NIH, for example, "new" firms—not previously funded by the NIH SBIR program—accounted for about 60 percent of applicant firms in recent years, and for about 40 percent of Phase I winners.

Data from NSF are quite similar, indicating not only that the percentage of new winners ranges from 41 percent to 63 percent, but that the number has increased from 1999 to 2005, even though the number of past winners in the pool of potential applicants continues to increase every year.

TABLE 3-6 New Phase I Applicants and Winners at NIH

	Percent of Applicants Not Previously Funded	Percent of New Firms Among Funded Firms
2000	63.0	47.1
2001	58.0	39.8
2002	61.8	42.2
2003	63.7	42.8
2004	63.5	41.7
2005	61.0	35.8

SOURCE: National Institutes of Health.

TABLE 3-7 Percent of Winning Companies that Had Not Previously Received an SBIR Grant from NSF, 1996-2003

Fiscal Year	Previous Winners	New Winners	Percent New Winners
1996	98	109	52.7
1997	99	116	54.0
1998	83	100	54.6
1999	113	81	41.8
2000	97	108	52.7
2001	92	95	50.8
2002	99	133	57.3
2003	138	236	63.1

SOURCE: National Science Foundation.

DoD's SBIR Program Office reports that 37 percent of awards on average go to firms that have not previously won a DoD SBIR award.[6]

To conclude, the data on new winners strongly makes clear that the various SBIR programs are open to new ideas from new firms, even though in some cases individual firms have been able to win a significant number of awards, remaining active participants in a growing program of innovative research. A detailed assessment of this issue is provided in Chapter 5.

[6]See National Research Council, *An Assessment of the SBIR Program at the Department of Defense*, Charles W. Wessner, ed., Washington, DC: The National Academies Press, 2009; Chapter 2: Awards.

4

SBIR Program Outputs

4.1 INTRODUCTION

Congress has tasked the National Academies to assess whether and to what extent the Small Business Innovation Research (SBIR) program has met the congressionally-mandated objectives for the program, and to suggest possible areas for improvement in program operations. Congress has, over the years, identified a number of objectives for the program, and these mandated objectives can be summarized as follows:

- Supporting the commercialization of federally funded research.
- Supporting the agency's mission.[1]
- Supporting small business and, in particular, woman- and minority-owned businesses.
- Expanding the knowledge base.

Congress has not prioritized among the four objectives, although report language and discussions with congressional staff suggest that commercialization has become increasingly important to Congress. Still, it remains important to assess each of the four objectives, and each should therefore be taken as equally important in evaluating the achievements and challenges of the SBIR program. These four objectives help to define the structure and content of this chapter.

---

[1]Each of the SBIR-funding agencies has a different mission, described in the agency volumes of this study. For a review of the different ways the program is operated at the five agencies under review, see National Research Council, *SBIR: Program Diversity and Assessment Challenges*, Charles W. Wessner, ed., Washington, DC: The National Academies Press, 2004.

Assessing program outcomes against these four objectives entails numerous methodological challenges. These challenges are discussed in detail in the National Research Council's (NRC) Methodology Report.[2]

4.1.1 Compared to What?

Assessment usually involves comparison—comparing programs and activities, in this case. Three kinds of comparison seem possible: with other programs at each agency, between SBIR programs at the various agencies, and with early-stage technology development funding in the private sector, such as venture capital activities. Yet, the utility of each of these three types of comparison is limited.

Other award programs at the agencies have fundamentally different objectives, such as promoting basic research through grant programs (at the National Institutes of Health [NIH] and the National Science Foundation [NSF]), developing capacity (awards for university infrastructure), or training. There are often no other dedicated programs for innovative small businesses. And no other programs for small businesses have as a primary goal the commercial exploitation of research. This fundamental difference in objectives makes it difficult to usefully compare an SBIR program with other programs at the relevant agency.

Comparisons between SBIR programs at different agencies appear superficially more useful, but must be regarded with considerable caution. As discussions in Chapter 1 of this volume and in the separate agency volumes indicate, the widely differing agency missions have shaped the agency SBIR programs, focusing them on different objectives and on different mechanisms and approaches. Agencies whose mission is to develop technologies for internal agency use via procurement—notably the Department of Defense (DoD) and the National Aeronautics and Space Administration (NASA)—have a quite different orientation from agencies that do not procure technology and are instead focused on developing technologies for use outside the agency.

Finally, SBIR might be compared with venture capital (VC) activities, but there are important differences. VC funding is typically supplied later in the development cycle when innovations are in, or close to, market—most venture investments are made with the expectation of an exit from the company within three years. VC investments are typically larger than SBIR awards—the average investment made by VC firms in a company was over $7 million in 2005,

[2]National Research Council, *An Assessment of the Small Business Innovation Research Program: Project Methodology*, Charles W. Wessner, ed. Washington, DC: The National Academies Press, 2004, pp. 20-21. For a broader discussion of the scope and limitations of surveys by the University of Michigan Survey Research Center, see Robert M. Groves et al., *Survey Methodology*, Wiley-IEEE, 2004.

BOX 4-1
Multiple Sources of Bias in Survey Response

Large innovation surveys involve multiple sources of bias that can skew the results in both directions. Some common survey biases are noted below.[a]

- **Successful and more recently funded firms are more likely to respond.** Research by Link and Scott demonstrates that the probability of obtaining research project information by survey decreases for less recently funded projects and it increased the greater the award amount.[b] Nearly 40 percent of respondents in the NRC Phase II Survey began Phase I efforts after 1998, partly because the number of Phase I awards increased, starting in the mid-1990s, and partly because winners from more distant years are harder to reach. They are harder to reach as time goes on because small businesses regularly cease operations, are acquired, merge, or lose staff with knowledge of SBIR awards.
- **Success is self reported.** Self-reporting can be a source of bias, although the dimensions and direction of that bias are not necessarily clear. In any case, policy analysis has a long history of relying on self-reported performance measures to represent market-based performance measures. Participants in such retrospectively analyses are believed to be able to consider a broader set of allocation options, thus making the evaluation more realistic than data based on third party observation.[c] In short, company founders and/or principal investigators are in many cases simply the best source of information available.
- **Survey sampled projects at firms with multiple awards.** Projects from firms with multiple awards were underrepresented in the sample, because they could not be expected to complete a questionnaire for each of dozens or even hundreds of awards.
- **Failed firms are difficult to contact.** Survey experts point to an "asymmetry" in their ability to include failed firms for follow-up surveys in cases where the firms no longer exist.[d] It is worth noting that one cannot necessarily infer that the SBIR project failed; what is known is only that the firm no longer exists.
- **Not all successful projects are captured.** For similar reasons, the NRC Phase II Survey could not include ongoing results from successful projects in firms that merged or were acquired before and/or after commercialization of the project's technology. The survey also did not capture projects of firms that did not respond to the NRC invitation to participate in the assessment.
- **Some firms may not want to fully acknowledge SBIR contribution to project success.** Some firms may be unwilling to acknowledge that they received important benefits from participating in public programs for a variety of reasons. For example, some may understandably attribute success exclusively to their own efforts.
- **Commercialization lag.** While the NRC Phase II Survey broke new ground in data collection, the amount of sales made—and indeed the number of projects that generate sales—are inevitably undercounted in a snapshot survey taken at a single point in time. Based on successive data sets collected from NIH SBIR award recipients, it is estimated that total sales from all responding projects will likely be on the order of 50 percent greater than can be captured in a single survey.[e] This underscores the importance of follow-on research based on the now-established survey methodology.

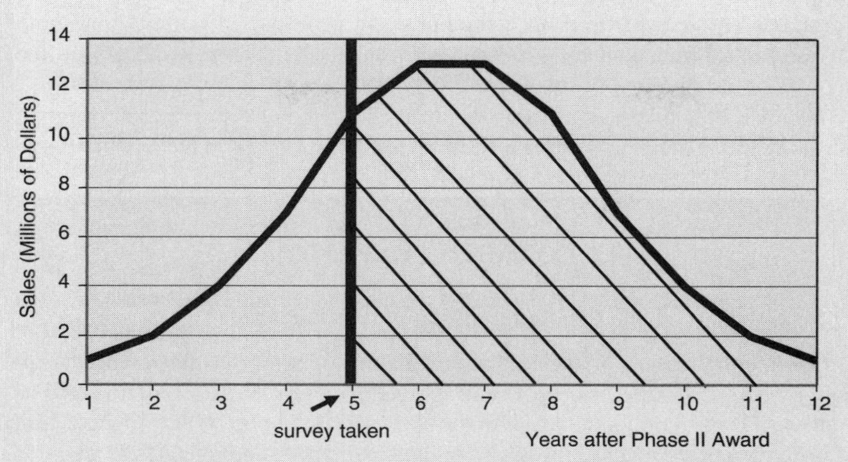

FIGURE B-4-1 Survey bias due to commercialization lag.

These sources of bias provide a context for understanding the response rates to the NRC Phase I and Phase II Surveys conducted for this study. For the NRC Phase II Survey, of the 4,523 firms that could be contacted out of a sample size of 6,408, 1,916 responded, representing a 42 percent response rate. The NRC Phase I Survey captured 10 percent of the 27,978 awards made by all five agencies over the period of 1992 to 2001. See Appendix A and B for additional information on the surveys.

---

[a]For a technical explanation of the sample approaches and issues related to the NRC surveys, see Appendix A.

[b]Albert N. Link, and John T. Scott, *Evaluating Public Research Institutions: The U.S. Advanced Technology Program's Intramural Research Initiative*, London: Routledge, 2005.

[c]While economic theory is formulated on what is called "revealed preferences," meaning individuals and firms reveal how they value scarce resources by how they allocate those resources within a market framework, quite often expressed preferences are a better source of information especially from an evaluation perspective. Strict adherence to a revealed preference paradigm could lead to misguided policy conclusions because the paradigm assumes that all policy choices are known and understood at the time that an individual or firm reveals its preferences and that all relevant markets for such preferences are operational. See {1} Gregory G. Dess and Donald W. Beard, "Dimensions of Organizational Task Environments," *Administrative Science Quarterly*, 29: 52-73, 1984; {2} Albert N. Link and John T. Scott, *Public Accountability: Evaluating Technology-Based Institutions*, Norwell, MA: Kluwer Academic Publishers, 1998.

[d]Albert N. Link, and John T. Scott, *Evaluating Public Research Institutions: The U.S. Advanced Technology Program's Intramural Research Initiative*, op. cit.

[e]Data from NIH indicates that a subsequent survey taken two years later would reveal very substantial increases in both the percentage of firms reaching the market, and in the amount of sales per project. See National Research Council, *An Assessment of the SBIR Program at the National Institutes of Health*, Charels W Wessner, ed., Washington, DC: The National Academies Press, 2009.

compared to less than $1 million for SBIR over a two to three year cycle.[3] VC investments are also focused on companies, not projects, and often come both with substantial management support and influence (e.g., through seats on the company's board). None of this is true for SBIR.

The lack of available comparators means that we must assess each program in terms of the benchmarks developed to review the program in the Methodology report described below.[4]

4.2 COMMERCIALIZATION

4.2.1 Challenges of Commercialization

Commercialization of the technologies developed under the research supported by SBIR awards has been a central objective of the SBIR program since its inception. The program's initiation in the early 1980s, in part, reflected a concern that American investment in research was not adequately deployed to the nation's competitive advantage. Directing a portion of federal investment in R&D to small businesses was thus seen as a new means of meeting the mission needs of federal agencies, while increasing the participation of small business and thereby the proportion of innovation that would be commercially relevant.[5]

Congressional and Executive branch interest in the commercialization of SBIR research has increased over the life of the program.

A 1992 GAO study[6] focused on commercialization in the wake of congressional expansion of the SBIR program in 1986.[7] The 1992 reauthorization specifically "emphasize[d] the program's goal of increasing private sector com-

---

[3]2005 saw some 3,000 deals worth an average of $7.35 million according to data from National Venture Capital Association. See National Venture Capital Association Web site, *<http://www.nvca.org/ffax.html>*.

[4]For a discussion of this and related methodological challenges, see, National Research Council, *An Assessment of the Small Business Innovation Research Program: Project Methodology*, op. cit.

[5]A growing body of evidence, starting in the late 1970s and accelerating in the 1980s indicates that small businesses were assuming an increasingly important role in both innovation and job creation. See, for example, J. O. Flender and R. S. Morse, *The Role of New Technical Enterprise in the U.S. Economy*, Cambridge, MA: MIT Development Foundation, 1975, and David L. Birch, "Who Creates Jobs?" *The Public Interest*, 65:3-14, 1981. Evidence about the role of small businesses in the U.S. economy gained new credibility with the empirical analysis by Zoltan Acs and David Audretsch of the U.S. Small Business Innovation Data Base, which confirmed the increased importance of small firms in generating technological innovations and their growing contribution to the U.S. economy. See Zoltan Acs and David Audretsch, "Innovation in Large and Small Firms: An Empirical Analysis," *The American Economic Review*, 78(4):678-690, September 1988. See also Zoltan Acs and David Audretsch, *Innovation and Small Firms*, Cambridge, MA: The MIT Press, 1990.

[6]U.S. General Accounting Office, *Federal Research: Small Business Innovation Research Shows Success But Can be Strengthened*, GAO/RCED-92-37, Washington, DC: U.S. General Accounting Office, 1992.

[7]PL 99-443, October 6, 1986.

mercialization of technology developed through federal research and development."[8] The 1992 reauthorization also changed the order in which the program's objectives are described, moving commercialization to the top of the list.[9]

The term "commercialization" means "reaching the market," which some agency managers interpret as "first sale"—that is, the first sale of a product in the market place, whether to public or private sector clients. This definition, however, misses significant components of commercialization that do not result in a discrete sale. It also fails to provide any guidance on how to evaluate the scale of commercialization, an important element in assessing the degree to which SBIR programs successfully encourage commercialization. The metrics for assessing commercialization can also be elusive,[10] and it is important to understand that it is not possible to completely quantify all commercialization from a research project:

- The multiple steps needed *after* the research has been concluded mean that a single, direct line between research inputs and commercial outputs rarely exists in practice; cutting edge research is only one contribution among many leading to a successful commercial product.
- Markets themselves have major imperfections, or information asymmetries so high quality, even path-breaking research, does not always result in commensurate commercial returns.
- The lags involved in the timeline between an early stage research project and a commercial outcome mean that for a significant number of the more recent SBIR projects, commercialization is still in process, and sales—often substantial sales—will be made in the future. The current "total" sales are in this case just a "snapshot half way through the race," and will require updating as the full impact of the award becomes apparent in sales.
- Yet the impact of SBIR awards also needs to be qualified. Research rarely results in stand-alone products. Often, the output from an SBIR project is combined with other technologies. The SBIR technology may provide a critical element in developing a winning solution, but that commercial impact—the sale of the larger combined product—is not captured in the data.

---

[8]PL 102-564 October, 28, 1992.

[9]These changes are described by R. Archibald and D. Finifter in "Evaluation of the Department of Defense Small Business Innovation Research Program and the Fast Track Initiative: A Balanced Approach" in National Research Council, *The Small Business Innovation Research Program: An Assessment of the Department of Defense Fast Track Initiative*, Charles W. Wessner, ed., Washington, DC: National Academy Press, 2000.

[10]See National Research Council, *An Assessment of the Small Business Innovation Research Program: Project Methodology*, op. cit.

- In some cases, the full value of an "enabling technology" that can be used across industries is difficult to capture.

All this is to say that commercialization results must be viewed with caution, first because our ability to track them is limited (indeed it appears highly likely that our efforts at quantification of research awards may understate the true commercial impact of SBIR projects) and because an award, and a successful project, cannot lay claim to all subsequent commercial successes, though it may contribute to that success in a significant fashion.

These caveats notwithstanding, it is possible to deploy a variety of assessment techniques to measure commercialization outcomes.

4.2.2 Commercialization Indicators and Benchmarks

This report uses three sets of indicators to quantitatively assess commercialization success:

(1) **Sales and Licensing Revenues** ("sales" hereafter, unless otherwise noted). Revenues flowing to a company from the commercial marketplace and/or through government procurement constitute the most obvious measure of commercial success. They are also an important indicator of uptake for the product or service. Sales indicate that the result of a project has been sufficiently positive to convince buyers that the product or service is the best available solution.

Yet if there is general agreement that sales are a key benchmark, there is no such agreement on what constitutes "success." Companies, naturally enough, focus on projects that contribute to the bottom line—that are profitable. Agency staff provide a much wider range of views. Some view any sales a substantial success for a program focused on such an early stage of the product and development cycle, while others seem more ambitious.[11] Some senior executives in the private sector viewed only projects that generated cumulative revenues at $100 million or more as a complete commercial success.[12]

Rather than seeking to identify a single sales benchmark for "success," it therefore seems more sensible to simply assess outcomes against a range of benchmarks reflecting these diverse views, with each marking the transition to a greater level of commercial success:

[11] Interviews with SBIR program coordinators at DoD, NIH, NSF, and DoE.

[12] Pete Linsert, CEO, Martek, Inc., meeting of the NRC Committee on Capitalizing on Science, Technology, and Innovation: An Assessment of the Small Business Innovation Research Program, June 5, 2005.

(a) *Reaching the market*—a finished product or service has made it to the marketplace
(b) *Reaching $1 million in cumulative sales* (beyond SBIR Phases I and II)—the approximate combined amount of standard DoD Phase I and Phase II awards
(c) *Reaching $5 million in cumulative sales*—a modest commercial success that may imply that a company has broken even on a project
(d) *Reaching $50 million in cumulative sales*—a full commercial success.

(2) **Phase III Activities Within DoD.** As noted above, Phase III activities within DoD are a primary form of commercialization for DoD SBIR projects. These activities are considered in Section 4.3 (Agency Mission) of this report and Chapter 5 (Phase III Challenges and Opportunities) of the DoD report.[13]

(3) **R&D Investments and Research Contracts.** Further R&D investments and contracts are good evidence that the project has been successful in some significant sense. These investments and contracts may include partnerships, further grants and awards, or government contracts. The benchmarks for success at each of these levels should be the same as those above, namely:
(a) Any R&D additional funding
(b) Additional funding of $1 million or more
(c) Additional funding of $5 million or more
(d) Funding of $50 million or more

(4) **Sale of Equity.** This is a clear-cut indicator of commercial success or market expectations of value. Key metrics include:
(a) Equity investment in the company by an independent third party
(b) Sale or merger of the entire company

4.2.3 Sales and Licensing Revenues

The most basic of all questions on commercialization is whether results from a project reached the marketplace. The NRC Phase II Survey[14] indicates that just under half (47 percent) of respondents had generated some sales, and that a further 18 percent still expected sales, though they had none at the time of the survey. In addition, 5 percent were still in the research stage of the project.

---

[13]National Research Council, *An Assessment of the SBIR Program at the Department of Defense*, Charles W. Wessner, ed., Washington, DC: The National Academies Press, 2009.

[14]Much of the primary data in this section of the report was derived from the NRC's Phase II Survey. See Appendix A for additional information about the NRC's Phase II Survey, including response rates.

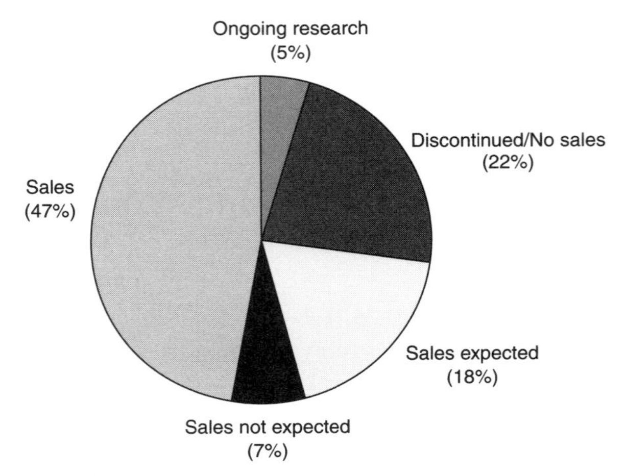

FIGURE 4-1 Sales from Phase II projects.
SOURCE: NRC Phase II Survey.

There is variation among the agencies, but these data are consistent with program objectives.[15]

4.2.3.1 Distribution of Sales

Research on early-stage financing strongly suggests a pronounced skew to the results, and this turns out to be the case. Most projects that reach the market generate minimal revenues. A few awards generate substantial results, and a small number bring in large revenues.

Of the 790 SBIR Phase II projects reporting sales greater than $0, average sales per project were $2,403,255. Over half of the total sales dollars were due to 26 projects (1.4 percent of the total), each of which had $15,000,000 or more in sales. The highest cumulative sales figure reported was $129,000,000.

This distribution is reflected in Figure 4-2.

Almost three quarters of the projects reporting sales greater than zero had $1 million or less in sales; two projects reported sales greater than $100 million. The latter by themselves accounted for 16.5 percent of all the revenues reported; together, the 1.7 percent of respondents reporting sales greater than $20 million accounted for 43.7 percent of all revenues reported.

These distributions are similar to those reported from other SBIR data

[15]See the section on venture capital in National Research Council, *An Assessment of the SBIR Program at the National Institutes of Health*, Charles W. Wessner, ed., Washington, DC: The National Academies Press, 2009.

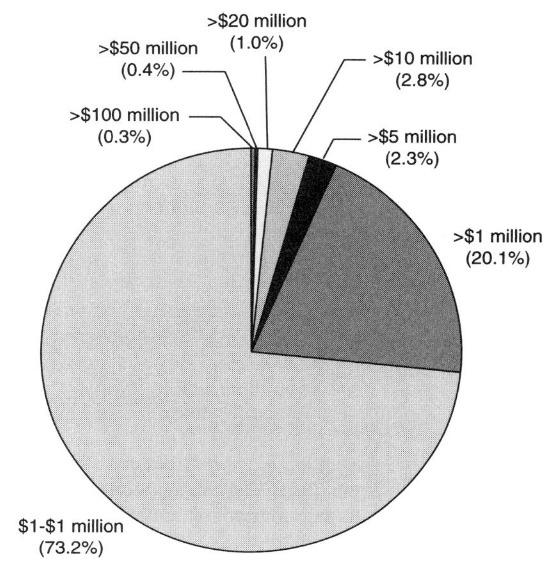

FIGURE 4-2 Distribution of projects with sales >$0.
SOURCE: NRC Phase II Survey.

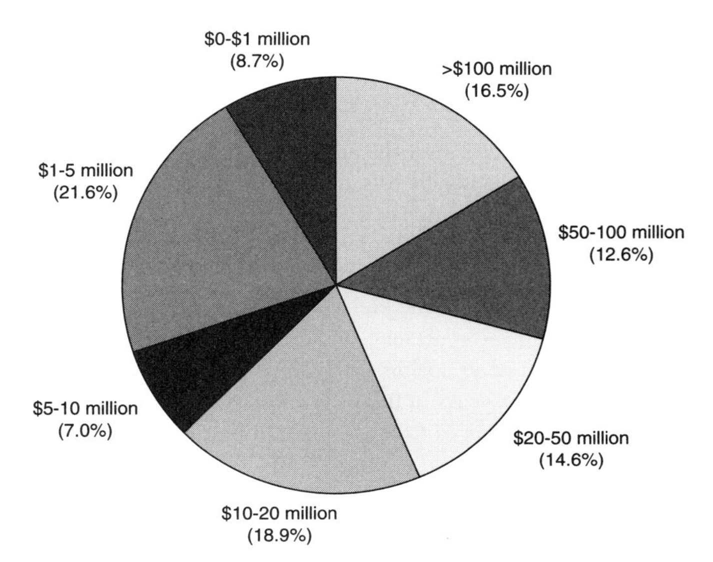

FIGURE 4-3 Distribution of sales, by total sales (percent of total sales dollars).
SOURCE: NRC Phase II Survey.

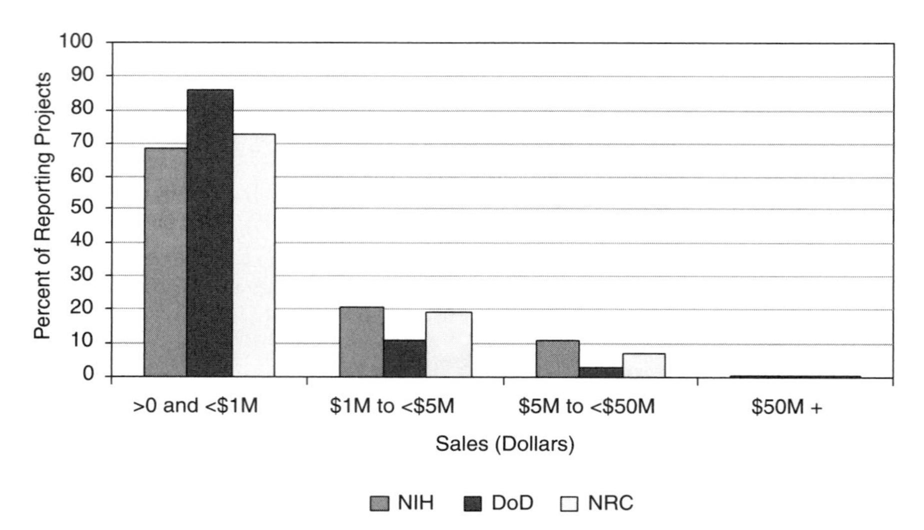

FIGURE 4-4 NIH sales by sales range (percent of projects in each range). Total for 1992-2002.
SOURCE: NIH data, NRC Phase II Survey, and DoD Commercialization database.

sources.[16] For example, at NIH, sales data are available from the NRC survey, a previous NIH survey, and the DoD commercialization database.[17] These data in Figure 4-4 show that about 10 percent of NIH projects that report any revenues report more than $5 million.

4.2.3.2 Sales Expectations

Because it may take years after the end of a Phase II for a commercial product to reach the market, many projects that do not yet report sales still expect them eventually. About 36 percent of NRC survey respondents with no sales (19 percent of all projects) still expected sales in the future. Analysis elsewhere suggests that these expectations are often optimistic, and that a considerably smaller number of these projects will, in the end, reach the market.

However, it is equally important to note that a complete accounting of all sales from the projects funded during 1992-2001 (the focus of the NRC survey) will be possible only some years in the future. Many projects have only recently reached the market, so the bulk of their sales will be made in the future and are

[16]See National Research Council, *An Assessment of the SBIR Program at the National Institutes of Health*, op. cit., Chapter 4, where data from the DoD Commercialization database and the NIH Phase II Survey are compared with the NRC data.

[17]The DoD database captures commercialization data from NIH projects where the firm subsequently applied for a DoD SBIR award.

TABLE 4-1 Most Important Customer (Percent of Responses)

Most Important Customer	Percent of Responses
Domestic private sector	35
Department of Defense (DoD)	32
Prime contractors for *DoD or NASA*	10
NASA	2
Other federal agencies	1
State or local governments	4
Export markets	14
Other	2

SOURCE: NRC Phase II Survey.

not captured in these survey data, which effectively capture initial sales (see Box 4-2).

4.2.3.3 Sales by Sector

The NRC Phase II Survey asked respondents to identify the customer base for the products. There are substantial differences between agencies. Only 1 percent of NIH respondents, for example, reported sales to DoD, in contrast to 38 percent for DoD respondents.[18]

4.2.3.4 Licensee Sales and Related Revenues

The indirect impact of licensee sales, where survey respondents report sales not made by their own company, is an important measure of success. These data are important as an indicator of the extended effects of SBIR beyond the immediate awardee company. However, they should be treated with an additional degree of caution, as respondents do not necessarily have as accurate information about another company as they have about their own.

Licensing activity within the program is significant: Just over 35 percent reported a finalized licensing agreement or ongoing negotiations towards one.[19]

Just over 5 percent of respondents reported licensee sales greater than $0, with three licenses reporting more than $70 million in sales each and accounting for more than half of all reported licensee sales. One project alone reported more than $200 million in licensee sales.

As the case study in Box 4-3 indicates, in some cases licensing has been pivotal in the commercialization of a successful technology: In some industries, there are no alternatives to using established channels.

[18]See Chapter 4 in National Research Council, *An Assessment of the SBIR Program at the National Institutes of Health*, op. cit.

[19]Eighteen percent of responding projects (242 firms) reported finalized licensing agreements, and 19 percent (249 firms) reported ongoing negotiations. NRC Phase II Survey, Question 12.

BOX 4-2
Underestimating Commercial Outcomes from the SBIR Program: The Impact of Systematic Characteristics of Survey Analysis

Among the SBIR agencies, only DoD requires that firms enter commercialization data into a database when applying for subsequent awards. This detailed database is a powerful source of information, primarily about DoD-oriented firms and projects. We would recommend that other agencies consider making the same requirement, and utilizing the existing DoD database for this purpose to minimize costs.

In the absence of such data, analysis of commercialization at the other SBIR agencies continues to rely on survey data. These data have important strengths and weaknesses.

The NRC survey and the NIH survey instruments were sent to all SBIR Phase II recipients from 1992 onward. This is the first effort to generate responses from the entire population of winning firms. The data generated are the best available for these agencies. However, there are two key sets of limitations, both of which have the effect of understating—perhaps very substantially—the amount of commercialization achieved. These two limitations can be called the "multiple awards effect" and the "snapshot effect."

The Multiple Awards Effect

Because some firms have received many awards, it is not feasible or reasonable to expect them to answer a similar questionnaire about each award that they received. As a result, both the NIH and NRC surveys limited the number of questionnaires sent to multiple winners. NIH sent one survey per firm, allocating one randomly to a specific project at multiple award firms. The NRC sent questionnaires about all projects to firms that had won three awards or less. It sent questionnaires to a sampling of the awarded projects for firms with more than three awards.

In both cases (somewhat less so for the (NRC), the effect has been to bias survey responses away from firms with multiple awards. This matters when there are systemic differences between the results provided by these different groups of firms. And the NRC survey indicates that firms with multiple awards are, in fact, likely to generate significantly higher levels of commercialization than are firms with smaller numbers of awards.

Using data from the DoD commercialization database to test this hypothesis, we found that firms receiving more than 15 awards generated an average of $1.39 million in sales per project; firms with fewer than 15 awards generated only $0.75 million per project. Firms with more awards generated on average 85 percent more sales per project.

Thus, the response bias in both NIH and NRC Phase II Surveys appear likely to have a significant downward impact on commercialization estimates.

The Snapshot Effect

Well-designed surveys provide an important insight into outcomes from SBIR projects. Necessarily, however, they provide a view of outcomes at the moment that the survey was completed.

For almost all products and services, sales follow some form of bell-shaped curve:

relatively slow sales as they begin market penetration, growth in sales until the market is saturated or competing products emerge, and decline until the product has been superseded. The shape of the curve differs between products, of course, and the entire curve can be completed in a matter of months for some software sales, or in decades for niche products in extremely long cycle industries (e.g., weapons platforms).

The survey, however, essentially takes a cross-section of the bell curve. It asks about levels of commercialization at a particular point in time. In essence, it asks about past sales, but can generate little reliable data on future sales.

Thus the average sales data generated by surveys reflects average sales *to date*. Using some simple analytic techniques, it is possible to suggest that on average, the NIH and NRC surveys excluded approximately 50 percent of the total lifetime sales of the products and services generated from SBIR awards.

And this hypothesis is at least supported by recent data from NIH, where the first resurvey of firms was done in 2005, three years after the initial 2002 survey. Results from the survey indicate that the number of firms with some sales increased from 29 percent of surveyed firms to 63 percent (this reflects the number of firms still in pre-commercialization at the time of the first survey).

The Recent Awards Effect

The snapshot effect is further complicated by the distribution of responses to the surveys. Two factors help tilt responses toward awards from more recent years.

First, the number of awards has been rising rapidly, especially at NIH and DoD since the late 1990s. As a result, a larger number of awards are concentrated in recent years.

Second, firms with awards from many years ago are harder to find, and are less likely to respond to surveys. As one commentator notes, "there are no SBIR shrines" at SBIR recipient companies—no one may remember receiving an award 10 years ago; the company may be out of business; the principal investigator (PI) may have left. As a result, more recent awards generate a higher percentage response rate.

The results of the factors are clear. At the NIH, of the original 758 survey respondents, 258—34 percent of all respondents—reached the market *after* the date of the first survey. The first survey captured less than half of the projects that had reached the market three years later, in 2005.

This is unsurprising (although it is important). Responses come preponderantly from projects where awards were made relatively recently—and where the snapshot effect is particularly important, as many are still in the early commercialization or even late development (precommercialization) phases of their projects. Thus the recent awards effect too tends to reduce commercialization estimates.

Conclusions

It is at this stage not possible to provide accurate estimates for the impact of these effects on commercialization estimates drawn from surveys. The limited evidence available to date, notably the analysis of the NIH second snapshot discussed above, suggests that the effect may be to substantially reduce commercialization estimates.

This analysis strongly suggests that follow-up surveys will be especially important, as they provide critical data for making precisely the assessments and modifications to the analysis that will be necessary to improve accuracy in the future.

BOX 4-3
Licensing Case Study

Applied Health Science and the Wound and Skin Intelligence System™

Applied Health Science (AHS) received a Phase I award to validate and automate the Pressure Sore Status Tool, a standardized assessment instrument for use in field settings for describing and tracking status changes in chronic wounds (e.g., pressure ulcers).

The WSIS (Wound and Skin Intelligence System™ or WSIS™) provides clinicians with the ability to assess risk and request a "case specific" prevention plan for reducing the probabilities that a wound will develop. The system tracks prevention and treatment outcomes over time, and relates these outcomes to individual risk and wound profiles and interventions employed. Thus, the system has the capacity to "learn" from its own experience.

The product was commercialized through the sale of rights to ConvaTec, a wholly owned unit of Bristol Myers-Squibb that is the largest wound products company in the world. ConvaTec funded commercialization, receiving in exchange licensing rights. This merged AHS technology and research capabilities with ConvaTec's global marketing power—the brand has a presence in about 80 countries worldwide. ConvaTec subsequently bought all rights to the software. AHS retained the worldwide data "pipelines," warehouse, and analytical functions. AHS also has a right-to-first-review for any elaborations of, or changes in the system.

AHS has announced current projections of $30 million in annual sales from the U.S. market alone, and expects to add one employee for each 75 users of the system. AHS and ConvaTec are also forming a series of strategic alliances with companies prepared to supply or develop add-on capabilities.

SOURCE: National Institutes of Health, *<http://grants1.nih.gov/grants/funding/sbir_successes/sbir_successes.htm>*.

4.2.4 Additional Investment Funding

Further investment in a recipient company related to the SBIR award project is another indication that the project work is of value. On average, SBIR projects received almost $800,000 from non-SBIR sources, with over half of respondents (51.6 percent) reporting some additional funds for the project from a non-SBIR source.[20]

Focusing just on the 839 projects that reported receiving more than $0 in additional funding, these projects reported average additional funding of $1,538,438, with almost $260,000 from federal agencies themselves, mostly from DoD. Three hundred eighteen projects reported some federal funding, averag-

---

[20]This included some 989 respondents.

TABLE 4-2 Sources of Additional Investments in SBIR Projects

Source	Average Amount ($)
a. Non-SBIR federal funds	259,683
b. Private Investment	
(1) U.S. venture capital	164,060
(2) Foreign investment	40,682
(3) Other private equity	125,690
(4) Other domestic private company	64,304
c. Other sources	
(1) State or local governments	9,329
(2) College or universities	1,202
d. Not previously reported	
(1) Your own company	113,454
(2) Personal funds	15,706
Total	$794,110

SOURCE: NRC Phase II Survey.

ing just under $1.6 million per project. At DoD, just over 30 percent of projects (205) reported additional federal funding outside SBIR, averaging just over $1.6 million per project.

About $165,000 per project came from venture capital sources. However, in those cases where venture funding was present, the amounts of funding were substantial: For the 50 responding projects with VC funding, the average per project was just under $8.3 million.

These figures reflect the well-known concentration of venture funding on a few, highly desirable projects. They also show that SBIR supports a wide range of projects which do have commercial prospects (as well as other possible benefits) that are not likely to be funded by venture capital.

Focusing more closely on venture funding at NIH, initial research indicates that approximately 50 of the 200 companies that won the most Phase II awards at NIH have received some venture funding.[21] This is reflected in Figure 4-5.

Total VC investment is approximately $1.59 billion in these 50 companies, a total that dwarfs the $272 million NIH SBIR investment in these companies.

In addition, the NRC Firm Survey determined that 15 firms had had initial public offerings, and that a further six firms planned such offerings for 2005-2006. SBIR firms also generate a significant number of new companies. Fourteen percent of responding firms indicated they had formed a spin-off company, with a total of 242 spin-off companies reported.

In contrast to the tightly concentrated distribution of venture funding, inter-

[21]Venture funding of awardees at NIH is discussed in more detail in Chapter 4 of the NIH report, National Research Council, *An Assessment of the SBIR Program at the National Institutes of Health*, op. cit.

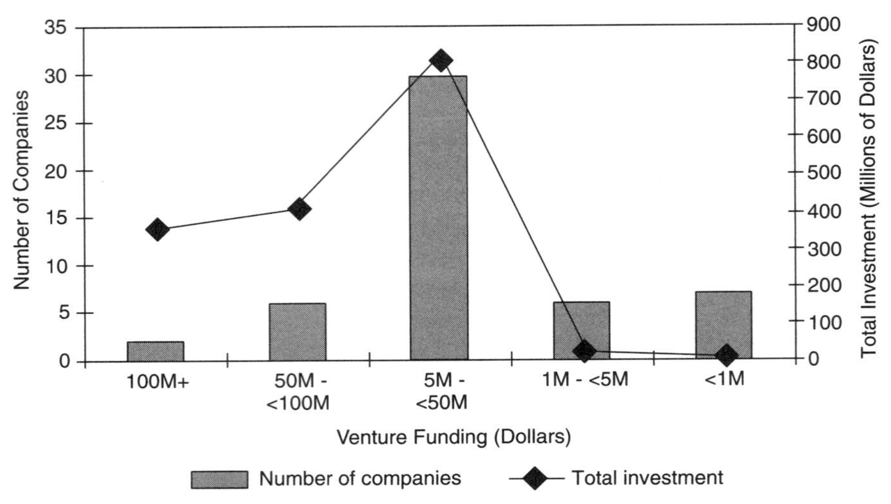

FIGURE 4-5 Distribution of venture funding for NIH top 200 Phase II winners.
SOURCE: VentureSource and other venture capital databases; National Institutes of Health awards database. See also National Research Council, *An Assessment of the SBIR Program at the National Institutes of Health*, Charles W. Wessner, ed., Washington, DC: The National Academies Press, 2009.

nal funding was by far the most widespread form of support, being reported by almost 50 percent of all respondents. Average internal funding ($113,000), was much lower than average venture funding.

About 7 percent of all respondents reported receiving "other private equity." This funding averaged $1.9 million per project for those projects receiving this kind of funding. Investments from government and academic sources were relatively scarce (less than 5 percent of respondents), and provided relatively small amounts (just over $225,000 per respondent receiving these funds).

4.2.5 Additional SBIR Funding

Aside from providing non-SBIR funds, the federal government in many cases makes further investments via the SBIR program itself. The NRC surveys asked respondents how many additional Phase I and Phase II awards followed each initial award, related to the original project.

About 40 percent of respondents reported receiving at least one additional related Phase II award, and slightly over half (53.8 percent) reported at least one additional Phase I award.

Relatively few projects received many related awards: Only 8.4 percent of respondents reported at least 5 related Phase I awards, and 9.7 percent received at least three related Phase IIs. This suggests that the "clustering" hypothesis—that

TABLE 4-3 Related SBIR Awards

Phase I			Phase II		
Number of Awards	Number of Respondents	Percent	Number of Awards	Number of Respondents	Percent
0	799	47.2	0	1006	59.5
1	351	20.7	1	335	19.8
2	214	12.6	2	187	11.1
3	122	7.2	3	60	3.5
4	64	3.8	4	44	2.6
5	41	2.4	5	32	1.9
6	25	1.5	6	8	0.5
7	13	0.8	7	6	0.4
8	17	1.0	8	1	0.1
9	7	0.4	9	4	0.2
10	8	0.5	11	1	0.1
11	4	0.2	12	2	0.1
12	14	0.8	28	6	0.4
15	2	0.1			
19	1	0.1	Total 96	1,692	
21	2	0.1			
30	1	0.1			
44	5	0.3			
65	1	0.1			
82	1	0.1			
Total 354	1692				

NOTE: Overall percentages use total responses plus missing responses as denominator.
SOURCE: NRC surveys.

SBIR type projects often require multiple awards, sometimes looping back to Phase I, before reaching the market—only applies to a limited number of projects, although case studies (e.g., SAM Technologies[22]) indicate that there are also important cases where such clustering does occur. In short, the data suggest that some companies and projects do attract a cluster of SBIR awards, but most do not. It may be, however, that this concentration of clustered awards reflects the limited number of commercial successes as well. Just under one half of respondents (47 percent) reported no additional related Phase I awards, and 60 percent reported no related Phase II.

4.2.6 SBIR Impact on Further Investment

Both the NRC and NIH Surveys sought additional information about the impact of the SBIR program on company efforts to attract third party funding—the

[22]See National Research Council, *An Assessment of the SBIR Program at the National Institutes of Health*, op. cit.

"halo effect" mentioned by some interviewees, who suggested that an SBIR award acted as a form of validation for external inventors.[23]

The fact that two-thirds of SBIR respondents did not attract outside funding, and that only 3.5 percent received venture funding, suggests that receiving a Phase II SBIR award does not in itself guarantee external funding. Survey responses from other surveys did, however, paint a more positive picture of these effects: 78 percent of NIH respondents said that they believed that additional capital had "resulted from" their SBIR participation—a strong statement.[24]

Case study interviews provided mixed views on this, with some interviewees strongly supporting the view that SBIR helps to attract investment and others suggesting that SBIR awards had had relatively little impact, although these views obviously reflect individual company experiences.

4.2.7 Small Company Participation and Employment Effects

Employment is another indicator of commercial success and also that the program is supporting small business.

The median size of a company receiving SBIR awards is relatively small—far lower than the 500-employee limit imposed by the SBA (see Figure 4-6).

The program focuses the bulk of its awards on very small companies. More than a third of awardees had between one and five employees at the time of award. A very substantial number (seventy percent) of respondent companies had 20 employees or fewer at the time of the Phase II award.

The NRC Survey sought detailed information about the number of employees at the time of the award and at the time of the survey and about the direct impact of the award on employment. Overall, the survey data showed that the average employment gain at each responding firm from the date of the SBIR award to the time of the survey was 29.9 full-time equivalent employees. Of course, very few of the companies that went out of business responded to the survey, so this question is particularly skewed toward firms that have been at least somewhat successful.

Most responding companies have expanded since the date of the Phase II award. The NRC Phase II Survey also shows that respondents enjoyed strongly positive employment growth after receiving a Phase II award. Table 4-4 shows that the percentage of companies with at least 50 employees more than doubled,

[23]For a discussion of the "halo effect" from awards by the Advanced Technology Program, see Maryann Feldman and Maryellen Kelley "Leveraging Research and Development: The Impact of the Advanced Technology Program," in National Research Council, *The Advanced Technology Program: Assessing Outcomes*, Charles W. Wessner, ed., Washington, DC: National Academy Press, 2001.

[24]See the Chapter 4 in National Research Council, *An Assessment of the SBIR Program at the National Institutes of Health*, op. cit.

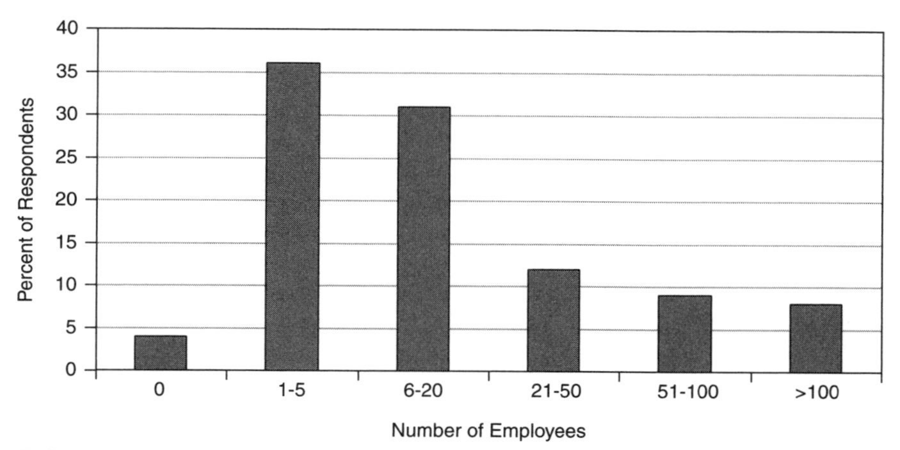

FIGURE 4-6 Distribution of companies, by number of employees at time of award.
SOURCE: NRC Phase II Survey.

from 16.5 percent to 35.4 percent of all respondents. Overall, survey respondents reported gains of 57,808 full time equivalent employees, with the top five respondents accounting for 18.4 percent of the overall net gain.

The NRC survey also sought to directly identify employment gains that were the direct result of the award. Respondents estimated that specifically as a result

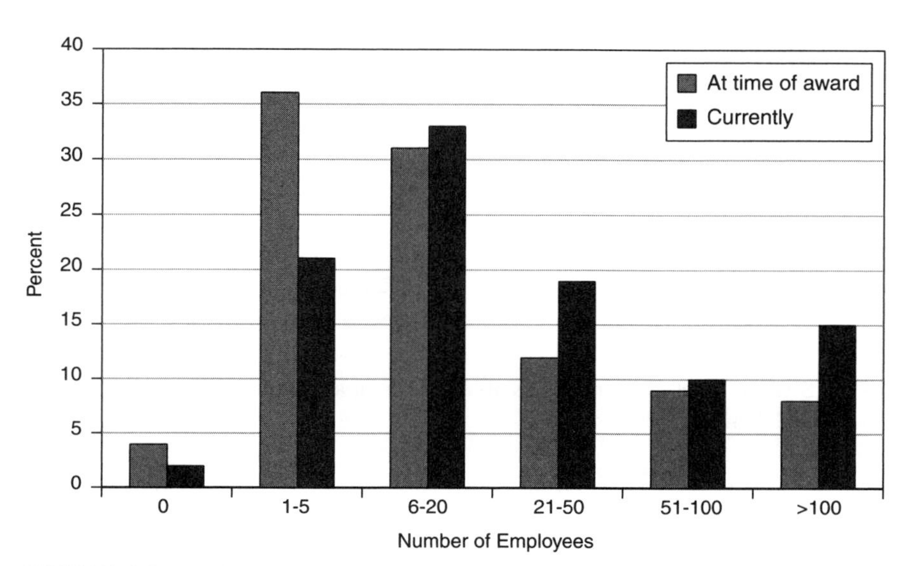

FIGURE 4-7 Employment distribution at responding companies, at time of award and currently.
SOURCE: NRC Phase II Survey.

TABLE 4-4 Employment at Phase II Respondent Companies, at the Time of Award and Currently

At Time of Award			Currently		
Number of Employees	Number of Responses	Percent	Number of Employees	Number of Responses	Percent
0	73	4.3	0	39	2.3
1-5	609	35.9	1-5	357	20.9
6-10	273	16.1	6-10	274	16.1
11-20	257	15.1	11-20	282	16.5
21-50	207	12.2	21-50	327	19.2
51-100	144	8.5	51-100	165	9.7
>100	135	8.0	>100	263	15.4
Total	1,698	100	Total	1,707	100

SOURCE: NRC Phase II Survey.

TABLE 4-5 Company-level Activities

	U.S. Companies/Investors		Foreign Companies/Investors	
Activities	Finalized Agreements (percent)	Ongoing Negotiations (percent)	Finalized Agreements (percent)	Ongoing Negotiations (percent)
Licensing agreement(s)	16	16	6	6
Sale of company	1	4	0	1
Partial sale of company	2	4	0	1
Sale of technology rights	5	9	1	3
Company merger	0	3	0	1
Joint Venture agreement	3	8	1	2
Marketing/distribution agreement(s)	14	9	8	4
Manufacturing agreement(s)	5	7	2	2
R&D agreement(s)	14	13	3	4
Customer alliance(s)	11	13	4	2
Other *Specify*__________	2	2	0	1

SOURCE: NRC Phase II Survey.

of the SBIR project, their firm was able to hire an average of 2.4 employees, and to retain 2.1 more.[25]

4.2.8 Sales of Equity and Other Company-level Activities

Company-level activities offer another set of indicators for measuring commercial activity, as these suggest that something of commercial value is being

[25]NRC Phase II Survey, Questions 16.

developed, even if no sales have as yet resulted. For example, although Neurocrine, an NIH SBIR awardee has yet to produce a product, it achieved an IPO that raised more than $100 million.[26]

The NRC Phase II Survey explored whether SBIR awardees had finalized agreements or ongoing negotiations on various company-level activities. The data show that marketing-related activities were most widespread. Activities with foreign partners were lower than similar activities with U.S. partners. Note, however, that the question asked specifically for outcomes that were the "result of the technology developed during this project"—a tight description.[27]

The impact of these activities is hard to gauge using quantitative assessment tools only. Box 4-4 illustrates how research conducted using SBIR funding seeded an entire generation of spin-off companies and joint ventures in a technology of potential significance for homeland security.

4.2.9 Commercialization: Conclusions

While accepting the view that there is no single, simple metric for determining the commercial success of an early stage R&D program such as SBIR, numerous metrics do provide the basis for making a broad determination of commercial outcomes at SBIR.

These data, taken together, strongly support the view that the program has a strong commercial focus, with considerable efforts to bring projects to market, with some success. Even though the number of major commercial successes has been few, that is normal for early stage high-risk projects, and the overall commercialization effort is substantial. Products are coming to market quickly, significant licensing and marketing efforts are under way for many projects, and approximately 30-40 percent of projects generate products that do reach the marketplace. These data all paint a picture of a program that is successful in commercializing innovative technologies in a variety of ways.

4.3 AGENCY MISSION

Each agency with an SBIR program has a different mission, and the contribution of SBIR to each specific mission must be assessed individually. These assessments, found in the individual agency reports, conclude that SBIR is indeed supporting each agency's pursuit of its specific mission.

Some more general observations can be made, however. An assessment of the extent to which SBIR supports agency mission can be divided into two areas:

---

[26]See Neurocrine case study in National Research Council, *An Assessment of the SBIR Program at the National Institutes of Health*, op. cit.

[27]NRC Phase II Survey, Question 12.

- *procedural alignment*—the extent to which the procedures of the agency SBIR program are aligned with the needs of the agency
- *program outcomes*—the extent to which outcomes from the program have the effect of supporting the agency mission.

It is important to note that the different missions of the agencies mean that some agencies define agency mission support more narrowly, or at least have much tighter metrics for assessing this element of the program. In particular, the procurement agencies—primarily DoD and NASA—assess contribution to agency mission primarily against the extent to which the agency itself uses outputs from the SBIR program. In contrast, the non-procurement agencies—NIH, NSF, and, to a great extent, the Department of Energy (DoE)—see support for

BOX 4-4
Intelligent Optical Systems: Intelligent, Distributed, Sensitive Chemical and Biochemical Sensors and Sensor Networks

Intelligent Optical Systems has developed a system for using the entire length of a specially designed fiber-optic cable as a sensor for the detection of toxins and other agents. This bridges the gap between point detection and standoff detection, making it ideal for the protection of fixed assets.[a]

SBIR-supported research has been followed by a focus on the development of subsidiaries and spin-offs at Intelligent Optical Systems. This activity has generated private investments of $23 million in support of activities oriented toward the rapid transition to commercially viable products.

Since January 2000, IOS has formed two joint ventures, spun out five companies to commercialize various IOS proprietary technologies, and finalized licensing/technology transfer agreements with companies in several major industries.

Optimetrics manufactures and markets active and passive integrated optic components based on IOS-developed technology for the telecommunication industry. Maven Technologies was formed to enhance and market the Biomapper technologies developed by IOS. Optisense manufactures and distributes gas sensors for the automotive, aerospace, and industrial safety markets, and will be providing H2 and O2 optical sensor suites designed to enhance the safety of NASA launch operations. Optical Security Sensing (OSS), which is IOS's newest spin-off company, was formed to commercialize chemical sensors for security and industrial applications.

IOS currently employs 40 scientists, and almost 80 percent of its revenues come from non-SBIR sources. The company currently holds 13 patents, with an additional 13 applications pending.

---

[a]*Point detection* means the contaminant comes into physical contact with the sensor and it is analyzed. In *standoff detection,* the sensor sees the contaminant at a distance and recognizes it, but the contaminant never comes in contact with the sensor.

mission much more broadly: For NIH, for example, support for mission can be construed as anything that improves medical knowledge or public health.

4.3.1 Procedural Alignment of SBIR Programs and Agency Mission

A procedural assessment reviews the steps taken by each agency program to ensure that the design and procedures of their SBIR program are aligned with the needs of the agency.

Agencies do this in different ways, but the following areas of analysis are broadly shared by all the agencies.

4.3.1.1 Topics and Solicitations

SBIR proposals are received by the agencies in response to published solicitations for proposals. These solicitations are the primary vehicle through which the agency expresses its areas of research interest to the scientific and technical community of small businesses. Within each solicitation, specific subject areas of interest are defined by individual topics. Topics can be focused tightly on a specific problem or requirement, or they may broadly outline an area of technical interest to the agency.

Aside from NIH, which expressly indicates that its topics are guidelines, not mandatory limits or boundaries on research that could be funded, all the agencies use topics to specify boundaries. In doing so, they are specifically delimiting areas of technical interest to the agency.

This is prima fascia confirmation that the SBIR programs support agency mission: Unless there is evidence that agencies are generating topics that are *not* aligned with the agency mission—and our analysis and interviews with staff and awardees found no trace of this—the use of topics and solicitations indicates that agencies are working to ensure that awards are aligned with the stated scientific and technical needs of the agency.

4.3.1.2 Topic Development Process—DoD

The agencies—notably DoD but, to a lesser extent, all the SBIR agencies—have an elaborate process for developing proposals for topics, sifting and assessing these proposals, and then finally deciding which topics should be published in the solicitation.

Agency procedures differ, and are described in the agency volumes. At DoD, for example, the topic selection process includes multiple levels of review. While this process has been criticized for adding considerable time between the initial identification of an agency's needs and the first dollars flowing to those needs, the process also seeks to ensure that all of the stakeholders feel that they have had a say. This is particularly important for DoD because it has a number of different stakeholder communities.

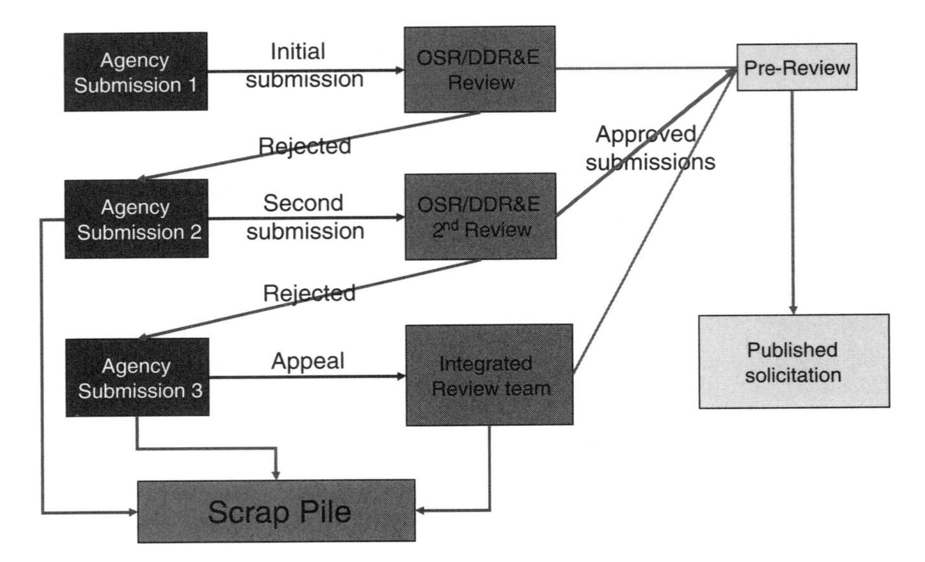

FIGURE 4-8 Topic review process.
SOURCE: NRC chart developed from interviews with Department of Defense staff.

Ultimate decision authority on the inclusion of topics in a DoD solicitation lies with the Integrated Review Team, which contains representatives from each of the awarding components. Topics are reviewed initially at DDR&E and are then returned to the agencies for correction of minor flaws, for revision and re-submission, or as discards.

DoD has also made considerable efforts to improve the topic selection process. In the late 1990s, DoD determined that topics were not being linked closely enough to the users of SBIR technologies in the acquisition community, and a conscious effort was made to ensure that more topics were "owned" by that community (with considerable success—see DoD report, Chapter 5: Program Management, for details[28]).

4.3.1.3 Topic Development Process—NSF

Under the 2005 NSF strategic plan, SBIR solicitation topics fit into three broad areas: (A) investment business focused topics, (B) industrial market driven topics, and (C) technology in response to national needs. (A, B, and C below). The topics list given in the strategic plan includes seven topics:

[28]National Research Council, *An Assessment of the SBIR Program at the Department of Defense*, op. cit.

A. Investment Business Focused Topics
   1. Biotechnology (BT)
   2. Electronics Technology (EL)
   3. Information Based Technology (IT)
B. Industrial Market Driven Topics
   1. Advanced Materials and Manufacturing (AM)
   2. Chemical Based Technology (CT)
C. Technology in Response to National Needs
   1. Security Based Technology (ST)
   2. Manufacturing Innovation (MI)

It appears that areas A and B above are expected to be relatively stable, and that area C is expected to have more frequent changes in specific topics as national needs change.

Even at NIH, where topics are viewed as guidelines, not boundaries for permissible research, the topic selection process is designed to ensure that area specialists within the agency's Institutes and Centers have substantial input. Area specialists are regularly encouraged to ensure that their wider research agendas are reflected in their selection of SBIR topics for publication.

Overall, agencies appear to be well aware that the topic/solicitation process is the primary mechanism through which the SBIR program is aligned with the agency's S&T objectives and its overall agency missions. Topics appear to be aligned with agency missions.

4.3.1.4 Award Selection Process

The selection of awards can also support an agency's mission, to the extent that the process reflects the agencies' priorities. A wide range of awards procedures are used at the various agencies, and these may differ substantially even between components of the same agency. For example, within DoD, Army, and Navy use different approaches, staff, and methodologies for selecting awardees.

All of these agency procedures share some basic characteristics: They assign considerable weight to the technical merit of proposals; They seek to ensure that selectors are, to some extent, independent of sponsoring agency components; and they have a set of written standards against which proposals are supposed to be measured. They differ in the nature of the reviewers (internal or external), the type of scoring (quantitative or qualitative), the extent of appeal and resubmission for nonawarded proposals, the degree to which scores are binding, and the discretionary powers of agency staff to make full or supplementary awards.

It is also in general fair to say that the point of reference with regard to agency mission reference at the point of award is not the agency's overall mission statement, but the specific expression of that statement in the form of the topic. Topics are used as expressions of the agency's mission. For example, DoE's overall mission is to "advance the national, economic, and energy security of the United States; to promote scientific and technological innovation in support of that mission; and to ensure the environmental cleanup of the national nuclear weapons complex."[29] Specific SBIR awards are made in relation to topics that support that mission—for example, to generate better battery cell technology. A high-quality, effective selection process, one that effectively identifies and funds projects that will help meet agency needs expressed via the topic, is therefore a process that supports the agency mission overall.

While selection procedures could potentially be improved, our analysis concludes that at every agency, the current selection processes largely succeed in aligning the technologies chosen with the expressed needs of the agency.[30]

4.3.2 Program Outcomes and Agency Mission

In contrast to the discussion above, the program outcomes for agency mission are both more difficult to assess and also more specific to individual agencies. All of the methodological difficulties in assessing outcomes discussed at the beginning of this chapter apply here; moreover, (unlike commercialization) there are no obvious and widely understood benchmarks that apply across agencies.

All agencies maintain a list of "success stories," describing SBIR awards that meet congressional goals. Some of these are focused on agency mission. However, the stories themselves, while illustrative of the power of the program to help develop new technologies, are of variable quality. Some agencies, such as NIH, use success stories written by the company in question without validating them. Even agencies that take a more systematic approach to case studies do not appear to have clear and transparent criteria for determining what should count as a "success story."

Each agency has found different ways to explore and describe outcomes related to agency mission. As a result, our overview must reflect these agency differences. Below, we provide a brief summary of agency approaches to identifying and measuring ways in which the SBIR program supports the agency mission.

4.3.2.1 NIH

For NIH, the primary mission is the development of fundamental knowledge and its application for improving health.

---

[29]Department of Energy Web site, accessed at <*http://www.energy.gov/about/index.htm*>.

[30]More details about each agency's selection procedures can be found in the individual agency volumes.

The NIH mission is science in pursuit of fundamental knowledge about the nature and behavior of living systems and the application of that knowledge to extend healthy life and reduce the burdens of illness and disability.

SOURCE: National Institutes of Health, *<http://www.nih.gov/about/ndex.html#mission>*.

A more detailed assessment of outcomes related to these objectives can be found in Chapter 3 of the NIH Report.[31] Data on patents and peer-reviewed publications resulting from awards can be used as an indicator of the development of fundamental knowledge. Commercialization measures are another possible indicator of the transfer of knowledge, as commerce marks a transaction that reflects that transfer.

Still, more direct measures would be useful. NIH has tried to develop these in several ways. Like other agencies, NIH maintains a Web page filled with "success stories." Unfortunately, these are entirely self-posted by companies, and reflect no selection or even verification by staff.

The NIH SBIR program has also led the way at NIH in developing metrics through the implementation of a recipient survey, with the first survey being deployed in late 2003. In this survey, the agency has sought to identify the populations targeted by SBIR projects. Figure 4-9 shows the distribution of projects by sector, for both projects that have already reached the market and those still being commercialized.[32]

The NIH survey also sought to quantify the number of people affected by a technology, asking respondents to place their project within ranges of affected populations by size. Such information, if accurate, would be useful. Unfortunately, respondents have wide latitude in answering, and are given no guidelines that might help to explain how to structure their answers. The responses are essentially guesses—perhaps biased guesses, as respondents may have a tendency to overestimate their projects' importance.

Still, the data categories show the wide variety of mission-related areas into which SBIR projects can be categorized—and all clearly fall within the broad mission definition of NIH. Responses can also indicate the distribution of projects between mission areas and can be used as a proxy for the degree of NIH interest in these different areas.

[31]National Research Council, *An Assessment of the SBIR Program at the National Institutes of Health*, op. cit.

[32]Note that percentages do not add up to 100 percent, as respondents are permitted to select more than one affected population.

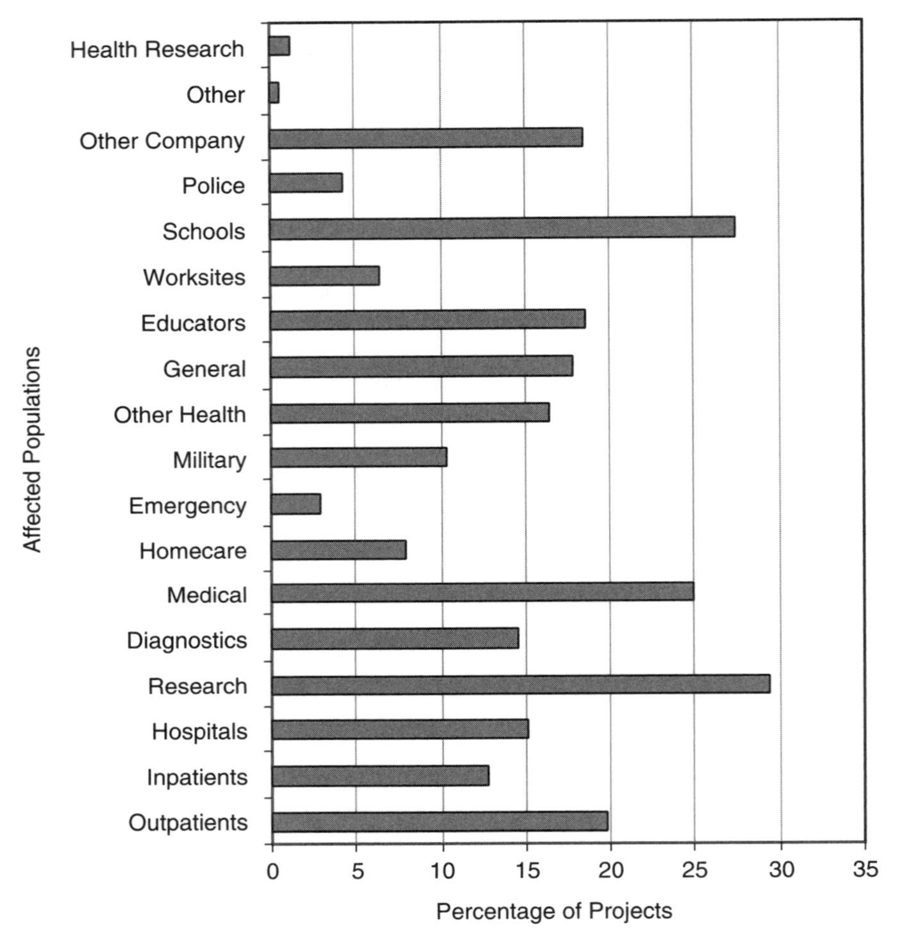

FIGURE 4-9 Distribution of NIH projects, by type of affected population.
SOURCE: National Institutes of Health, *National Survey to Evaluate the NIH SBIR Program: Final Report*, July 2003.

4.3.2.2 DoD

The DoD SBIR program's primary goal is the provision of technologies that are employed as part of defense systems being developed, acquired, or maintained to meet DoD mission needs. In effect, that acquisition stream defines, for operational purposes, agency needs, and the extent to which SBIR is, in fact, part of the acquisition process has become an important indicator of support for agency mission.

The mission of the Department of Defense is to provide the military forces needed to deter war and to protect the security of our country.

SOURCE: Department of Defense, *<http://www.defenselink.mil/admin/about.html>*.

This was not always the case. As discussed in the DoD volume, the program's objectives have evolved since the late 1990s. Some elements of the management have recently focused on finding ways to more closely align SBIR projects and programs with the acquisition process of the Services.

Much of the discussion that follows will focus on acquisitions and on Phase III at DoD. However, it is still important to remember that, even at DoD, not all outcomes that support the mission result in acquisitions and Phase III. For example, agency staff have indicated that SBIR can be an important way of assessing technologies that in the end do not pan out. The awards thus act as a low-cost probe of the technological frontier. So, even when no acquisition occurs, SBIR can still provide valuable support for the agency mission in terms of information on technological dead ends, promising technological options, or use resulting from the award itself.

Still, even with these multiple functions, as participants at the NRC's Phase III Symposium affirmed, Phase III and acquisitions are regarded by DoD as the core focus of the SBIR program, and the key indicator for measuring success in supporting agency mission. Many speakers at the NRC's Phase III Symposium made clear that the take-up of SBIR-funded technologies into the DoD acquisitions program was the benchmark against which DoD's SBIR program should be judged.

DoD has made a more conscious effort to measure SBIR impacts on agency performance than any of the other agencies under study. Because DoD is a procurement agency, there is also one clear set of indicators which, if properly captured, could provide critical benchmarks and feedback to the agency.

DoD has two tools for measuring follow-on Phase III contracts. The first is the DD350 contracting forms, which is supposed to capture whether a contract is a follow-on to an SBIR award. However, only well-trained contracting officers who understand SBIR are likely to correctly fill out the form. Internal assessment of the DD350 by DoD suggests that not all contracting officers use the form correctly. Consequently, at many DoD components, SBIR follow-on contracts are only erratically reported, if at all.[33] The other tool for measuring follow-on Phase III contracts is the DoD SBIR/STTR Commercialization Database. This database is used to calculate firms' Commercialization Achievement Index and

[33]See more extended discussion of this issue in Chapter 6 of National Research Council, *An Assessment of the SBIR Program at the Department of Defense*, op. cit.

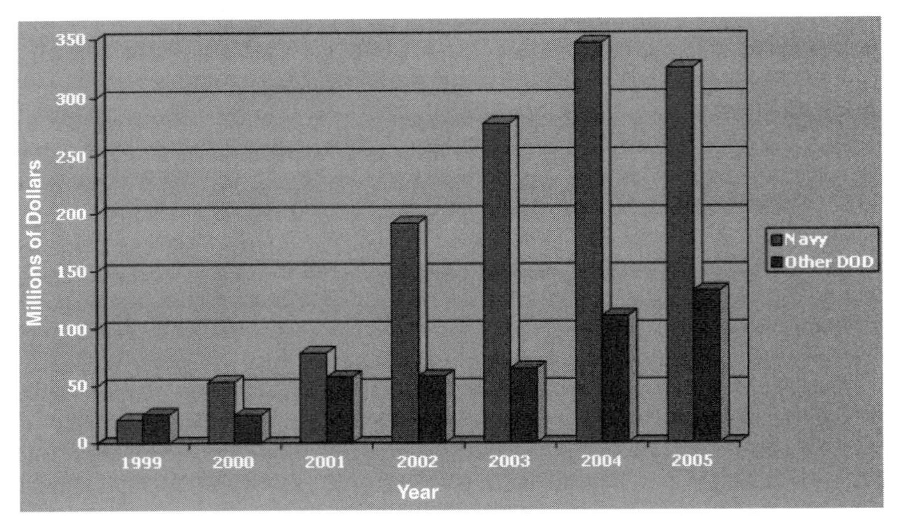

FIGURE 4-10 Phase III awards, annual totals, 1999-2005.
SOURCE: DoD SBIR/STTR Commercialization Database, provided by John Williams, U.S. Navy, April 7, 2005.

to generate the Company Commercialization Reports that are packaged with proposals to evaluators.

The Navy has recently made efforts—including the investment of external program dollars—to improve the quality of DD350 data. This effort is reflected in recent DD350 results, which show Navy as accounting for more than 70 percent of all DoD Phase III awards.[34]

The DD350 results do show that the amount of Phase III contracts generated have been climbing steadily in recent years, particularly at the Navy (see Figure 4-10).

These data also show that the amount of Phase III contracts being identified is greater than the amount of funding expended on SBIR, once commercialization has been appropriately lagged.

A fuller discussion of DoD's efforts regarding Phase III can be found in Chapter 5 of the study of SBIR at DoD.[35] Leaving complexities aside, however, it is possible to draw the following general conclusions:

---

[34]While these results are impressive, it is likely that DD350 data at other components will undercount Phase III successes unless those components undertake a similar effort to improve the quality of their data.

[35]See National Research Council, *An Assessment of the SBIR Program at the Department of Defense*, op. cit.

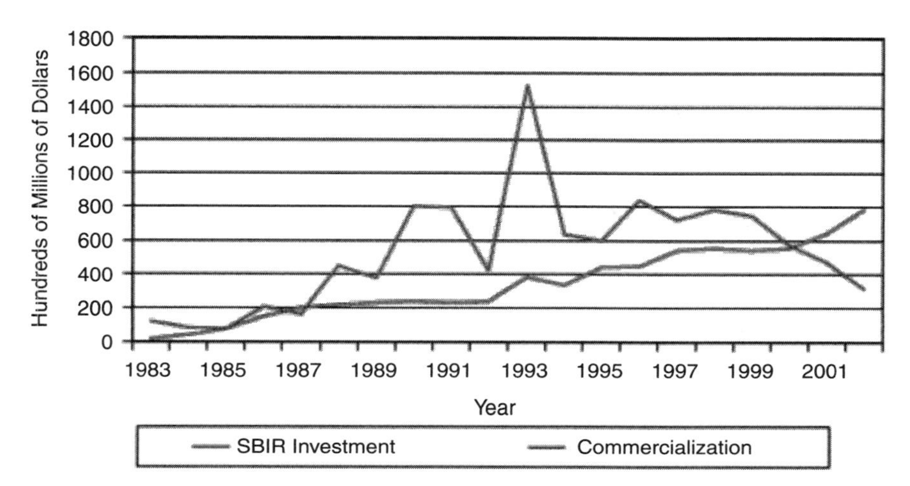

FIGURE 4-11 Reported commercializations versus SBIR budget.
SOURCE: Michael Caccuitto, DoD SBIR/STTR Program Administrator, Presentation to SBTC "SBIR in Rapid Transition" Conference, Washington, DC, September 27, 2006.

- DoD is increasingly making an effort to measure support for agency mission.
- Part of this assessment is increasingly quantitative, with the Navy leading the way in utilizing data from the DD350 forms. Nonetheless, there are barriers and difficulties that tend to reduce the number and amount of Phase III contracts counted in the DD350 tracking forms.
- The demonstration effect of the Navy program and growing awareness of SBIR's potential has increased senior management's awareness of SBIR's contributions to the Defense mission.[36]
- The growing interest and dedicated management attention by Prime contractors is enhancing the potential for insertion of SBIR outputs into the acquisition process.
- The reluctance of other agencies to provide management funding at the level of the Navy's effort remains a constraint on maximizing the return on the nation's SBIR investment.

In short, recent improvements in DoD activities related to Phase III do suggest a slow but significant change in attitude toward SBIR within DoD. SBIR is seen less as a tax imposed by Congress on otherwise useful R&D activities, and

[36]See statements by Charles Holland, Deputy Under Secretary of Defense for Science and Technology in National Research Council, *SBIR and the Phase III Challenge of Commercialization*, Charles W. Wessner, ed., Washington, DC: The National Academies Press, 2007.

more as an opportunity to bring innovative ideas and products to meet mission objectives.

4.3.2.3 DoE

DoE has, to a considerable extent, relied on process and procedures to ensure that its mission is being supported by the SBIR program. The agency has developed topic and award selection procedures that ensure that the primary driver of the program will be R&D managers within the agency, rather than either external peer reviewers or SBIR program managers. The latter play a minimal role in topic and award selection.

Mission
Discovering the solutions to power and secure America's future

Vision
The Department's vision is to achieve results in our lifetime ensuring: Energy Security; Nuclear Security; Science-Driven Technology Revolutions; and One Department of Energy—Keeping our Commitments.

SOURCE: Department of Energy, *<http://www.cfo.doe.gov/strategicplan/mission.htm>*.

In FY2005, this process resulted in 49 technical topics, from within the 12 program areas (see Table 4-6).

TABLE 4-6 DoE Phase I Awards by Program Area

Program Area	Number of Topics	Number of Grant Applications Submitted	Number of Phase I Awards
Fossil Energy	7	247	29
Advanced Scientific Computing Research	3	47	9
Basic Energy Sciences	9	247	56
Biological and Environmental Research	6	182	47
Environmental Management	0	0	0
Nuclear Physics	4	47	25
High Energy Physics	5	111	46
Fusion Energy Sciences	3	80	18
Nuclear Energy	1	11	3
Energy Efficiency and Renewable Energy	6	470	38
Nonproliferation and National Security	3	61	11
Electric Transmission and Distribution	2	48	8
TOTALS	49	1,551	290

SOURCE: Department of Energy Web site. Accessed at *<http://www.science.doe.gov/sbir/awards_abstracts/sbirsttr/statisticsinfo.htm>*.

There does not appear to be a systematic effort under way at DoE to determine whether the SBIR program supports agency mission. As with all other agencies, to the extent that agency mission involves the development of technical knowledge, there are indicators for whether SBIR companies are producing this knowledge.

4.3.2.4 NSF

NSF is similar to DoE in its reliance on procedures to ensure that agency mission is supported. As with DoE, technical managers are involved in the development of SBIR topics, and there is a strong focus on ensuring that awards are made only within defined topic areas. Topic development also aims to involve technical area program managers.

The National Science Foundation (NSF) is an independent federal agency created by Congress in 1950 "to promote the progress of science; to advance the national health, prosperity, and welfare; to secure the national defense . . . "

SOURCE: National Science Foundation, *<http://www.nsf.gov/about>*.

The SBIR program has taken on the specific role and mission of being the commercialization arm of NSF; the NSF program manager has been explicit in focusing the program on commercialization, arguing that the remaining 97.5 percent of NSF is focused on basic science and other primarily noncommercial aspects of scientific inquiry. While at one level this is correct, most of the rest of the NSF budget is largely focused on university research and is not open to small company-based research.

Thus in one sense, support for agency mission is best measured in terms of commercialization (discussed in Section 4.2). In another sense, NSF is indeed focused on support for the generation of new scientific knowledge. Here traditional metrics include patents and peer-reviewed publications. These aspects of mission support for NSF are discussed below. Both topics, and the program as a whole, are covered in more detail in the NSF volume.

4.3.2.5 NASA

NRC used a unique approach to address the question of support for NASA's agency mission. NRC surveyed agency technical managers (COTRs), who are in charge of the research areas within which SBIR awards are made. The survey sought to measure the *quality* of the research from the perspective of technical staff who managed both SBIR and non-SBIR programs.

NASA's mission is to pioneer the future in space exploration, scientific discovery, and aeronautics research.

SOURCE: National Aeronautics and Space Administration, *<http://www.nasa.gov/about/highlights/what_does_nasa_do.html>*.

The research found that the COTRs believed that 68 percent of surveyed projects generated useful information and found that results from 58 percent of the projects were sufficiently positive to encourage the COTR to seek out additional funding, within or outside the SBIR program.

It is especially encouraging that over 30 percent of projects were sufficiently positive that technical managers sought non-SBIR funding within NASA for further development.

Agency mission at NASA can also be measured in terms of agency take-up of SBIR-funded technologies. This may be especially important at NASA, as the commercial market for space-related technologies is likely to be small. NASA does maintain a set of success stories, but Phase III activities do not appear to be tracked in any coordinated fashion.

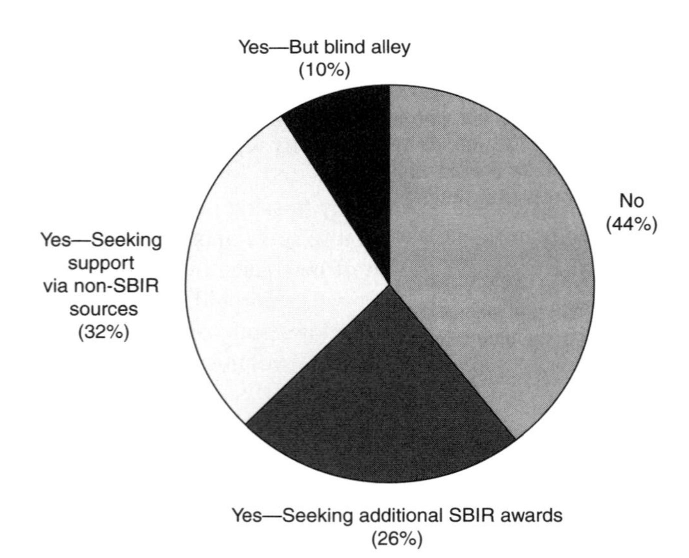

FIGURE 4-12 NASA staff perspectives on SBIR awards.
SOURCE: NRC Project Manager Survey.

4.4 SUPPORT FOR SMALL, WOMAN-OWNED, AND DISADVANTAGED BUSINESSES

4.4.1 Support for Woman- and Minority-owned Firms

Support for women and disadvantaged persons is one of the four primary congressional objectives for the SBIR program. Unfortunately, there are significant concerns regarding the collection of data related to the participation of woman- and minority-owned firms in the program. More importantly, the data that does exist raises questions about whether agencies adequately meet the needs of such firms.

4.4.1.1 Data Concerns

There are problems with the collection of data related to woman- and minority-owned firms at the two largest SBIR agencies. In both cases, these difficulties make it hard to determine how well each agency is meeting this element of the congressional mandate.

At DoD, the problem lies in the collection of data about applications. Even though DoD publishes an annual report which includes data on the number of applications and the success rates of women- and minority-owned firms in winning Phase I and Phase II awards, these data suffer significant deficiencies. For example, the demographic status of applications is entirely self-determined by applicants, and it is likely that in some cases this data is not reported accurately. BRTRC (a data contractor) has identified 53 firms that listed minority or women ownership on some, but not all of the proposals they submitted during FY2005. Looking across years, firms were identified that showed women ownership some years, then no status, then women ownership again. One firm that submitted about 10 proposals annually first listed itself as minority-owned, then for several years claimed no special ownership, then claimed to be woman-owned.

After awards are made and moved to a separate database table, DoD works to correct errors in the demographic status, so the awards data are much more accurate in this regard. Since firms get no preference in selection for being minority- or woman-owned, they may not be motivated to ensure that this part of the application is correct.

Because the applications data are not accurate, conclusions about success rates are also uncertain. And, as discussed below, it is hard to determine what the data mean—and why certain outcomes occur.

At NIH, the NRC study identified anomalies which, on closer inspection, indicated that for some years, NIH was not capturing women- and minority-ownership data accurately. Following discussion with the NRC, NIH made a significant effort to recalculate the data for women and minority participation in the NIH SBIR program. However, apparent anomalies in the NIH data for 2001 and 2002 could not be resolved by the time of publication of this report.[37]

---

[37]This is a correction of the text in the prepublication version released on July 27, 2007.

In addition, it would be helpful if all agencies captured and regularly reported data about the demographics of principal investigators (PIs), as well as company ownership, as many company founders have prior experience as PIs, so PI demographics may be a useful leading indicator of minority- and women-owned businesses in the program.

4.4.1.2 Award Patterns

Of the five agencies studied by the NRC, one—NASA—has no apparent issues in relation to awards to woman- and minority-owned firms. That is to say, their pattern of awards matches those for other firms, and their share of awards is in line with the average for all agencies, and shows no obvious negative trends in any area.

This is not the case for the remaining four agencies. At DoD, there has been a substantial decline in the award shares of minority-owned firms. The share of awards to woman-owned firms has been relatively low at NSF, and the application success rates for woman-owned firms have been lower than those for other businesses. At DoE, success rates for both woman- and minority-owned firms have been lower than for other small businesses. At NIH, issues focus on award shares for minority-owned firms, and on the discordance between the award share to woman-owned firms and the number of female scientists and engineers working in the life sciences. And, as noted the lack of quality data suggest that this issue is not adequately monitored or analyzed, reflecting a need for greater management attention.

DoD

While Phase I awards to woman-owned firms have continued to increase as a percentage of all Phase I awards, the percentage of Phase I awards being made to minority-owned firms has declined since the mid-1990s. The percentage fell below 10 percent for the first time in 2004, and is down by a third since the early 1990s.

The absence of reliable data on the demographics of applications makes it impossible at this point to determine why minority-owned businesses are getting a declining share of Phase I awards. We do not know whether minority-owned business applications are down, whether the success rate of those applications has fallen, or both.

NSF

At NSF, the existence of good applications data makes it relatively easy to determine why woman-owned firms have been doing less well in recent years. Woman-owned businesses have been less successful in getting applications approved than have all applicants in every year since 1994. In half the years, the success rate of woman-owned businesses was less than 70 percent the rate of

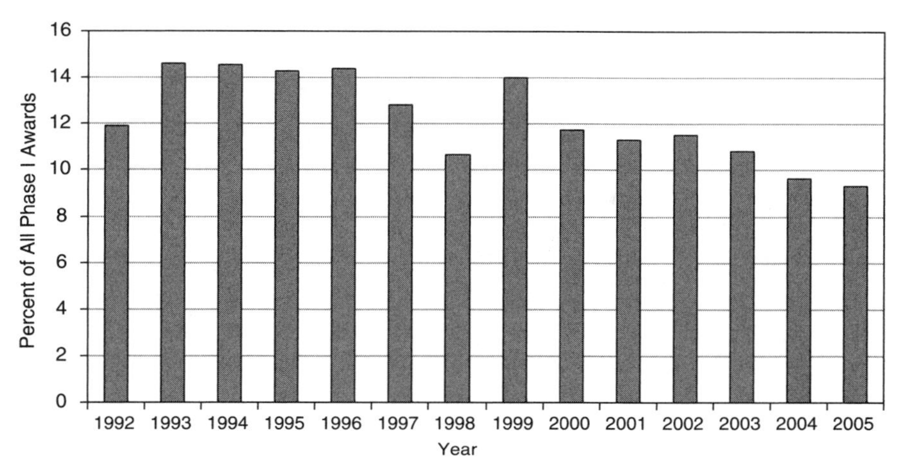

FIGURE 4-13 Minority-owned business shares of Phase I awards at DoD, 1992-2005.
SOURCE: Department of Defense Awards Database.

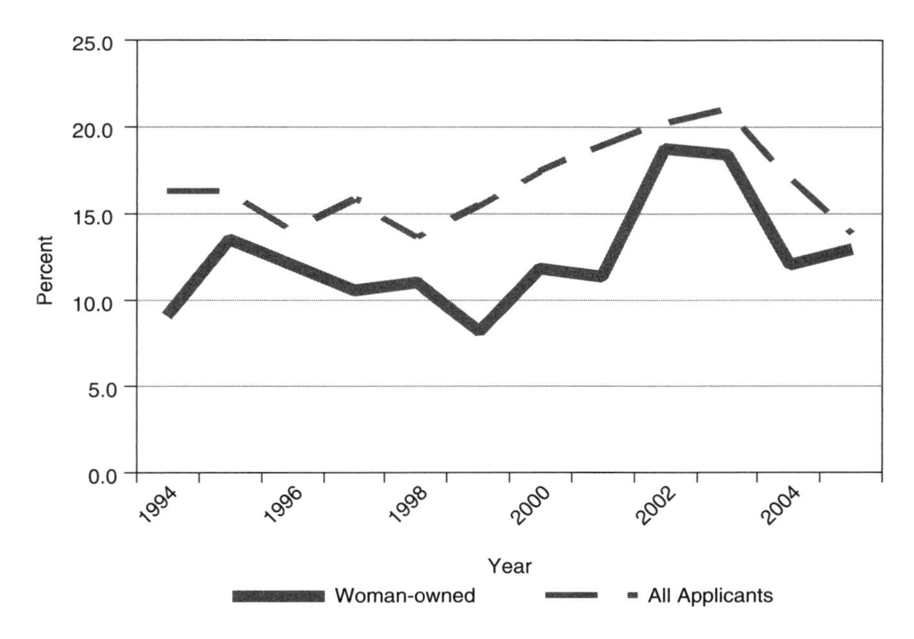

FIGURE 4-14 NSF: Comparative success rates for woman-owned and for all applicants in having their Phase I applications approved, 1994-2005.
NOTE: This is a correction of the prepublication version released on July 27, 2007.
SOURCE: Developed from data provided by the NSF SBIR program.

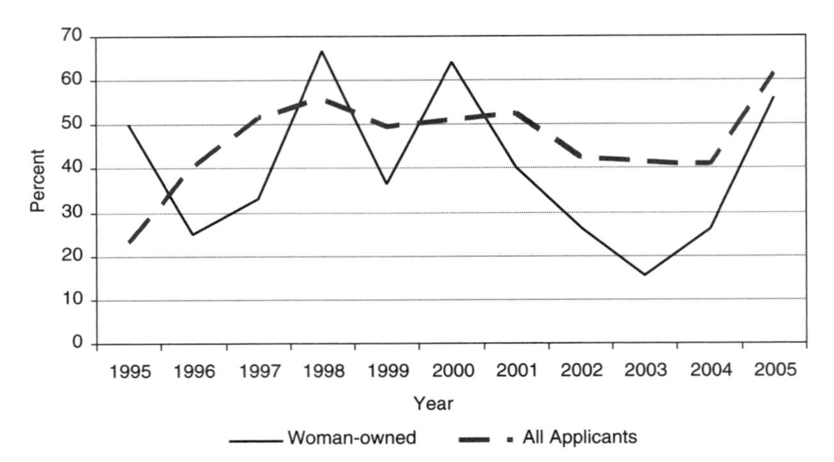

FIGURE 4-15 NSF: Comparative Phase II success rates for woman-owned and for all applicants, 1995-2005.
NOTE: This is a correction of the prepublication version released on July 27, 2007.
SOURCE: Developed from data provided by the NSF SBIR program.

all applicants. Over the period 1994-2005, woman-owned businesses accounted for 12 percent of all Phase I applications but received less than 10 percent of all Phase I awards.

Low success rates occur during Phase II as well. In all except three of the ten years from 1995 through 2005, the success rate of woman-owned businesses fell below that of all businesses in having their Phase II applications funded. Woman-owned businesses contributed 9 percent of all Phase II applications submitted from 1995 to 2003 (203/2,299) and received 7.5 percent of all Phase II grants (79/1,059). From 2002 to 2003, the success rate of woman-owned businesses in getting Phase II grants was particularly low but recently has recovered significantly.

DoE

At DoE, success rates are also an area of interest. Woman-owned businesses accounted for almost 8 percent of all DoE Phase I awards during 1992-2003, and for 10.5 percent of Phase I applications. During the same period, woman-owned businesses accounted for 7.8 percent of Phase II applications in 1995-2003 and for 6.7 percent of actual Phase II awards.

Data on minority-owned firms was similar. They submitted 16.5 percent of all Phase I applications in 1992-2003 and received 13.2 percent of all Phase I awards. Between 1995 and 2003, they accounted for 13.7 percent of Phase II applications and 11.8 percent of Phase II awards.[38]

[38]Data are unreported by woman- and minority-ownership status for 1992-1994.

DoE is the only agency where minority-owned firms apply more and receive more awards than woman-owned firms.

NASA

The relatively steady trends for both Phase I and Phase II awards at NASA for both woman- and minority-owned businesses are shown below (see Figures 4-16 and 4-17).

The absence of detailed applications data for woman- and minority-owned businesses means that we currently are not able to determine whether these trends are the result of a faster increase in these firms' number of proposals than other small businesses, improved success rates, or a combination of both.

4.4.2 Small Business Support

At one level, the SBIR program obviously provides support for small business, in that it gives funding only to businesses with no more than 500 employees—the SBA definition of a small business. However, it has been less clear whether SBIR has provided additional support for small business, or simply aggregates existing small business research dollars under the program's umbrella.

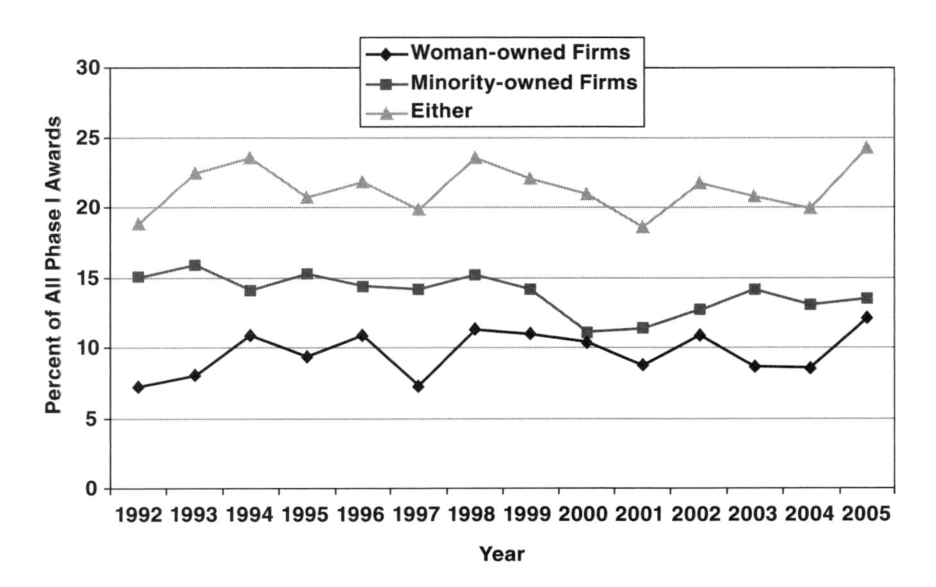

FIGURE 4-16 Phase I awards at NASA: woman- and minority-owned businesses' share of all awards.
NOTE: This is a correction of the prepublication version released on July 27, 2007.
SOURCE: National Aeronautics and Space Administration Awards Database.

FIGURE 4-17 Phase II awards at NASA: woman- and minority-owned businesses' share of all awards.
NOTE: This is a correction of the prepublication version released on July 27, 2007.
SOURCE: National Aeronautics and Space Administration Awards Database.

The most direct way to address the question is to review the amount of R&D funding going to small business at the agency as a whole, and then compare that with trends in SBIR funding. Unfortunately, this comparison is publicly available only for one agency, NIH, which publishes separate data on small business shares of research funding.

NIH data (Figure 4-18) show that the share of all NIH small business funding being disbursed through the SBIR program has fallen steadily since soon after the 1983 inception of the NIH program. After peaking at 90 percent of all small business research funding in the mid-1980s, SBIR's share fell steadily to about 72 percent, before falling further in 2004 (the most recently available data).

At NIH, the rapid growth of SBIR in recent years has supplemented, not supplanted, small business funding through other mechanisms at NIH. Increasing amounts—and shares—of small business research funding are available outside the SBIR program.

4.4.3 Project-level Impacts

One way of measuring SBIR's impact is to ask awardees whether their projects would have been implemented without SBIR program funding. Data from the

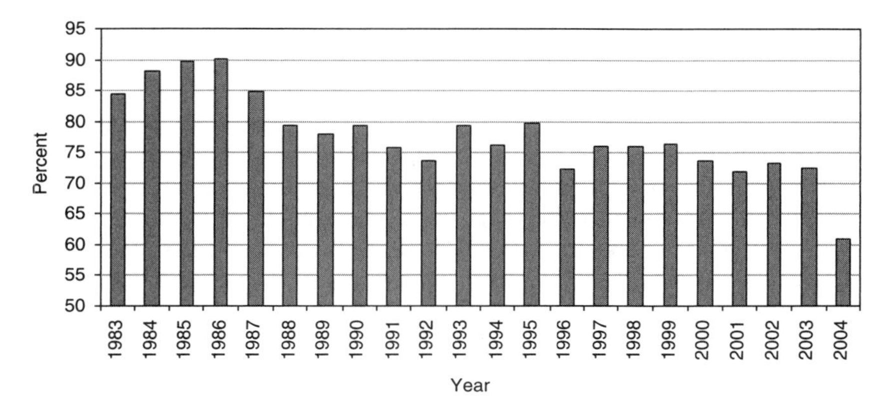

FIGURE 4-18 SBIR share of small business research funding at NIH
SOURCE: National Institutes of Health Awards database.

NRC Phase II Survey shown in Figure 4-19 strongly suggest that SBIR provides funding that plays a determinant role to most of the projects that receive it.

According to the respondents, more than 70 percent of projects would likely not have proceeded at all without SBIR. This finding reflects the known difficul-

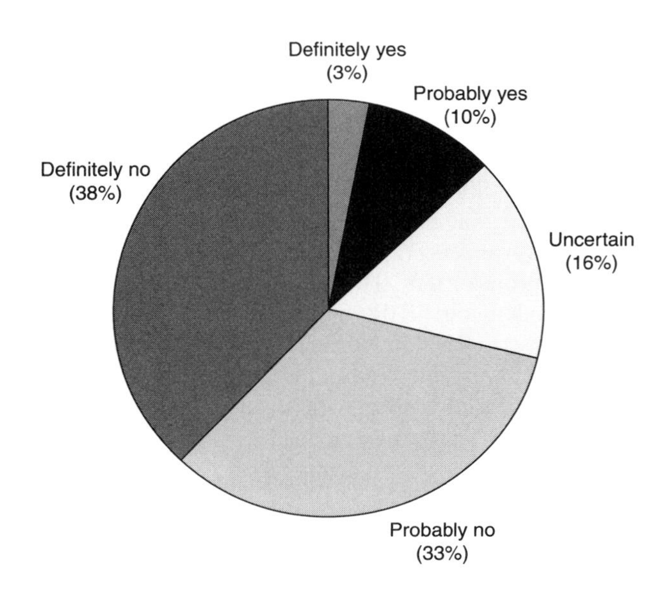

FIGURE 4-19 Would the project proceed without SBIR funding? (Percent of respondents.)
SOURCE: NRC Phase II Survey, Question 13.

ties in funding high-risk early-stage research in all scientific fields. SBIR seems to provide critical funding necessary to fund many early stage projects.

Respondents also indicated that many of the 13 percent of projects that were "definitely" or "likely" to have continued in the absence of SBIR funding would have had significant delays and other changes. More than half of these respondents (54 percent) noted that the project's scope would have been narrower. Sixty-two percent of projects that would have continued would have been delayed, and 47 percent expected the delay would have been at least 12 months.[39]

4.4.4 SBIR Impacts Different Types of Companies in Different Ways

Professor Irwin Feller has proposed a typology of companies supported by SBIR funding, capturing the critical differences in company capabilities and aspirations. This section describes, using both data and case studies, how these five different kinds of companies (or different stages of company development) have been supported by the NIH SBIR program.

Feller's typology differentiates between five types of SBIR-supported company:

(1) **Start-up Firm.** This is a new firm, typically without marketable products, and usually with minimal funding and limited personnel resources.

(2) **R&D Contractor.** As described by Cramer, these firms make a strategic choice to specialize in the performance of R&D rather than in marketing products or services.[40]

(3) **Technology Firm.** These firms have developed a core technology, which is then deployed into products and services.

(4) **Scientific Firm.** These businesses are described by Cramer as "firms [that] are generally small and were founded by scientists to explore whether a particular research area can generate ideas or products that might attract investors."[41]

(5) **Transformational Firm.** These companies start out as highly (or partially) dependent on SBIR, or on other government R&D contracts, which they use to develop a product that turns out to have considerable commercial value. This leads the company to become a production-oriented commercial vendor, with a concomitant decrease in the role of SBIR on the firm's progression.

---

[39]NRC Phase II Survey, Questions 14 and 15.

[40]See Reid Cramer, "Patterns of Firm Participation in the Small Business Innovation Research Program in Southwestern and Mountain States," in National Research Council, *The Small Business Innovation Research Program: An Assessment of the Department of Defense Fast Track Initiative*, op. cit.

[41]Ibid.

TABLE 4-7 SBIR Awards and Company Foundation

	Firms Indicating Role of SBIR in Their Foundation	
	Number of Responses	Percent of Responses
No	908	79
Yes	92	8
Yes, in part	149	13
Total	1,149	100.0

SOURCE: NRC Firm Survey.

All of these firm types have been identified through case studies as benefiting from SBIR awards. However, SBIR impacts on company formation are an especially important component of the program's overall support for companies; in addition, the role of R&D contractor companies has been a matter of controversy within the program for some time.

4.4.4.1 Start-ups

Responses to the NRC Firm Survey indicate that just over 20 percent of firms were founded entirely or in part as a result of SBIR awards. These data are supported by a number of the cases examined for this study. One of them is summarized in Box 4-5.[42]

While supporting company formation is not a direct objective of the program, it is easily construed as part of the mandate to support small businesses, and the program does, in fact, meet this important objective.

4.4.4.2 R&D Contractors

R&D contractor-type companies have been the subject of pointed criticism from some observers of the SBIR program. These critics have noted that some of these companies receive substantial SBIR program funds, but produce little in the nature of commercial goods and services. There is also minimal "take-up" of their R&D into the acquisition streams of the procurement agencies. These companies have been derogatively referred to as SBIR "mills," living off SBIR awards.

This issue is discussed in more detail in the section in Chapter 2 on multiple-award winners. Here, however, we should simply note that this criticism often misses important changes at these companies. The last point has more general application: reliance on SBIR awards tends to decline as the size of the company

[42]See Sociometrics case study in Appendix D for a more detailed review of SBIR at this company.

BOX 4-5
Company Formation: The Case of Sociometrics, Inc.

Sociometrics is a successful winner of SBIR awards that has also successfully commercialized some of its SBIR-generated technologies. Over the past twenty years, it has received more than 15 Phase II awards, and has become a premier publisher of educational and software materials related to behavioral change.

The founder of Sociometrics, Dr. Josefina Card, was a university researcher when she came into contact with Ms. Mary Baldwin, then a program officer at the National Cancer Institute. According to Dr. Card, Ms. Baldwin encouraged her to start a company that could seek NIH funding via the new SBIR mechanism then being developed. Dr. Card formed the company, then applied for and won Phase I and Phase II awards.

According to Dr. Card, not only was SBIR funding absolutely critical to the company's survival during its early years, the company would never have been founded without access to the seed capital represented by SBIR.

grows, and most companies that survive do grow. This observation is supported by data from the NRC Phase II Survey.

Table 4-8 shows, for NIH awardees, how responding firms' SBIR focus compares with their revenue at the time of the survey (2005). The data show that as companies get larger, their reliance on SBIR revenues tends to decline. Of the 36 NIH respondents with at least $5 million in revenues, 30 (78.9 percent) reported no more than a 10 percent focus on SBIR. Conversely, of the 102 firms reporting at least 80 percent focus on SBIR, 100 reported annual firm revenues of no more than $1 million. The data also show that there are a number of companies with revenues of $1-5 million where SBIR accounts for more than half of the revenues.

4.4.4.3 Transitioning Firms

A recurring theme in the case studies is that contract research is often used as a bridge to commercialization. For example, Polymer Technologies—an NIH awardee—was originally a contract research house, but it now manufactures a wide range of high technology medical devices at its Berkeley, California plant. It provides OEM services to major device manufacturers around the world.

Similarly, T/J Technologies—an NSF awardee—currently obtains most of its revenue from contract research, but its longer term strategy is to develop partnerships for commercializing its material technologies.

Firms' strategies evolve, and in particular, companies that now only do con-

TABLE 4-8 NIH Phase II SBIR Awards by Overall Company Revenue and Percentage Dependence on SBIR as a Source of Revenue

	Number of Phase II Awards by Percent Dependence on SBIR						
Firm Revenues ($)	0	1-10	11-25	26-50	51-75	76-100	Total
0	31	41	2	2	2	14	61
<100k	12	15	7	16	14	29	81
100k–<500k	10	16	7	9	18	25	75
500k–<1M	28	33	17	28	21	32	131
1M–<5M	13	26	3	10	20	2	61
5M–<20M	2	21	5	2	2	0	30
20M–<100M	0	3	0	1	0	0	4
100M+	2	2	0	0	0	0	2

SOURCE: NRC Firm Survey.

tract research may have ambitions or find it necessary to move to other models. Agencies' push to commercialization can encourage this process.

4.4.5 Growth Effects

No existing data sets measure the effect of SBIR on company growth. However, NRC survey respondents did provide their own estimates of SBIR impacts on their companies' growth.

Almost 55 percent of respondents indicated that more than half of the growth experienced by their firm was directly attributable to SBIR. This is evidence of

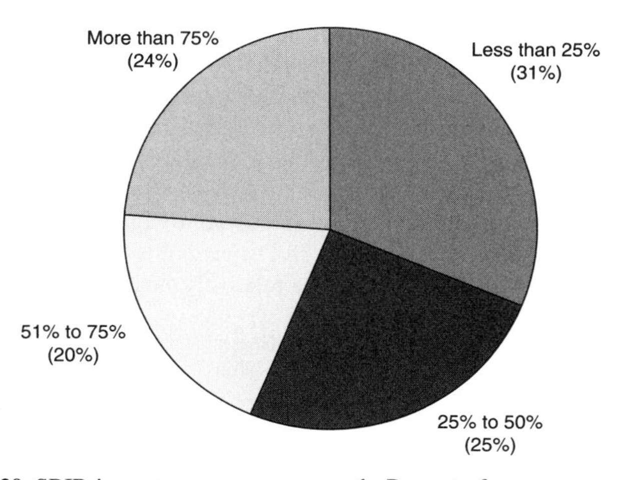

FIGURE 4-20 SBIR impacts on company growth: Percent of company growth attributable to SBIR awards.
SOURCE: NRC Firm Survey.

the powerful impact of SBIR on the future development of firms winning SBIR awards.

4.4.6 Conclusions

SBIR supports small high technology businesses at a time when other sources of financial support are especially difficult to find. Businesses use these funds for a variety of purposes, in pursuit of several distinct strategies, as captured in Box 4-6.

4.5 MULTIPLE-AWARD WINNERS AND NEW PROGRAM ENTRANTS

In this section, we will review the data about both the incidence of firms winning multiple SBIR awards and the extent to which the program is open to new companies that have not previously won SBIR awards.

4.5.1 The Incidence of Multiple-Award Winners

The absence of a high quality unified database of award winners makes it difficult to track MAWs. Because there is no single ID for SBIR firms, MAWs must be tracked using firm names, and this method is not reliable: not only do firms change names, even the smallest change in the spelling of a firm name on the application cover sheet will create a second record for that firm.

4.5.2 Commercialization and Multiple-Award Winners

While the DoD commercialization database does not contain information on all companies or all awards,[43] it does provide the best data on overall outcomes from awards to multiple SBIR-award winners, partly because it is specifically designed to do so. DoD has been concerned to ensure that companies winning more than four or five awards (the number changed in 2005) are generating results that support DoD's mission.[44]

The DoD data covers firms and awards from all agencies, reported within the DoD system (by firms that apply for DoD awards). It is therefore incomplete. However, it does show both the fact that a relatively small number of companies

[43]According to BRTRC, which manages the DoD database, the data collected from the agencies on Phase II awards made from 1992 to 2001 identified 2,257 firms that had received at least one Phase II, but were not in the DoD database, and were therefore not included in the table. Of these 2,257, only six had received 15 or more Phase II during the ten years for which BRTRC received award data. Although inclusion of pre-1992 and post-2001 awards would have increased that number, it seems safe to conclude that the firms in the DoD data represent a large majority of multiple winners.

[44]Although the database covers all agencies, some agencies are underrepresented owing to the focus on DoD-oriented firms. NIH awardees, for example, account for only 7 percent of entries in the database.

BOX 4-6
SBIR as an Enabler and Lifeline for Some High-Tech Companies

Faraday Technologies, Inc. The SBIR program enabled the company to undertake research that otherwise it would not have done. It sped the development of proofs of concept and pilot-scale prototypes, opened new market opportunities for new applications, led to the formation of new business units in the company, and enabled the hiring of key professional and technical staff. The SBIR "is well structured to allow taking on higher risk . . ."[a]

Immersion Corporation. SBIR grants gave Immersion the ability to grow its intellectual property portfolio, the core of its commercial success. The company leveraged the government funding to attract investment funding from private sources.

ISCA Technology, Inc. The SBIR program was essential to the survival of the company after it hit a major financial setback on its initial path. "The NSF SBIR gave us lots of prestige; it gave us credibility." The company used SBIR funding to bring advanced technology to a predominantly low-tech area.

Language Weaver. ". . . the STTR/SBIR from NSF created Language Weaver and what we are today. Without that we would have shelved the technology."

MicroStrain, Inc. The company found the NSF SBIR program's "more open topics" particularly helpful in the early stages when the company was building capacity.

MER Corporation. The SBIR program allowed the company to steadily improve and advance its R&D capabilities. It also helped MER's owners to keep control of their company.

National Recovery Technologies, Inc. "Without the SBIR program, NRT wouldn't have a business. We couldn't have done the necessary technical development and achieved the internal intellectual growth. . . . SBIR saved our bacon."

[a]See Faraday Technology case study in National Research Council, *An Assessment of the SBIR Program at the National Science Foundation*, Charles W. Wessner, ed., Washington, DC: The National Academies Press, 2008.

account for a larger share of awards, and also that there are now very many firms that win a great many awards.

The top 1 percent of winners—27 firms—accounted for 16.8 percent of all recorded Phase II projects. The top 7 percent of award winners—200 firms—accounted for 46.5 percent of all awards.

Of course, these data alone do not determine whether any of these firms are "mills"—whether they do not commercialize, and whether they are SBIR-dependent for revenues.

TABLE 4-9 DoD Data on Multiple Award Winners, All Agencies, 1992-2005.

Number of Phase II SBIR per Firm	Number of Firms	Number of Projects in CCR Database	Percent of Firms	Percent of Projects
≥125 projects	5	941	0.2	6.2
≥75 and <110 (no firms had between 111 and 124 projects)	5	485	0.2	3.2
≥50 and <75	17	1,067	0.6	7.0
≥25 and <50	77	2,692	2.6	17.8
≥15 and <25	101	1,858	3.5	12.3
>0 and <15	2,715	8,101	93.0	53.5
Total	2,920	15,144	100	100

SOURCE: Department of Defense Commercialization Database.

The first question can be answered, at least in aggregate. As noted, data from the DoD commercialization database indicates that the largest award winners are in fact—on average—stronger commercializers than firms with few awards.

The 2,920 firms reporting through the commercialization database reported a total of $20.3 billion in commercial outcomes, including $13.2 billion in sales. The five companies with more than 124 projects generated approximately 60 percent greater commercial returns on average than did the projects with fewer than 15 awards, and accounted for $1.4 billion in sales.

The 7 percent of firms with the highest number of awards—which accounted for 46.5 percent of awards, accounted for 60 percent of the commercial outcomes. This does provide a rebuttal to the claim that multiple winners are simply mills: At the most basic level, the data show that the multiple winners commercialize more than firms receiving fewer awards.

Thus the criticism of MAWs seems in general to be misplaced. It results from an overly negative reading of limited data focused too tightly on raw commercialization data, and perhaps on assessments of the program in its first years of operation.[45]

A fuller assessment of the role of multiple winners reveals multiple dimensions, and multiple contributions:

- **Active Commercialization.** Aggregate data from the DoD commercialization database indicates that the companies winning the most awards are, on average, more successful commercializers than those winning fewer awards.
  - While data from this source are not comprehensive, they do cover the vast majority of MAWs—at least at DoD—and the data indicate that,

[45]See, for example, U.S. General Accounting Office, *Federal Research: Small Business Innovative Research Shows Success But Can be Strengthened*, GAO/RCED-92-37, op. cit.

TABLE 4-10 Commercialization Results from Multiple Award Winners

Number of Phase II SBIR per Firm	Average Commercialization of Projects with Award Years prior to 2004 ($)	Average Sales of Projects with Award Years Prior to 2004 ($)	Average Funding of Projects with Award Years Prior to 2004 ($)
≥125 projects	2,067,719	1,384,571	683,148
≥75 and <110 (no firms had between 111 and 124 projects)	1,117,325	526,623	590,703
≥50 and <75	4,103,125	3,586,611	516,514
≥25 and <50	1,710,140	1,048,787	661,354
≥15 and <25	1,375,061	863,310	511,750
>0 and <15	1,300,886	751,418	549,468

SOURCE: Department of Defense Commercialization Database.

on average, firms with the largest number of awards commercialize as much or more than do all other groups of awardees; in the aggregate, there is failure of multiple-award winners to commercialize.

- ○ Case studies also show that some of the biggest award winners have successfully commercialized, and have also met the needs of sponsoring agencies in other ways.

- **Declining SBIR Revenues.** For some MAWs at least, even though they continue to win a considerable number of awards, the contribution of SBIR to overall revenues has declined: At the commonly cited "mill" firm Radiation Monitoring Devices, for example, SBIR now only accounts for 16 percent of total firm revenues.
- **Graduation.** Some of the biggest Phase II winners have graduated from the program either by growing beyond the 500 employee limit or by being acquired. Foster-Miller, for example, was bought by a foreign-owned firm. Legislating to solve a problem with companies that are, in any event, no longer eligible seems inappropriate.
- **Agency Research.** Contract research can be *valuable* in and of itself. Agency staff indicate that SBIR fills multiple needs, many of which do not show up in sales data. For example, some agency staff have noted that they use SBIR awards to conduct low-cost probes of the technological frontier. Awards are conducted on time and on budget, and can effectively test technical hypotheses, potentially saving extensive time and resources later.
- **Spin-off Companies.** Some MAWs spin off companies to commercialize innovations. Creating new firms is a valuable impact and a valuable contribution of the program.

- **Rapid Response.** MAWs have provided the highly efficient and flexible capabilities needed to solve pressing problems rapidly. For example, Foster Miller Inc. responded to needs in Iraq by developing and the manufacturing add-on armor for Humvees.[46]

All these points suggest that while there have been companies that depend on SBIR as their primary source of revenue for a considerable period of time, and there are some who fail to develop commercial results, such behavior is limited and does not constitute a significant impediment to SBIR program goals.

There is one final point to consider. Given the fact that SBIR awards meet multiple agency needs and multiple Congressional objectives, it is difficult to see how the program might be enhanced by the imposition of an arbitrary limit on the number of applications per year, as is currently the case at NSF. The evidence—outlined above—supports the conclusion that there is no multiple winner problem per se, although there may be "problem," i.e., underperforming, firms—a judgment the agency management must make. To the extent agencies continue to see issues in this area, we would strongly suggest that they consider adopting some version of the DoD "enhanced surveillance" model, in which multiple winners are subject to enhanced scrutiny in the context of the award process. The exclusion model used by NSF seems simply inappropriate.

4.5.3 The Incidence of New Entrants

Just as analysis of multiple winners provides a view of award concentration, analysis of new entrants provides a view of award dispersion. The data unequivocally show that the agency SBIR programs are open to significant numbers of new entrants.

4.5.3.1 National Science Foundation

In every year except 1999, the percentage of NSF award-winning companies that have not previously won Phase I awards at NSF is over 50 percent, and the percentage increased by more than half since 1999, to 63.1 percent in 2003.

[46]Foster-Miller's Last Armor TM, which uses Velcro-backed tiles to protect transport vehicles, helicopters and fixed wing aircraft from enemy fire, was developed on two Phase I SBIRs and a DARPA Broad Agency Announcement. The technology helped improve the safety of combat soldiers and fliers in Bosnia and Operation Desert Storm. Access at <*http://www.dodsbir.net/SuccessStories/fostermiller.htm*>. (This is a correction of the text in the prepublication version released on July 27, 2007.)

4.5.3.2 National Institutes of Health

NIH is unique among the agencies in providing data on new applicants—companies not previously funded at NIH—and on new awards to those companies.

The trendline shows that the share of applications from previously unfunded companies has actually increased somewhat, and, in any event, remains within fairly narrow boundaries (58-63.5 percent of applications come from previously unfunded companies). Given that previously funded companies have already passed a quality screen, it is not surprising that the share of awards going to previously unfunded companies is lower.

Nonetheless, at least 35 percent of awards have gone to previous nonwinners in each year since 2000, although that share has declined from 47 percent in 2000 to just over 35 percent in 2005. This decline might partly reflect the fact that the number of previous winners has increased sharply during this period, and that many of these new "previous winner" firms continue to apply for more awards.

4.5.3.3 Department of Defense

At Defense, the data appear to show similar award patterns for Phase II awards. According to the SBIR program office at DoD, 37 percent of FY2005 Phase II awards went to companies that had not previously won a Phase II SBIR award from DoD.

This steady infusion of new firms constitutes a major strength of the program. It underscores the positive impact of SBIR awards, as they encourage innovation activity across a broad spectrum of firms, create additional competition among suppliers for the procurement agencies, and provide agencies with new mission-oriented research and solutions.

4.6 SBIR AND THE EXPANSION OF KNOWLEDGE

Quantitative metrics for assessing knowledge outputs from research programs are well-known, but far from comprehensive. Patents, peer-reviewed publications, and, to a lesser extent, copyrights and trademarks, are all widely-used metrics, and are discussed in detail below.

However, these metrics do not capture the entire transfer of knowledge involved in programs such as SBIR. Michael Squillante, Vice-President for Research at Radiation Monitoring Devices, Inc., points out that there may be benefits from the development and diffusion of knowledge that are not reflected in any quantitative metric:

> For example, our research led to a reduction in the incidence of stroke following open-heart surgery. Under an NIH SBIR grant we developed a tool for medical

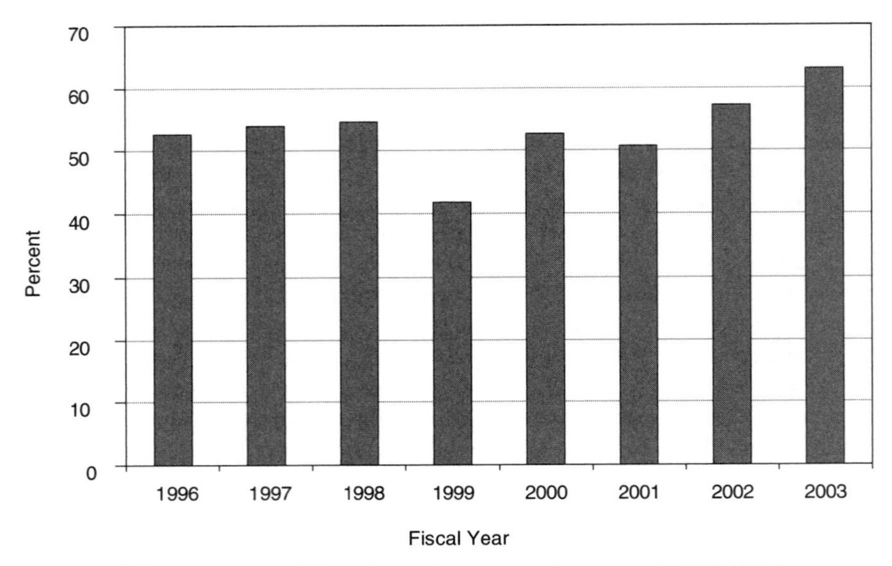

FIGURE 4-21 Percentage of Phase I awards to companies new to the NSF SBIR program, 1996-2003.
SOURCE: National Science Foundation.

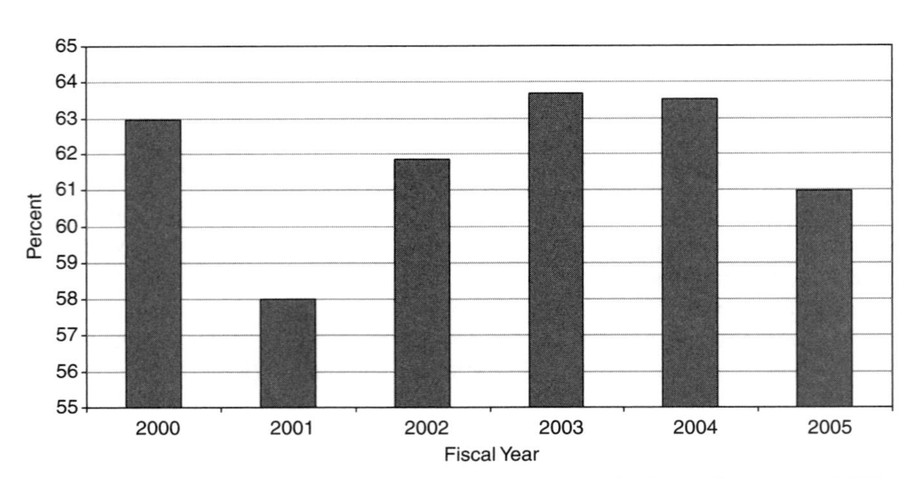

FIGURE 4-22 "New" Phase I applicants (percent of all applicants) at NIH, FY2000-2005.
SOURCE: National Institutes of Health data.

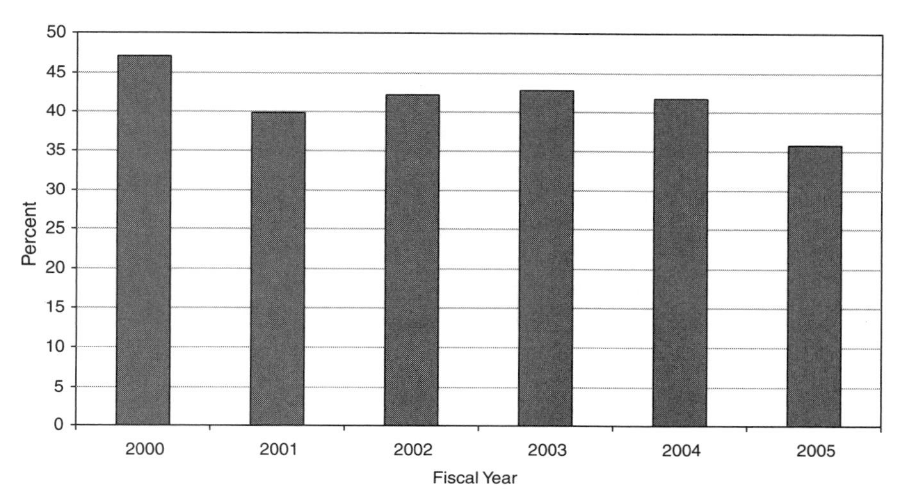

FIGURE 4-23 Percentage of Phase I awards to companies new to the NIH SBIR program, FY2000-2005.
SOURCE: National Institutes of Health data.

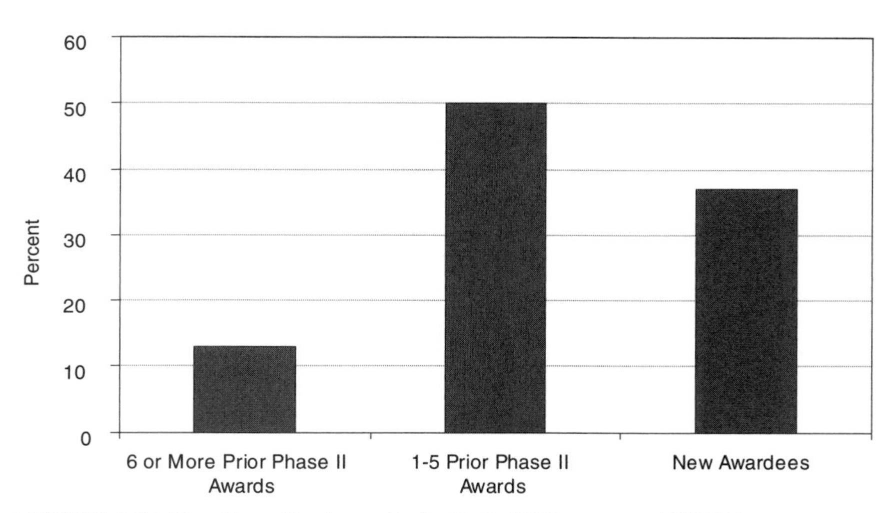

FIGURE 4-24 New Phase II winners in the DoD SBIR program, FY2005.
NOTE: Firm data as of March 2006. The "New Awardees" category includes prior Phase I award winners that have not won a Phase II award.
SOURCE: Michael Caccuitto, DoD SBIR/STTR Program Administrator, and Carol Van Wyk, DoD CPP Coordinator, Presentation to SBTC "SBIR in Rapid Transition Conference," Washington, DC, September, 27 2006.

researchers who were examining the causes of minor and major post-operative stroke occurring after open-heart surgery.[47]

This was not a project whose success could be measured in commercial terms; the market (stroke researchers) is far too small to generate large revenues. However, the impact of the research was profound.

It is, therefore, quite important to understand that the quantitative metrics discussed below are an indicator of the expansion of knowledge; they reflect that expansion but do not fully capture it. In particular, they say little about the impact of that knowledge. In the case illustrated above, for example, there was no patent, no commercial sales, no impact on the company's bottom line—but the research, nevertheless, made a remarkable improvement in outcomes for stroke patients undergoing surgery.

4.6.1 Patents

According to the Small Business Administration, small businesses produce 13 to 14 times more patents per employee than do large patenting firms. These patents are twice as likely as are large firm patents to be among the one percent most cited.[48]

With respect to SBIR, the NRC survey data indicate that about 30 percent of respondents received patents related to their SBIR research. About two-thirds of projects generated at least one patent application, and as with most other SBIR metrics, the distribution of outcomes is highly skewed.

The data show that three companies account for almost 20 percent of all patent applications related to the surveyed SBIR award. However, they were not especially successful, as no company reported more than 20 successful patents granted. Overall, 2.2 percent of the responding companies applied for more than five patents, and 1.3 percent received more than five.

More detailed case study analysis also indicates the importance of intellectual property to firm strategies. At NSF, a number of the case studies showed this clearly.

---

[47]Presentation to NRC Research Team by Michael Squillante, vice president, Radiation Monitoring Devices, Inc. June 11, 2004.

[48]Accessed on May 16, 2007 at <*http://app1.sba.gov/faqs/faqindex.cfm?areaID=24*>. Drawing on seminal empirical research, Acs and Audretsch found that small business have comparatively higher rates of innovation—specifically, that "the number of innovation increases with increased industry R & D expenditures but at a decreasing rate. Similarly, while the literature has found a somewhat ambiguous relationship between concentration and various measures of technical change, our results are unequivocal—industry innovation tends to decrease as the level of concentration rises." See Zoltan J. Acs and David B. Audretsch, "Innovation in Large and Small Firms: An Empirical Analysis," op. cit.

4.6.2 Scientific Publications

The NRC Phase II Survey determined that slightly more than half of the respondents had published at least one related scientific paper (45.4 percent). About one-third of those with publications had published only a single paper, but one company had published 165 papers on the basis of its SBIR research, and several others had published at least 50 papers.

These data fit well with case studies and interviews, which suggested that SBIR companies are proud of the quality of their research. Publications are featured prominently on many Web sites, and companies like Advanced Brain Monitoring, SAM Technologies, and Polymer Research all made the point during interviews that their work was of the highest technical quality as measured in the single measure that counts most in the scientific community—peer-reviewed publication.

Publications therefore fill two important roles in the study of SBIR programs:

- First, they provide an indication of the quality of the research being conducted with program funds. In this case, more than half of the funded projects were of sufficient value to generate at least one peer-reviewed publication.
- Second, publications are themselves the primary mechanism thorough which knowledge is transmitted within the scientific community. The existence of the articles based on SBIR projects is therefore direct evidence that the results of these projects are being disseminated widely, which in turn means that the congressional mandate to support the creation and dissemination of scientific knowledge is being met.

We note that no agency seems to have in place evaluation programs for determining whether similar knowledge effects are being generated at the same or at a different rate outside the SBIR program. Likewise, there is insufficient tracking to determine citation rates for these articles.

4.6.3 SBIR and Universities

According to agency and congressional staff, one of the implicit, though not formal objectives, of the SBIR program is that it supports the transfer of knowledge from universities to the commercial marketplace.

This transfer can happen in many different ways. SBIR funding:

- Helps university scientists form companies;
- Enables small firms to use university faculty and to employ graduate students as specialized consultants;
- Permits the use of university laboratory facilities; and
- Encourages other less formal types of collaboration.

TABLE 4-11 Distribution of Patents

Patent Applications					Patent Awards				
Number of Patents	Number of Companies	Total Number of Patents	Percent of Applicants	Percent of Applications	Number of Patents	Number of Companies	Total Number of Patents	Percent of Recipients	Percent of Patents Received
0	1,060	0	62.3	0.0	0	1,202	0	70.7	0.0
1	378	378	22.2	23.6	1	319	319	18.8	34.0
2	124	248	7.3	15.5	2	99	198	5.8	21.1
3	60	180	3.5	11.3	3	36	108	2.1	11.5
4	25	100	1.5	6.3	4	14	56	0.8	6.0
5	17	85	1.0	5.3	5	8	40	0.5	4.3
6	11	66	0.6	4.1	6	11	66	0.6	7.0
7	4	28	0.2	1.8	7	3	21	0.2	2.2
8	2	16	0.1	1.0	10	2	20	0.1	2.1
9	2	18	0.1	1.1	11	2	22	0.1	2.3`
10	6	60	0.4	3.8	15	2	30	0.1	3.2
11	4	44	0.2	2.8	17	1	17	0.1	1.8
12	1	12	0.1	0.8	20	2	40	0.1	4.3
13	2	26	0.1	1.6	Total	1,701	937	100	100
14	1	14	0.1	0.9					
25	1	25	0.1	1.6					
100	3	300	0.2	18.8					
Total	1,701	1,600	100	100					

SOURCE: NRC Phase II Survey, Question 18.

BOX 4-7
Intellectual Property and Company Strategy Among NSF Awardee Case Studies

A number of NSF case studies showed extensive use of patenting:

- Faraday Technology has had 23 patents issued in the U.S. and three foreign issued, since its founding in 1991. Historically, this amounts to 1.4 issued patents per employee. Patents and the fees they generate are the central focus of Faraday's business strategy, and the company investigates citations by other companies of its patents to obtain knowledge about potential customers.
- Immersion Corporation has more than 270 patents issued in the U.S. and another 280 pending in the U.S. and abroad. According to Immersion, its patent portfolio is at the heart of its wealth-generation capacity.
- ISCA has received a trademark on its most recent insect lure technology, which is expected to generate substantial growth for the company.
- Language Weaver has more than 50 patents pending worldwide, which underpin its commercialization approach. MER has had more than a dozen patents granted.
- MicroStrain reported patenting to be "very important" to its commercialization strategy.
- NVE reported 34 issued U.S. patents and more than 100 patents issued worldwide.
- T/J Technologies has had seven patents granted and has a number of others pending.

Over a third (36.5 percent) of all NRC Phase II Survey respondents indicated that there had been involvement by university faculty, graduate students, and/or a university itself in developed technologies. Based on the responses of all survey respondents, this involvement took a number of forms (see Table 4-13).

The wide range of roles played by university staff and students underscores the multiple ways in which SBIR projects enhance the knowledge base of the nation. Involvement in these projects provides different opportunities for university staff than those available within academia.

Many of the companies interviewed for this study noted that their university connections were extremely important. Radiation Monitoring Devices said that they spent more than $1 million annually on contracted research at universities under their SBIR awards.[49] Advanced Targeting Systems, in San Diego, has forged an extended and successful research partnership with a senior scientist at

[49]Interview with Michael Squillante, vice president, Radiation Monitoring Devices, June 10, 2005.

TABLE 4-12 Publications

Submitted					Published				
Number of Publications	Number of Respondents	Total Publications	Percent of Respondents	Percent of Total Publications	Number of Publications	Number of Respondents	Total Publications	Percent of Respondents	Percent of Total Publications
0	902	0	53.0	0.0	0	929	0	54.6	0.0
1	253	253	14.9	8.4	1	252	252	14.8	8.7
2	213	426	12.5	14.1	2	208	416	12.2	14.4
3	116	348	6.8	11.5	3	108	324	6.3	11.2
4	65	260	3.8	8.6	4	65	260	3.8	9.0
5	63	315	3.7	10.5	5	57	285	3.4	9.9
6	15	90	0.9	3.0	6	11	66	0.6	2.3
7	5	35	0.3	1.2	7	4	28	0.2	1.0
8	16	128	0.9	4.2	8	18	144	1.1	5.0
9	4	36	0.2	1.2	9	3	27	0.2	0.9
10	18	180	1.1	6.0	10	16	160	0.9	5.6
11	3	33	0.2	1.1	11	4	44	0.2	1.5
12	7	84	0.4	2.8	12	5	60	0.3	2.1
13	2	26	0.1	0.9	13	2	26	0.1	0.9
15	3	45	0.2	1.5	15	3	45	0.2	1.6
17	1	17	0.1	0.6	17	1	17	0.1	0.6
20	1	20	0.1	0.7	20	1	20	0.1	0.7
23	1	23	0.1	0.8	22	1	22	0.1	0.8
25	1	25	0.1	0.8	25	1	25	0.1	0.9
30	2	60	0.1	2.0	30	2	60	0.1	2.1
33	1	33	0.1	1.1	33	1	33	0.1	1.1
37	1	37	0.1	1.2	37	1	37	0.1	1.3
40	1	40	0.1	1.3	40	1	40	0.1	1.4
45	1	45	0.1	1.5	41	1	41	0.1	1.4
50	1	50	0.1	1.7	50	3	150	0.2	5.2
52	2	104	0.1	3.5	60	1	60	0.1	2.1
60	1	60	0.1	2.0	75	1	75	0.1	2.6
75	1	75	0.1	2.5	165	1	165	0.1	5.7
165	1	165	0.1	5.5	Total	1,701	2,882	100	100
Total	1,701	3,013	100	100					

SOURCE: NRC Phase II Survey, Question 18.

TABLE 4-13 University Involvement in SBIR Projects

2%	The Principal Investigator (PI) for this Phase II project was a faculty member.
3%	The Principal Investigator (PI) for this Phase II project was an adjunct faculty member.
22%	Faculty or adjunct faculty member(s) work on this Phase II project in a role other than PI, e.g., consultant.
15%	Graduate students worked on this Phase II project.
13%	University/College facilities and/or equipment were used on this Phase II project.
3%	The technology for this project was licensed from a University or College.
5%	The technology for this project was originally developed at a University or College by one of the percipients in this phase II project.
17%	A University or College was a subcontractor on this Phase II project.

SOURCE: NRC Phase II Survey.

the University of Utah, which they described as "critical to the development and testing of their products."[50]

These stories and the many others like them that we encountered suggest that the flow of information and funding between small businesses and universities working within the SBIR framework is not simple or unidirectional. For example, the constant flow of feedback, testing, and insights between university researchers and Advanced Targeting Systems staff helped to move the company forward toward product deployment into new research areas while providing real-life problems for research to address.

What is now clear from this research—and could certainly be the subject of further analysis—is the extent to which universities themselves see SBIR as a mechanism for technology transfer, commercialization, and additional funding for university researchers.

4.6.3.1 University Faculty and Company Formation

Data from the NRC Firm Survey strongly support the hypothesis that SBIR has encouraged university faculty to form companies to commercialize their inventions and technologies. Two-thirds of all responding companies had at least one academic founder, and more than a quarter had more than one.

The same survey found that about 36 percent of founders were most recently employed in an academic environment before founding the new company.

These data and evidence from case studies strongly suggests that SBIR has indeed encouraged academic scientists to bring their work to a more commercial environment.

[50]See Advanced Targeting System case study in National Research Council, *An Assessment of the SBIR Program at the National Institutes of Health*, op. cit.

4.7 CONCLUSIONS

The extended discussion of metrics and indicators in the Methodology Report show quite conclusively that there is no single simple metric that captures "results" from the program. Each of the four congressional mandates is best assessed separately, and within each, there are a multiple issues to be addressed.

Bearing all these points in mind, it is still possible to summarize the results of our research in straightforward terms.

Commercialization. Approximately 30-40 percent of Phase II projects produce innovations that reach the market, with a few generating substantial returns. Other indicators of commercialization, such as licensing activities, marketing partnerships, and access to and utilization of further investments from both private and public sources, all confirm that while returns are highly skewed, the results in general are positive, and are consistent with other public and private sources of early-stage financing.

Agency Mission. We found many examples of funded projects that clearly made a major contribution to the mission of the awarding agency. To an increasing extent, SBIR is becoming more firmly integrated into the overall R&D strategy of each agency. At the procurement agencies—DoD and NASA—we found that data from the agencies and from NRC surveys indicated that many SBIR projects did result in utilization by the agencies. We also determined that the agencies had made important efforts to improve the alignment between agency needs and SBIR implementation, and that procedures for encouraging or ensuring alignment were in place at all agencies. Traditionally hostile views of the program among many acquisition and R&D managers have begun to shift in a much more positive direction. Many projects made a very valuable contribution to agency mission, and these contributions were often far out of proportion to the commercial returns involved.[51]

Support for Small Business, and Woman- and Minority-owned Businesses. SBIR significantly supports small high technology businesses in general, and the NRC research team discovered that SBIR had an important catalytic effect in terms of company foundation—providing the critical seed money to fund a company's first steps. SBIR also had strongly influenced companies' decisions to initiate individual projects: more than 65 percent of NRC Phase II Survey respondents at every agency believed that their projects would not have gone forward without SBIR, and, of the remainder, most believed that the projects

[51]The use of Savi RFID tracking systems reportedly enabled large reductions in shipping requirements for U.S. force deployments. The potential savings far outweighs the commercial value of the tracking devices. See the Savi Technology case study in Appendix D of National Research Council, *An Assessment of the SBIR Program at the Department of Defense*, op. cit.

would have been delayed and/or would have had a reduced scope. While some efforts have been made to increase participation by woman- and minority-owned businesses, more needs to be done. At DoD and NIH data on these businesses were incomplete or missing. At all of the agencies except NASA, what data there was suggested that there were areas of concern in the program (different areas at different agencies), and it did not appear that the agencies had focused sufficient resources and management attention on these areas.

Support for the Advancement of Scientific and Technical Knowledge. The program funds cutting edge research, as it was designed to do. Most of the agencies use "technical innovation" in some form as a critical selection parameter. Outcomes in the form of patents and peer-reviewed scientific publications are encouraging, with about two-thirds of recipients publishing at least one related scientific paper. Surveys of program managers and technical points of contact at DoD, NASA, and DoE indicated that they saw SBIR projects as approximately equivalent in research quality to non-SBIR projects.[52] Moreover, case studies indicate that knowledge generated in SBIR-funded research can be picked up indirectly in many ways, and often continues to be productive long after the original project has been concluded.

It is therefore appropriate to conclude that the program is meeting all four of the congressional objectives.

[52]After results were normalized to eliminate the most favorable and most negative responses.

5

Program Management

5.1 INTRODUCTION

Each agency's Small Business Innovation Research (SBIR) program operates under legislative constraints that set out program eligibility rules, the program's three-phase structure, and (to a considerable extent) phase-specific funding limitations. Beyond these basic structural characteristics, programs have flexibility to decide how they select, manage, and track awards.

The similarities and differences across the agencies are discussed in detail in the following sections.

5.2 TOPIC GENERATION AND UTILIZATION

All applications for SBIR funding are made in response to published documents—termed solicitations—that describe what types of projects will be funded and how funding decisions will be made. Each agency publishes its solicitation separately. Subject areas eligible for SBIR funding are referred to as "topics" and "subtopics," and each agency chooses its topics and subtopics in a different fashion with different goals in mind.

We have identified three distinct models of topic use at the SBIR agencies:

(1) **Acquisition-oriented Topic Procedures.** The Department of Defense (DoD) and the National Aeronautics and Space Administration (NASA) solicit for topics that channel applications toward areas that are high acquisition priorities for the awarding agency. For these agencies, the goal is to target R&D spending towards projects that will provide technolo-

gies that can eventually be acquired by the agency for agency use. Applications are not accepted outside topic areas defined in the solicitation.

(2) **Management-oriented Topic Procedures.** At the National Science Foundation (NSF) and the Department of Energy (DoE), topics are used to serve the needs of agency management, not as a method of acquiring technology. NSF primarily focuses on ensuring that the research agendas of the various NSF directorates (divisions) are closely served. This enhances agency staff buy-in for SBIR. At DoE, topics are also used as a screen to reduce the number of applications so that the agency's limited SBIR staff is able to manage the program effectively. As a result, neither agency will accept applications outside topic areas defined in the solicitation.

(3) **Guideline-oriented Topic Procedures.** The National Institutes of Health (NIH) is the only agency that uses topics as guidelines and indicators, not hard delimiters. NIH issues annual topic descriptions, but emphatically notes that it primarily supports investigator-driven research. Topics are defined to show researchers areas of interest to the NIH Institutes and Centers (ICs), not to delimit what constitutes acceptable applications. Applications on any topic or subject are therefore potentially acceptable.

5.2.1 Acquisition-oriented Approaches

Acquisition-oriented approaches are designed to align long range research with the likely needs of specific agencies and programs. How well this works depends to a great degree on the specific topic development mechanisms used at each agency.

5.2.1.1 Topic Development

In DoD, topics originate in Service Laboratories[1] or in Program Acquisition Offices. Many awarding units within DoD do not have their own laboratories, and depend on the Service Laboratories for "in-house" expertise. They request topics from these experts, who thus become topic authors. These authors frequently become the technical monitors for the contracts that are awarded based on their topics.

Since 1999, DoD has made a number of efforts to ensure that topics are closely aligned with the needs of acquisition programs. For example, each major acquisition program has a SBIR liaison officer, who works with SBIR program managers within DoD, and with the SBIR contractor community. Their job is to provide a mechanism through which contractors can communicate with end cus-

[1]The laboratories develop technologies to meet long-term agency needs.

tomers in acquisition programs. The liaison officers may author topics, or cause them to be authored.

5.2.1.2 Topic Review

The topic review process at an acquisition-oriented agency is stringent and extensive. Alignment with mission and with anticipated technology needs is crucial. After each individual DoD component has completed its own internal topic review process, the results go through the department-wide DoD review process. This multistage topic approval process takes a considerable amount of time, sometimes more than a year.

Intracomponent review varies in duration, somewhat correlated by size (smaller components with fewer proposed topics take less time). The Army topic review is a centralized online process that takes 4 months, while Air Force review is less centralized (but also online) and takes up to 15 months.

5.2.1.3 Broad versus Narrow Topics

There is a tension between the need to write topics tightly to ensure that they align well with acquisition programs, and the desire to write them broadly enough to fund novel mechanisms for resolving agency problems. Components have tended to adopt one or other of these approaches. For example, the Missile Defense Agency (MDA) has until recently had only a small number of broad topics, evolving only gradually from year to year, which provides great flexibility to proposing firms.

DoD is addressing this tension by calling for all topics to be written to allow a bidder "significant flexibility" in achieving the technical goals of the topic.[2] For example, the Navy might need corrosion-resistant fastenings for use on ships. Instead of writing a topic that specifies the precise kinds of technology to be utilized, managers are now pushing for topics that simply state the problem and leave the technology itself up to applicant.

5.2.1.4 Quick Response Topics

Particularly since 9/11, and the start of the conflict in Iraq, DoD has had acquisitions needs with short timeframes. Quick response topics offer a new way to short-cut the often lengthy topic review process.

As part of the DoD's SBIR FY-2004.2 solicitation, the Navy added three special SBIR Quick Response Topics. The rules for these topics were slightly different: They offered a three- to six-month Phase I award of up to $100,000 (with no option), with successful Phase I firms being invited to apply for a Phase

[2]This is a correction of the text in the prepublication version released on July 27, 2007.

II award of up to $1,000,000.[3] The topics were released more rapidly because they did not go through the DoD level approval process.

5.2.1.5 Topic/Funding Allocation

The allocation of topics effectively controls the allocation of money. There are two basic models for agency funding allocation via topics:

- **Percentage of the Gross.** All SBIR programs struggle with the problem of agency buy-in. Many agencies have traditionally viewed SBIR as a "small business tax" on their research budgets. This is even more important where the agency is itself the most important market for SBIR products and services, and acquisitions officers are the key customers. SBIR is funded through a set-aside of 2.5 percent of agency extramural research, and some senior agency staff have claimed that they could find better ways to spend that money. To address these concerns, some agencies and some components have designed their SBIR programs to tighten the link between the results of the SBIR research and the agency unit that has been "taxed" to fund the SBIR program. One way to do this is by allotting a specific number of topics to each funding unit, which then effectively controls the content of the research funded by "its" SBIR money. For example, in the Army and Air Force, the number of topics is allocated to agency laboratories specifically on the basis of each lab's overall R&D budget—and hence its contribution to the SBIR program funding pool.[4]
- **Technology-driven.** Topic decisions can be more centralized or otherwise divorced from funding unit control. This allows more flexible administration and more rapid response to urgent needs (via easier reallocation of funding between technical areas), but risks reduced unit buy-in.

5.2.1.6 DoD Pre-release

Federal Acquisition Regulations prohibit any SBIR applicant contact with the relevant agency after a solicitation opens, other than through written questions to the contracting officer, who must then make the question and the answer available to all prospective bidders.

Under its pre-release program, DoD posts the entire projected solicitation on the Internet about two months before the solicitation is to open. Each topic includes the name and contact information for the topic author. Firms may contact

[3]The Navy's quick review approach is discussed further in National Research Council, *An Assessment of the SBIR Program at the Department of Defense*, Charles W. Wessner, ed., Washington, DC: The National Academies Press, 2009.

[4]The Air Force also allocates the number of funded Phase I projects.

the authors, and discuss the problem that the government wants to solve as well as their intended approach.

Often, a short private discussion with the topic author can help a company avoid the cost of a proposal, or give the company a better idea of how to approach the SBIR opportunity. Small firms can learn of possible applications of their technology in acquisition programs, and determine which primes are involved in those programs. Firms new to SBIR often get procedural information, and are steered to the appropriate DoD Web sites for further information.

5.2.1.7 Company Influence on Topic Development

In some sense, private company influence on topic development is both positive and sought after. Companies often understand cutting edge technologies better than the agencies, and may be able to suggest innovative lines of research. At DoD, involvement of prime contractors in topic development is one further way to help build linkages between the major acquirers of technology and the SBIR program.

There have been some criticisms that topics are "wired"—that they are written for a specific company. There is no simple way to determine whether topics are wired, at DoD or elsewhere. Since about 40 percent of DoD Phase I winners are new to the program does, however, suggest that new or unconnected firms do have significant opportunities to participate in the SBIR program. It is also worth noting that while firms with long connections to DoD are best placed to help design topics friendly to their specific capacities, those long connections are in many cases, according to agency staff,[5] the result of effective performance building to a strong track record in the course of previous SBIR awards.

5.2.2 Management-oriented Approaches to Topic Utilization

Management-oriented approaches differ from the acquisition-driven programs in that they do not evaluate applicants based on the government agency's acquisitions needs.

Instead, there are management priorities. The two agencies using this approach differ substantially in the origins of these priorities. At NSF, SBIR funding decisions are driven by the technical decisions of the major funding directorates. Each directorate provides one annual list of topics, and their objective is explicitly to direct SBIR funding into targeted research areas. DoE's priorities are described in the case study below.

[5]See National Research Council, *SBIR and the Phase III Challenge of Commercialization*, Charles W. Wessner, ed., Washington, DC: The National Academies Press, 2007.

BOX 5-1
Topics at DoE Technical Areas

For the 2003 competition, the 11 program areas at DoE were as follow:

- Defense Nuclear Nonproliferation (3 Topics: 82 applications received);
- Biological and Environmental Research (6: 140 apps.);
- Environmental Management (2 Topics 10-11: 38 apps);
- Nuclear Energy (1 Topic: 15 apps);
- Basic Energy Sciences (8 Topics: 222 apps);
- Energy Efficiency (6 Topics: 287 apps);
- Fossil Energy (6 Topics: 169 applications received);
- Fusion Energy Sciences (3 Topics: 66 applications received);
- Nuclear Physics (4 Topics: 56 applications received);
- Advanced Scientific and Computing Research (2 Topics: 62 applications received); and
- High Energy Physics (6 Topics: 87 applications received).

SOURCE: Department of Energy.

5.2.2.1 Case Study: DoE

DoE attempts to provide each of its technical program areas with a "return on investment" equal to its SBIR contribution. Each year, for each technical program area, the SBIR office attempts to ensure that the dollar amount of all awards (Phase I plus Phase II) awarded to proposals submitted to that program area's technical topics is equal to that program area's contribution to the SBIR set-aside. Although the dollar amount is the primary consideration, the number of awards each program area gets—in both Phase I and Phase II—is also roughly proportional to the funding it provides SBIR awards.

To generate a "fair" return, each program area is provided with a number of technical topics that is proportional to its share of the total contribution. For example, if Fossil Energy (FE), provided 10 percent of the DoE's set-aside, FE would receive 10 percent of the technical topics, and also would receive Phase I and Phase II awards (from proposals submitted to those topics) whose dollar value equaled 10 percent of the set-aside.

About six months before the fall publication of the annual program solicitation, the DoE SBIR Program Manager sends a "call for topics" memo to all Portfolio Managers responsible for technical program areas within DoE. This call also goes to all technical topic managers (TMs) from the prior year. The call includes the National Critical Technologies List,[6] as well as guidelines for topic development.

[6]As conveyed to DoE from the U.S. Small Business Administration.

Topics and subtopics are submitted to or generated by the portfolio managers, who determine which topics to forward to the SBIR Program Office. Portfolio managers develop a topic list that matches the amount of funding allocated to them, and this list is 100 percent funded by the SBIR Office, which performs no further review.

The DoE SBIR Program Manager then edits the topics into a common format, and may seek to narrow topics to limit the number of applications.[7] About 50 percent of DoE topics change from year to year.

5.2.3 Investigator-driven Approaches

NIH is proud of its investigator-driven approach. The agency believes its approach is the best way to fund the best science, because it substitutes the decisions of applicants and peer reviewers for the views of agency or program staff.

A small but growing percentage of NIH SBIR funding, however, goes to Program Announcements (PAs) and Requests for Applications (RFAs), which are Institute-driven requests for research proposals on specific subjects, covering the remainder. Discussions with agency staff suggest that this percentage may increase, perhaps substantially.[8]

5.2.3.1 Standard Procedure at NIH

The NIH Web site and the documents guiding SBIR applications are both replete with statements that investigators are free to propose research on any technical subject, and that the topics published in the annual solicitation are only guides to the current research interests of the Institutes and Centers.

Topics are developed by individual ICs for inclusion in the annual omnibus solicitation. Typically, the Program Coordinator's office sends a request to the individual program managers (SBIR Point of Contact) at each IC. These PMs in turn meet with division directors, the locus of research assignments within the IC.

Division directors review the most recent omnibus solicitation (with their staff), and suggest changes and new topics based on recent developments in the field, areas of particular interest to the IC, and agency-wide initiatives with implications within the IC. The revised topics are then resubmitted for publication by the Office of Extramural Research (OER), which does not appear to vet the topics further.

[7]Interview with Bob Berger, former DoE SBIR Program Manager, March 18, 2005.
[8]Interview with National Cancer Institute staff on March 6, 2007.

5.2.3.2 Procedures for Program Announcements and Requests for Proposals (RFAs)

PAs and RFAs occupy a position closer to management-oriented approaches. Essentially, they add a parallel structure within the NIH SBIR program through which the ICs and the agency as a whole can fund their own research priorities.

PAs/RFAs are announcements of research funding areas that the IC expects to prioritize. PAs are areas of special interest that are still evaluated within the broad applicant pool for SBIR. RFAs are evaluated separately, and its applicants compete for a separate pool of SBIR funding set aside by the awarding IC.

In both cases, the announcement is published by one or more ICs as a reflection of top research priorities at the IC. It offers applicants a more directed path, aligned with agency priorities. NIH does still try to ensure that while PAs and RFAs define a particular problem, they are written broadly enough to encompass multiple technical solutions to the problem.

PAs and RFAs appear to be an effort to develop a middle ground between topic-driven and investigator-driven research. Essentially, by layering PA/RFA announcements on top of the standard approach, NIH seeks to focus some resources on problems that it believes to be of especially pressing concern, while retaining the standard approach as well.

5.2.4 Topics: Conclusions

There are some obvious advantages to an investigator-driven approach:

- Applications are likely to respond to current concerns more rapidly (given the long lags involved in topics developed with other mechanisms).
- Multiple deadlines are easier because an official solicitation does not have to be adjusted.
- Better science may be attracted, as promising and truly innovative new technologies are not excluded.

Acquisition- and management-driven models have their own advantages:

- Agencies can much more easily impose their own research agendas.
- Agencies can better ensure that SBIR research is aligned with eventual agency acquisition needs.
- Narrow technical "windows" mean that the agency can limit the technologies being addressed by a particular solicitation.
- Narrower topic options mean fewer applications, limiting the amount staff and reviewer time required.

The case for better science may be the most important point here. By definition, if topics are designed to sharply limit/focus research on particular problems, or in some cases on particular technologies, they are excluding other kinds of research. And, simple mathematics would suggest, if possible applicants are excluded a priori, the average quality of successful applications is likely to fall.

This suggests the following conclusions:

- The acquisition-oriented model seems appropriate at the agencies where eventual technology acquisition by the agency is the primary objective.
- The management-oriented model seems much less immediately defensible.
    - It is difficult to see how the narrowing of potential topics at NSF best serves the interest of science: in conjunction with the annual funding cycle, this ensures that excellent science may have to wait for several years before its number comes up in the NSF topics lottery.
    - It is even more difficult to defend the narrowing of selection areas on the basis of administrative convenience, as at DoE. Limited resources mean that all agencies limit the number of SBIR awards that are made, but it is hard to understand how doing so through a deliberate decision to reduce the scope of possible applications makes much sense.
- The NIH investigator-driven model seems to be working effectively for NIH.
- Hybrid models are worth further assessment. NIH appears to be evolving toward a hybrid model, as the share of awards going to RFAs and PAs is growing. This suggests possible options with which other agencies could experiment. For example, DoD components might reserve a small percentage of funding for "open" topics that are essentially investigator-driven.

It is worth noting that in other countries, hybrid models are used. For example, in Sweden, approximately half of Vinnova's[9] research funding is distributed to agency programs focused on the research needs of specific industries. The other half is allocated to proposals for research initiated by companies.[10]

5.3 OUTREACH MECHANISMS AND OUTCOMES

5.3.1 Introduction

As the program has become larger and more important to early-stage funding for high-tech companies, outreach activities by the programs have increasingly come to be complemented by outreach initiatives at the state level.

At the different agencies, outreach programs seem to share three key objectives:

---

[9]VINNOVA is the Swedish national technology agency.

[10]VINNOVA Web site, accessed at <*http://www.vinnova.se*>.

- Attracting new applicants.
- Ensuring geographical diversity in applications by attracting new applicants from under-represented regions.
- Expanding opportunities for disadvantaged and woman-owned companies.

These objectives are implicit, rather than explicit, in the outreach activities of all the programs. With the growth in electronic access, it is unclear whether greater outreach efforts could be cost-effective.

5.3.2 Outreach Mechanisms

At the program level, all the agencies conduct outreach activities, though those efforts vary in kind and degree. Important types of outreach activities include:

- **National SBIR Conferences**, which twice a year bring together representatives from all of the agencies, often at locations far from the biggest R&D hubs (e.g., the Spring 2005 national conference took place in Omaha, Nebraska).
- **Agency-specific Conferences**, usually held annually, which focus on awardees from specific agencies (e.g., the NIH conference, which is usually held in Bethesda, Maryland).
- **Phase III Conferences**, focused on bringing together awardees, potential partners, such as defense industry prime contractors, and potential investors (e.g., the Navy Opportunity Forum, held annually in the Washington, DC area).
- **The SWIFT Bus Tour**, which makes an annual swing through several "under-represented" states, with stops at numerous cities along the way. Participants usually include some or all of the agency program managers/directors/coordinators.[11]
- **Web Sites and Listservs** are maintained by all of the agencies. These Web sites contain application and support information. Some, such as the DoD's site, are elaborate and comprehensive, and most allow users to sign up for a news listserv.
- **Agency Publications** are used to spread information about the SBIR program. NASA, for example, has a variety of print and electronic publications devoted to technology and research promotion. The other departments use this publicity vehicle to a lesser degree.
- **Other, non-SBIR-focused Publications** are sometimes used by the agen-

[11]Agencies have different titles for their primary SBIR program manager.

cies. For example, DoD publishes solicitation pre-publication announcements in its *Commerce Daily*.

- **Demographic-focused Outreach.** Agencies acknowledge the need to ensure that woman- and minority-owned businesses know about SBIR and are attracted to apply. They have organized a series of conferences to encourage participation in the program by woman- and minority-owed business, often run in conjunction with other groups. For example, the DoD Southeastern Small Business Council held a conference for Woman-owned Small Businesses in Tampa, Florida, in June 1999.

In general, these activities are the responsibility of the SBIR Program Coordinator at the agency, rather than other agency staff. They are also funded out of general agency funds, as the governing SBIR legislation does not permit programs to use SBIR funds for purposes other than awards (and a limited amount of commercialization assistance. See below.).

Generally, agencies do not provide much funding for outreach, though there are pronounced differences between the agencies. NIH, for example, has developed a popular annual SBIR conference, whereas DoE has no funds for such an initiative. Within DoD, the Navy has provided significant extra operating funds for its SBIR program, which has therefore been able to take on outreach and other initiatives not open to the programs at other DoD components.

5.3.3 Outreach Outcomes

5.3.3.1 New Winners

As SBIR programs have become larger, with longer track records and better online support, it is increasingly less likely that potential candidate companies will remain unaware of the program, or that such companies will have difficulty understanding how to apply.

Still, new applicants remain an important component of the applicant pool (see Figure 5-1, for example). This may suggest that the current outreach activities are having the intended effect. Much learning about new opportunities such as SBIR is conveyed directly. Another interpretation of the data is that outreach activities have become increasingly unnecessary in a world where a wide range of economic development institutions at the state and local level automatically point small high-tech companies toward SBIR as a funding source, and where public information about SBIR is fairly widespread and online.

Anecdotal evidence from the SWIFT bus tour participants suggests that outreach activities do encourage new firms to enter the program. The program coordinator for NIH has noted that there is usually an upturn in applications from

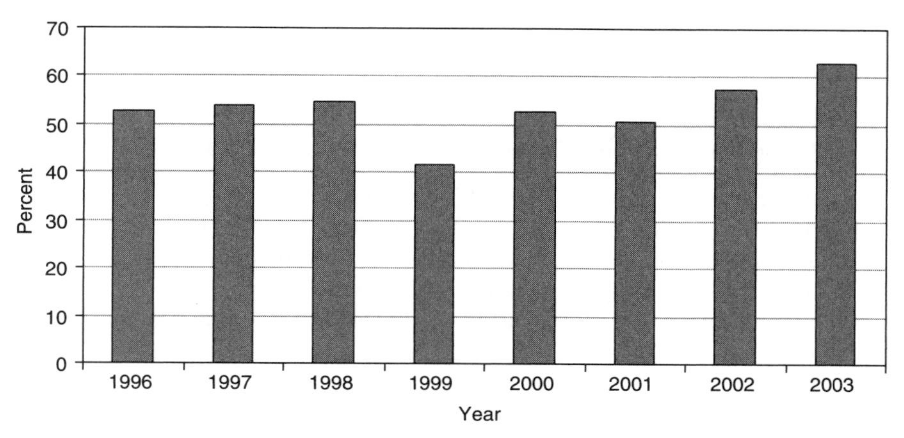

FIGURE 5-1 Phase I new winners at NSF (previously nonwinners at NSF).
SOURCE: National Science Foundation.

visited states after the tour, especially from states which have had relatively few applicants in the past.[12]

5.3.3.2 Applicants from Under-served States

The distribution of SBIR awards among states varies widely. California and Massachusetts consistently receive a large share of awards in federal R&D programs.[13] This is also true for SBIR. Conversely, about a third of the states receive few awards.

At NASA, for example:

- California and Massachusetts dominate, garnering 36.7 percent of Phase II awards.
- The top five states account for 54.4 percent of all awards.
- Nineteen states had ten or fewer phase II awards from FY1992 to FY2003, and 16 had less than four.

As yet, no systematic research has been conducted that would assess the distribution of SBIR awards relative to those of other research programs, or to other R&D indicators such as the distribution of scientists and engineers in the population.

---

[12]Jo Anne Goodnight, Personal Communication, May 15, 2004.

[13]See National Science Board, *Science and Engineering Indicators 2006*, Arlington, VA: National Science Foundation, 2006, Chapter 8: State Indicators. Accessed at *<http://www.nsf.gov/statistics/seind06/c8/c8.cfm>*.

This issue is discussed in more detail in Chapter 3. In general, the evidence suggests that while the broad pattern of awards has not changed much, more awards are going to states that previously received none or only a few.

5.3.3.3 Applications from Woman- and Minority-owned Businesses

Congress has mandated that one objective of the SBIR program is to increase support for woman- and minority-owned businesses. Agencies track participation by these demographic groups in their programs. Unfortunately, application data from DoD is insufficiently accurate for use in this area.

Our analysis in Chapter 3 utilized the SBA data, and generated Figure 3-13.

Two core questions emerge:

First, the decline in the share of awards to minority-owned businesses is, on the surface, an issue, given that support for these businesses is one of the four Congressional objectives for the program. At a minimum, it will be important for agencies to determine why this trend is occurring, and—if necessary—to identify ways to address it.

Second, even though the share of awards going to woman-owned businesses has been increasing, it is worth noting that women now constitute almost half of all doctorates in certain scientific and engineering fields—such as life sciences—and that award rates lag this considerably. Again, further assessment by the agencies is necessary to determine if there is a problem, and what should be done to address it.

5.3.4 Conclusions—Outreach

Information for potential applicants falls into two areas: (1) basic program information for potential applicants who have not heard of the program or have not yet applied and won awards; and (2) detailed technical information about specific solicitations and opportunities for those familiar with the program.

Interviews with agency staff and SBIR awardees indicate that the agencies have substantially upgraded their SBIR online information services in recent years. The Web has become the primary source of information for both existing awardees and potential new applicants.

While some of the Web sites are still not especially easy to navigate or utilize, collectively they represent a vast improvement in outreach capacity for the agencies. Moreover, online services help to ensure that potential applicants get up-to-date information.

For those with little information about the program, the states have increasingly come to play an important information-providing role. Almost all states have a technology-based economic development agency (TBED) of some kind. Many have TBEDs that are well-known and effective organizations. These organizations are aware of SBIR. They help to organize local and regional conferences with the SBIR agencies, they attend other SBIR events, and they are often in direct contact with agency staff.

Some TBEDs, such as the Innovation Partnership in Philadelphia, include SBIR as a key component. They center many of their activities around SBIR by helping local companies develop winning applications and integrate SBIR into a wider system for economic development. For example, the Innovation Philadelphia has a multiple funding model for early stage financing, which includes a role for SBIR.[14]

Overall, most companies with some reasonable likelihood of success in applying to the SBIR program have heard about SBIR and can find out how to fill in the appropriate application forms. To the extent that they want help in the process, there is a small cottage industry of SBIR consultants eager to provide that service.

Thus, it appears that the general outreach function historically fulfilled by the SBIR Agency Coordinators/Program Managers may now be changing toward a more nuanced and targeted role. Outreach can focus on enhancing opportunities for under-served groups and under-served states, or on specific aspects of the SBIR program such as the transition to the market after Phase II (e.g., the July 2005 DoD "National Phase II Conference—Beyond Phase II: Ready for Transition" held in San Diego), while relying on the Web and other mechanisms to meet the general demand for information.

5.4 AWARD SELECTION

5.4.1 Introduction

SBA Guidelines require that an agency should use the same selection process for its SBIR program as it does for its other R&D programs; as award selection processes differ across agencies, so do selection procedures for SBIR.

The selection process is critical for the long-run success of any government award program. For SBIR, the selection process is complicated by the fact that the program serves many masters and is aimed at many objectives. A successful selection process must therefore meet a number of quite distinct criteria. Discussions with agency staff and award winners, and with the other stakeholders, suggest that the following are key criteria for judging the quality of the SBIR selection processes:[15]

- **Fair.** Award programs must be seen to be fair, and the selection process

[14]Innovation Philadelphia Web site, accessed at <*http://www.ipphila.com*>.

[15]While these are in reality the implicit criteria against which agencies value their selection procedures, these criteria are not explicitly recognized in any agency, and the agencies balance them quite differently.

is a key component in establishing fairness. Despite numerous complaints about details of the NIH selection process, there is uniform acceptance among stakeholders that, in general, the process is fair.

- **Open.** A successful SBIR program must be open to new applicants. Agencies generally score high on this, with a third or more of Phase I awards at most agencies going to new winners.
- **Efficient.** The selection process must be efficient, making the best of use of the time of applicants, reviewers, and agency staff.
- **Effective.** The selection process must achieve the objectives set for it. In practice, these can be summarized as selecting the applications that show the most promise for:
  - Commercializing results, and
  - Helping to achieve the agency's missions,[16]
  - While expanding new knowledge, and
  - Providing support for woman- and minority-owned businesses.

The effectiveness of the agencies' SBIR programs is discussed elsewhere, in the Chapter 4: SBIR Program Outputs. The remaining components of the selection process are discussed in more detail below, along with some other selected topics, including the degree of centralization, and a more detailed review of the NIH approach based on outside reviewers.

5.4.2 Approaches to Award Selection

Key questions in discussing award selection concern the use of outside reviewers, and who makes final decisions on funding.

5.4.2.1 Use of External Reviewers

DoD. DoD components do not use external reviewers.[17] Two or three technical experts at the laboratory level review each proposal. Proposals are judged competitively on the basis of scientific, technical, and commercial merit, in accordance with the criteria listed in Box 5-2. Prior to the closing of the solicitation, the responsibility for each topic has been clearly established, so reviewers can access "their" proposals immediately after the closing. If a proposal is beyond the

[16]The other mandated goals of the program tend to have less impact on award selection, emerging from the implementation of awards.

[17]There are exceptions for support personnel working at government laboratories. The Air Force section of the solicitation states "Only government personnel and technical personnel from Federally Funded Research and Development Center (FFRDC), Mitre Corporation and Aerospace Corporation, working under contract to provide technical support to Air Force product centers (Electronic Systems Center and Space and Missiles Center respectively), may evaluate proposals."

expertise of the designated reviewers, the person with overall topic responsibility will obtain additional reviewers.

NIH. NIH relies exclusively on external reviewers to provide a technical and commercial assessment of proposals. A few applications received for specific technologies being sought by the agency (via the RFA mechanism) are evaluated at the proposing unit, but even then, outside reviewers provide much of the technical input.

NASA. NASA does not use outside reviewers. Initial technical reviews are done by a panel of technical experts convened at the relevant NASA Centers. Program staff at headquarters subsequently review the awards for alignment with NASA objectives and for other purposes.

DoE. DoE uses internal experts recruited for each specific proposal from across the agency to provide technical and commercial evaluations. After receiving the completed reviews, proposals are scored by the Technical Topic Manager (TTM). Only proposals scoring at least +2 (out of a possible score of +3) are eligible for selection. The SBIR office reviews for discrepancies between the average reviewer score and the TM's score, and resolves any conflict.

Portfolio Managers for technical program areas select the awardees from that program area. Different programs within DoE have differing philosophies and mechanisms for award selection. Some provide equal awards to each of the program area's topics. Others select applications with the highest score. DoE tries to decentralize decision making, so that those with the greatest knowledge of the technology make the decisions.

NSF. NSF uses at least three outside (non-NSF) reviewers for each application. Their ratings are then used to guide the Program Officer in making funding decisions. Final approval follows review by the Division of Grants and Agreements.

5.4.2.1.1 Outsider-dominated Processes: The NIH Approach

The peer review process at NIH is by far the most elaborate of all the agencies. It is operated primarily through the Center for Scientific Research (CSR), a separate IC at NIH which serves only the other ICs—it has no direct funding responsibilities of its own.

Study Sections. Applications are received at CSR and are assigned to a particular study section[18] based on the technology and science involved in the proposed research. Panels can either be permanent panels chartered by Congress,

[18]As review panels are known at the National Institutes of Health.

or temporary panels designated for operation by NIH. Most SBIR applications are assigned to temporary panels, many of which specialize in SBIR applications only. This trend appears to be accelerating, as the requirements for assessing SBIR applications—notably the commercialization component—are quite different from the basic research conducted under most NIH grants.

Special emphasis panels (SEPs) (temporary panels) are reconstituted for each funding cycle. Almost all SBIR applications are addressed by SEPs, which have a broader technology focus, and less permanent membership, than chartered panels. Many SEPs tend to draw most of their applications from a subset of ICs. For example, the immunology Integrated Review Group (IRG) covers about 15 ICs, but 50 percent of its work comes from the National Institute of Allergy and Infectious Diseases (NIAID), with a further 33 percent from the National Cancer Institute (NCI).

Procedures. Applications are assigned to a subset of the panel—two lead reviewers and one discussant. These panelists begin by separating out the bottom half of all applications. These receive a written review, but are not formally scored. The top 50 percent are scored by the three reviewers and then, after discussion, by the whole panel.

Panel Makeup. NIH guidelines are that at least one panelist should have small business background. One current panel, for example, had 13 small business representatives out of 25 panelists. This is a recent change; previous panels in this technical area had been dominated by academics. NIH guidelines mandate that panels have 35 percent female and 25 percent minority participation.

Scoring is based on five core criteria:

- Significance
- Approach
- Innovation
- Investigators
- Environment

No set point values are assigned to each. Scores are averaged (no effort is made to smooth results for example by eliminating highest and lowest scores), and range between 100 (best) and 500 (worst). Winning scores are usually in the 180-230 range or better, although this varies widely by IC and by funding year. Scores are computed and become final immediately.

Reviewers and Funding Issues. Reviewers are specifically not focused on the size of the funding requested; they are tasked to review whether the amount is appropriate for the research being proposed. As a result, the question of trade-offs between a single larger award against multiple smaller awards is not addressed. On the other hand, this approach permits applicants to propose projects of larger scope and value to the agency mission.

Reviewers and Outcomes of Prior Awards. If the proposing company has received more than 15 Phase II awards, it must note that on its applications. Otherwise, the application forms have no place to list previous awards. Companies with strong track records try to make sure that this is reflected in the text of their application. However, there is no formal mechanism for indicating the existence of, or outcomes from, past awards. Reviewers also do not know the minority or gender status of the PI or of the company's owners.

Positive and Negative Elements of Outside Review. On the positive side, strong outside review generates a range of benefits. These include:

- The strong positive endorsement for applications that comes from formal peer review.
- The alignment of the program with other peer-reviewed programs at NIH.
- The perception of fairness related to outside review in general.
- The absence of claims that awards are written specifically or "wired" for particular companies.
- The probability that reviewers will have technical knowledge of the proposed award area.

On the negative side, recent efforts to infuse commercialization assessment into the process have had mixed results at best. Problems include:

- Quality of reviews decreasing as workload increases.
- Perceptions of random scoring.
- Conflict of interest problems related to commercialization.[19]
- Substantial delays in processing.
- Growing questions about the trade-offs between different size awards.

Overall, outside review adds fairness but also adds complexity and possibly delay. The generic NIH selection process has not yet been substantially adjusted to address the needs of companies trying to make rapid progress in an increasingly competitive environment. Delays that might be appropriate at an institution focused on academic and basic research, may be less applicable to smaller businesses working on a much shorter development cycle.

[19]One example of conflict of interest may rise when the commercialization reviewer has a similar or competitive product planned for or already in the market.

5.4.3 Fairness

The fairness of selection procedures depends on several factors. These include:

- Transparency—is the process well known and understood?
- Implementation—are procedures always followed?
- Checks and balances—are outcomes effectively reviewed by staff with the knowledge and the authority to correct mistakes?
- Conflict of interest—are there procedures in place and effectively implemented to ensure that conflicts of interest are recognized and eliminated?
- Appeals and resubmissions—are there effective appeals and/or resubmission procedures in place?
- Debriefings—do agency staff give both successful and unsuccessful applicants good feedback about their proposals?

Each agency has strengths and weaknesses in these areas. They are summarized below.

5.4.3.1 NIH

Transparency. At NIH, the review process is well known to many applicants because it is almost the same as selection procedures for all other NIH awards. Interviews with awardees suggested that many had sat on NIH review panels and were familiar with review procedures.

Implementation. The NIH review procedures are highly formalized and are implemented by a professional and independent review staff at CSR.

Checks and Balances. NIH scores somewhat lower on this factor than do other agencies, because so much power is placed in the hands of independent reviewers. Resubmission (see below) replaces appeals for practical purposes at NIH, but adds substantial delay. Thus, there are few options available for applicants who believe that reviewers misunderstood or misjudged their proposals.

Conflict of Interest. NIH does have clear conflict of interest regulations in place for reviewers, and also has procedures in place that allow applicants to seek to exclude individual panel members from reviewing their application. However, the extent to which this works in practice is not clear, and may depend on individual CSR officers. Conflict of interest has been raised repeatedly by awardees as a problem.

Resubmissions are the standard mechanism for improvement and/or appeal at NIH, and about one-third of all awards are eventually made after at least one resubmission. Awardees noted in interviews that this capacity to resubmit enhances perceptions of fairness.[20]

5.4.3.2 DoD

DoD's SBIR program is regulation-driven (as is natural at a procurement-oriented agency). Its processes are governed by the Federal Acquisition Regulations (FAR), the Defense FAR Supplement (DFARS), and component specific supplements.[21] Interviewees among the companies presented few complaints that the program was unfair." Overall, DoD relies on the possibility of appeals, and on the self-interest of topic managers who are seeking to find technology that they can use, to ensure that the process is both fair and seen to be fair.

Transparency. DoD selection processes are less transparent than those at NIH. Information about selection criteria is available in the text of the solicitation, but there is no information there on selection processes—i.e., who will review the proposal, and how the review will be conducted. These procedures are agency specific, and in most cases minimal information is publicly available. For example, the Air Force has no information publicly posted on the agency's Web site concerning selection procedures.[22]

Implementation. There are no data concerning the degree to which DoD SBIR procedures are followed. As the procedures themselves are far from transparent, neither applicants nor researchers can easily evaluate whether the agency plays by its own rules.

Checks and Balances. While practices vary widely between agencies, DoD always gives applicants several layers of effective proposal review, and it is reasonable to conclude that checks and balances are formally in place. It is not clear how well these work in practice. Some components have more checks in place than others—the Army uses its lead scientists as gatekeepers for example, while other agencies have programs that are more decentralized.

Resubmission and Appeals. Resubmission is effectively impractical as topics change substantially between solicitations. The appeals process is much more important. GAO rules for appeals procedures are published on the DoD SBIR Web site, and appeals are utilized by aggrieved companies. In essence, the DoD system is the inverse of the NIH resubmission approach: at NIH, companies are free to propose any line of research they like, and can resubmit applications if

[20]For example, interview with Dr. Josephine Card, Sociometrics, Inc.

[21]This is a correction of the text in the prepublication version released on July 27, 2007.

[22]As of June 15, 2006.

BOX 5-2
DoD Evaluation Criteria
Evaluation Criteria—Phase I

The DoD Components plan to select for award those proposals offering the best value to the government and the nation considering the following factors.

a. The soundness, technical merit, and innovation of the proposed approach and its incremental progress toward topic or subtopic solution.
b. The qualifications of the proposed principal/key investigators, supporting staff, and consultants. Qualifications include not only the ability to perform the research and development but also the ability to commercialize the results.
c. The potential for commercial (government or private sector) application and the benefits expected to accrue from this commercialization as assessed utilizing the criteria in Section 4.4.

Where technical evaluations are essentially equal in merit, cost to the government will be considered in determining the successful offeror.

Technical reviewers will base their conclusions only on information contained in the proposal. It cannot be assumed that reviewers are acquainted with the firm or key individuals or any referenced experiments. Relevant supporting data such as journal articles, literature, including government publications, etc., should be contained or referenced in the proposal and will count toward the 25-page limit.

Evaluation Criteria—Phase II

The Phase II proposal will be reviewed for overall merit based upon the criteria below.

a. The soundness, technical merit, and innovation of the proposed approach and its incremental progress toward topic or subtopic solution.
b. The qualifications of the proposed principal/key investigators, supporting staff, and consultants. Qualifications include not only the ability to perform the research and development but also the ability to commercialize the results.
c. The potential for commercial (government or private sector) application and the benefits expected to accrue from this commercialization.

The reasonableness of the proposed costs of the effort to be performed will be examined to determine those proposals that offer the best value to the government. Where technical evaluations are essentially equal in merit, cost to the government will be considered in determining the successful offeror.

Phase II proposal evaluation may include on-site evaluations of the Phase I effort by government personnel.

SOURCE: Department of Defense SBIR Solicitation, FY2005.

they do not like the result of selection; at DoD, topics are restricted and resubmission is not permitted, so companies are directed strongly toward an appeals process instead. There are no public data about the extent of utilization.

5.4.3.3 NASA

Transparency. While the NASA SBIR Participation Guide[23] provides a general description of the selection procedure, it could be more helpful to applicants if it added more details about the process. The decentralized character of the program—with selection being conducted initially at the center level—makes transparency hard to achieve.

Implementation. It appears that all the NASA Centers operate the selection procedures in similar ways, as their end-products must be sent to HG for review by the SBIR Program Office and by representatives of the Mission Directorates.

Checks and Balances. There are multiple levels of checks and balances in the NASA selection system. Initially, proposals are reviewed by Center Committees, which collectively provide some balance against individual enthusiasm for a proposal. Tentative rankings are then submitted to fairly extensive review by SBIR program management and by other headquarters staff.

Conflicts of Interest. As with DoD, the absence of external reviews removes one potential source of such conflicts. Internally, champions of a proposal may have many reasons for supporting it, but the process is designed to balance out these justifications against objective criteria and other competing interests.

5.4.3.4 NSF

Transparency. All NSF SBIR reviewers are provided with instructions and guidance regarding the SBIR Peer Review process and are compensated for their time. NSF policy regarding the NSF review process and compensation can be found in the *NSF Proposal and Grant Manual.*

The NSF Report discusses the use of "additional factors." These may include the balance among NSF programs; past commercialization efforts by the firm; excessive concentration of grants in one firm or with one principal investigator; participation by woman-owned and socially and economically disadvantaged small business concerns; distribution of grants across the states; importance to science or society; and critical technology areas. However, there is no further explanation of how these factors might be applied or when they might be applied.

Implementation. The relatively small size of the NSF program and the centralized management structure ensure that implementation is uniform.

---

[23]National Aeronautics and Space Administration, "The NASA SBIR and STTR Programs Participation Guide," June 2005, accessed at <*http://sbir.gsfc.nasa.gov/SBIR/zips/guide.pdf*>.

Checks and Balances. Applications are reviewed by an outside panel, and then a final decision is made by the SBIR staff and the program manager. This indicates a limited level of review.

Conflict of Interest. NSF appears to take seriously applicant views on appropriate reviewers. It does not appear that reviewers sign a conflict of interest form.

5.4.4 Efficiency

For programs to work well, they must be efficient in term of their use of resources, both internal and external to the agency.

The efficiency objective must be balanced against other considerations. Hypothetically, having one awarding officer making instant decisions would be highly efficient, but would be neither fair nor necessarily effective.

Still, it is possible to construct indicators or measures of efficiency from the perspective of both the applicant and the agency. While many of the efficiency considerations listed in Box 5-3 are a factor in agency program management, it is also noteworthy that no agency has conducted an analysis of program efficiency from the perspective of applicants and award winners. Also, no agency has, to our knowledge, conducted any detailed surveys of customer satisfaction among these populations, focused on program improvement.

BOX 5-3
Possible Efficiency Indicators for the SBIR Selection Process

External: Efficiency for the Applicant

- Time from application to award
- Effort involved in application
- Red tape involved
- Output from application (not including award)
- Re-use of applications

Internal: Efficiency for the Agency

- Moves the money
- Minimizes staff resources
- Maximizes agency staff buy-in
- Minimizes appeals and bad feelings

Timelines are discussed in the section on timelines and gaps.

5.4.5 Other Issues

5.4.5.1 Electronic Submission

All agencies have now moved to electronic submission of SBIR applications, though DoE and NIH only did so relatively recently. Each agency uses its own home grown electronic submission system, even though the existing systems at DoD and NASA have been highly regarded by recipients.[24]

5.4.5.2 Degree of Centralization

It is somewhat misleading to talk about the degree of centralization. Programs handle selection so differently that the same functions are not performed at all agencies. However, it is possible to identify some shared functions (see Table 5-1).

5.4.6 Selection—Conclusions

We return to our initial conceptual framework of assessment. Are the selection procedures operated at the various agencies meeting the core requirements in this area? Are they:

- **Fair.**
- **Open.**
- **Efficient.**
- **Effective.**

Fairness. From discussions with applicants and staff, it would appear that fairness is substantially enhanced by two key characteristics:

- **Transparency.** The more applicants understand about the process, the fairer it appears. Generally, the debriefing programs provide sufficient feedback to meet this requirement, and the agency's procedures are reasonably well-known. NIH is unique in listing the names of outside panel members.
- **Conflict of Interest.** Conversely, the more outsiders are used for assessment, the more conflict of interest appears to be a potential problem. Claims of problems at NIH cropped up regularly in interviews, and it appears that procedures to address conflicts of interest could be strengthened there.

Openness. There have been comments in interviews that the DoD program

[24]Case study interviews.

TABLE 5-1 Comparing Selection Procedures Across Agencies

Procedure	NIH	NASA	NSF	DoE	Army	Navy	Air Force	MDA
Initial admin review	CSR	?	?	SBIR program	TPOC?			
Technical assessment	Outsiders	TPOC/ Centers	Outsiders	TTM/ Agency staff	TPOC/ agency staff	TPOC/ agency staff	TPOC/ agency staff	TPOC/ agency staff
Scoring mechanisms	Quant.	Qual.	Qual.	Quant.	Varies	Varies	Varies	Varies
Program alignment review	N/A	PM	PM	N/A	Lead Agency scientists			
Final rankings list	PM	SBIR dir.	PM	PM	Lead Agency scientists			
Effective funding approval	Div director	SBIR dir.	SBIR dir.	PM	SSA	SSA	SSA	PM/ SSA

SSA = designated contracts officer/program liaison
TPOC = technical point of contact
SBIR dir. = SBIR Program Coordinator or similar function
Outsiders = outside technical reviewers
Quant. = quantitative (numerical) scoring system
Centers = NASA Centers (e.g., Goddard Space Center)
PM = Program Manager (technical area manager, not SBIR manager)

favors companies that are known to the program. However, the data (see Chapter 2) indicate that at least one-third of awards go to new winners every year, and interviewees stated that the program was sufficiently open to new ideas and companies. These comments did not appear in relation to other agencies' SBIR programs

Efficiency. At most agencies, SBIR managers run the program with limited administrative funds. This limitation means lean program operations, but leanness does not necessarily mean that the needs of applicants and the agency are met as effectively as possible. In general, discussions with staff and awardees lead us to conclude that agency budgetary and efficiency concerns tend to trump the needs of applicants. For example, limiting the number of applications from a firm, or tightening topic definitions to reduce the number of applications, are both "efficient" in the sense that they allow agencies to process applications using the minimum amount of administrative funding. Yet both strategies run the risk of eliminating firms' and technologies' potential value.

Effectiveness. The effectiveness of the SBIR program is discussed in more detail in Chapter 4: SBIR Program Outputs. However, effectiveness can also be judged partly by the number of applications the program receives. Application numbers continue to rise.

5.5 FUNDING CYCLES AND TIMELINES

The agencies differ in the number of funding cycles they operate each year, ranging from four at DoD (in which only some components participate) to three at NIH and one at DoE. NSF operates two funding cycles, but any given technical topic will be available in only one of the two.

This has resulted in significant funding gaps between Phase I and Phase II, and after Phase II. The NRC Phase II Survey indicated that on average, 73 percent of companies experienced a gap; the average gap reported by those respondents with a gap was 9 months. Five percent reported a gap of two or more years.

Agencies also differ substantially in their efforts to close the funding gaps between Phase I and Phase II, and after Phase II. In general, it seems reasonable to posit the following models:

- "A standard annual award" model, used at NASA and DoE.
- A "gap-reducing model," favored at NIH and DoD.

NSF appears to primarily use the standard model, with some gap reducing features.

5.5.1 The Standard Model

Agencies using the standard model differ in the timing of milestones and the handling of administrative material. DoE, for example, uses a 9-month Phase I; NASA staff note that use of a sophisticated electronic tool for submission and program management has worked to reduce the gap between Phase I and Phase II.

Yet despite these differences, which can have an important impact on both awardees and program performance, the basic structure of the system can be captured (see Figure 5-2).

Key features of this model include:

- A single annual submission date (usually).
- No gap-funding mechanisms between Phase I and Phase II.
- No Fast Track program.
- No Phase IIB funding.

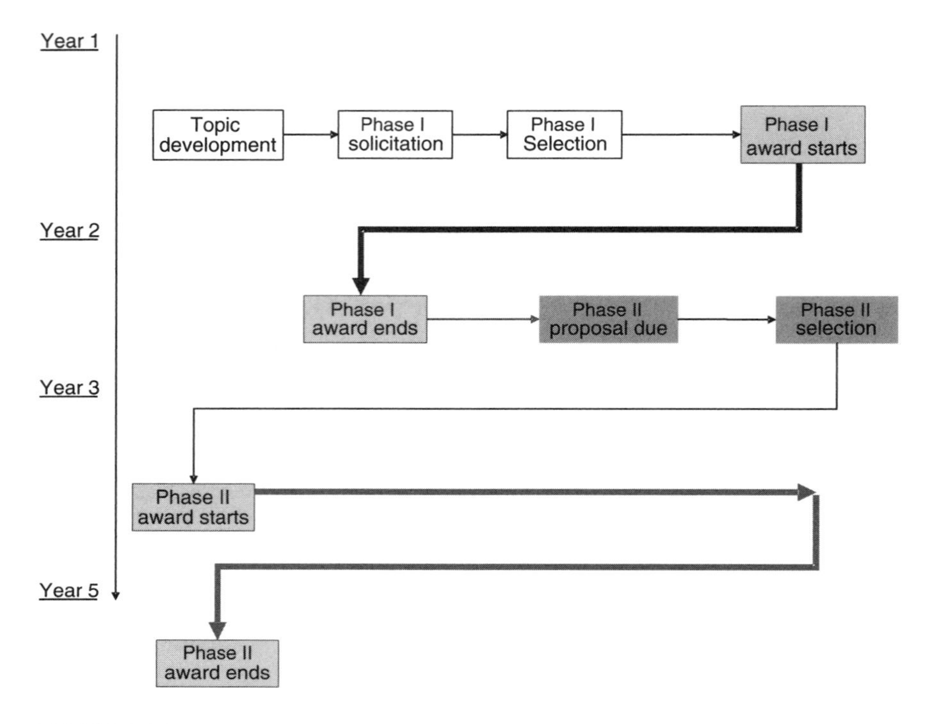

FIGURE 5-2 Standard model for SBIR award cycles.

5.5.2 The Gap-reduction Model

The gap-reduction model looks quite different. It includes some or all of a range of features designed to reduce gaps and improve the time performance from initial conception to final commercial product.

Drawing on the efforts made in this direction at the NIH and some components of DoD, as well in some respects at NSF, we can extract the following elements of a gap-reduction approach to award cycles.

5.5.2.1 Multiple Annual Submission Dates

More submission dates reduces the time that companies must wait before applying. Annual submission means a potential wait of up to a year.

NIH provides three annual submission dates for awards, in April, August, and December. DoD now offers a quarterly solicitation schedule, although few of the components participate in all four. SOCOM (Special Operations Command), for example, still participates in only two, and staff there are working to persuade

TABLE 5-2 Fixing the Phase I-Phase II Funding Gap

Agency Unit	Phase I-Phase II Gap Funding
NIH	3 mo. at own risk
NSF	None
DoE	None
NASA	None
DoD	
Army	50k after Phase II selection
Navy	30k after Phase II selection
Air Force	None
DARPA	None
DTRA	None
MDA	None
SOCOM	None
CBD	30k after Phase II selection
OSD	9 mo. for Phase I
NIMA	None

the agency's engineers and topic managers that more submission dates would result in better and more relevant applications.[25]

5.5.2.2 Topic Flexibility

Topics have important implications for company research timelines. A company that wishes to pursue a particular technology may have a problem at agencies with "tight" topic boundaries because the companies must wait until the topic shows up in a published program solicitation. If they miss that window of opportunity, that topic may not show up again for several years—in which case the effective gap between opportunities for that company is several years, regardless of whether the agency provides multiple annual award deadlines.

This suggests that an approach that writes broad topics, and offers some opportunity for "out-of-topic" projects, is one that best fits with the gap-reduction model.

5.5.2.3 Phase I—Phase II Gap Funding

The gap between the end of Phase I and the beginning of Phase II is potentially difficult for companies; smaller companies in particular may not have the resources or the other contracts in hand to keep working, and may even lose critical staff.

[25]Telephone interview with Special Operation Command (SOCOM) SBIR Program Manager, June 8, 2005.

Two mechanisms appear to have emerged to provide funds during the gap between the conclusion of Phase I and the start Phase II funding.

Several DoD components use some version of an option arrangement. They reduce the size of the initial Phase I award (usually to $70,000), and add an "option" for 3 months additional work for $30,000-50,000. Projects that win a Phase II award may become eligible for the option, which is designed to cover the funding gap. The "option" programs are activated only when the agency awards a Phase II contract, so it is unclear as yet whether these options are applied to all phase II winners, or how well they cover the gap, as the end of Phase I may well be some months before the start of a Phase II award at these components.

NIH does not use the option approach. Instead, companies that anticipate winning a Phase II can work for up to three months at their own risk, and the cost of that work will be covered if the Phase II award eventually comes through. If it does not, the company must swallow the cost. This may work reasonably well for companies that have sufficient cash flow to continue work, but less well for smaller companies with more limited resources.

5.5.2.4 Fast Track Models

Some agencies now offer some kind of Fast Track model to close the gap between Phase I and Phase II.

DoD

At DoD, starting in 1992, the Ballistic Missile Defense Organization (BMDO) developed a program called "co-investment," under which applicants who could demonstrate additional funding commitments, gained preferences during the Phase II application process, and were also offered more rapid transitions. This effectively eliminated the gap. In 1992, less than half of all BMDO awardees had such commitments; by 1996 this figure had risen to over 90 percent.[26]

In 1995, DoD launched a broader initiative. Under the Fast Track policy, SBIR projects that attract matching cash from an outside investor for their Phase II effort, can receive interim funding between Phases I and II, and are evaluated for Phase II under an expedited process.

Companies submit a Fast Track application, including a statement of work and cost estimates, between 120 and 180 days after the award of a Phase I contract, along with a commitment of third party funding for Phase II.

Subsequently, the company must submit its Phase I Final Report and its Phase II proposal no later than 210 days after the effective date of Phase I, and must certify, within 45 days of being selected for a Phase II award, that all matching funds have been transferred to the company.

[26]National Research Council, *The Small Business Innovation Research Program: An Assessment of the Department of Defense Fast Track Initiative*, Charles W. Wessner, ed., Washington, DC: National Academy Press, 2000, p. 27.

TABLE 5-3 NAVSEA Fast Track Calendar of Events

Responsible Party	Required Deliverable/Event	Due Date	Delivered to:
SBIR Company	Fast Track application	150 days after Phase I award	NAVAIR SBIR Program Office, TPOC, Navy Program Office, and DoD
SBIR Company	Five-to ten page electronic summary	150 days after Phase I award	NAVAIR SBIR Program Office
SBIR Company	Phase II proposal Phase I Final Report	181-200 days after Phase I award	Technical point of contact and NAVAIR Program Office
NAVAIR	Acceptance or rejection of Phase II proposal	201-215 days after Phase I award	SBIR Company
SBIR Company	Proof that third-party funding has been received by SBIR company	45 days after acceptance of Phase II proposal	Contract Specialist
SBIR Company	Final accounting of how investor's funds were expended	Include in final Phase II Progress report	Technical point of contact

SOURCE: NAVSEA Web site.

At NAVAIR, for example, a subcomponent of the Navy, Fast Track proposals will be decided and returned to the company within a maximum of 40 days from submission, which may be a month before the end of Phase I.[27] This provides "essentially continuous funding" from Phase I to Phase II. About 6-9 percent of DoD awards are Fast Track.[28]

Data from the NRC study indicates that Fast Track does reduce the gap between Phase I and Phase II. Survey data showed that for Fast Track recipients, more than 50 percent reported no gap at all, and the average gap was 2.4 months; for a control group of awardees, only 11 percent reported no gap, and the average gap was 4.7 month—about double that for Fast Track projects.[29]

However, recent awards data indicates that DoD components and companies are focusing more on Phase II+ arrangements, and less on Fast Track (see DoD Report: Chapter 2).[30]

[27]Department of Defense, "NAVSEA Fast Track Calendar of Events," accessed at <*http://www.navair.navy.mil/sbir/ft_cal.htm*>.

[28]National Research Council, *The Small Business Innovation Research Program: An Assessment of the Department of Defense Fast Track Initiative*, op. cit, p. 28.

[29]See National Research Council, *The Small Business Innovation Research Program: An Assessment of the Department of Defense Fast Track Initiative*, op. cit., pp. 66-67.

[30]See National Research Council, *An Assessment of the SBIR Program at the Department of Defense*, op. cit.

The DoD model is premised on the view that matching funding adds both legitimacy to a project and value in the form of additional research, and that this should therefore qualify the project for more rapid approval.[31]

NIH

The NIH Fast Track program is quite different from that at DoD—in fact, the similarity of names is highly confusing. Essentially, the NIH program allows applicants to propose a complete Phase I-Phase II program of research. No matching funds are required, and projects are reviewed and selected within the normal timeframe for review at NIH—i.e., they are not expedited.

The advantage to the applicant is that NIH Fast Track awards are designed to dramatically reduce uncertainty by awarding a Phase II at the same time as a Phase I, and also by reducing the time needed between awards.

Under normal NIH procedures, success with a Phase I will be followed by a Phase II application, which will be decided between 8 months and 12 months after the end of Phase I. Under Fast Track, Phase II can start immediately.

To date, there is little evidence about the impact of the program. Initial administrative difficulties have led to substantial confusion for some awardees.

5.5.2.5 Shorter Selection Cycles

Some agencies appear to manage the selection process in a shorter time period than others. Relatively minor changes to the process could make a substantial impact for companies. For example, NIH projects which do not receive funding typically receive their formal review comments too late for those comments to be applied to a resubmitted application in the next funding round—which means another 4 months delay until the next opportunity. NIH is aware of this and is initiating pilot programs to address this problem.

5.5.2.6 Closely Articulated Phase I-Phase II Cycles

Agencies that use different cycles for Phase I and Phase II can arrange them so that the end of Phase I is timed to meet the Phase II application process quite closely. DoD, NSF, and DoE already do this, but NIH does not. At NIH, Phase II applications have the same deadlines as Phase I.

---

[31]To address the efficacy and impact of the DoD Fast Track program, the SBIR program management has asked the Academies to update the 2000 study. That study found that the program was having a positive effect on commercialization. The new study will survey participants in the Fast Track program, in order to determine subsequent outcomes. For the initial Academy review, see National Research Council, *The Small Business Innovation Research Program: An Assessment of the Department of Defense Fast Track Initiative*, op. cit.

5.5.2.7 Phase II+ Programs

Phase II+ programs are designed to help bridge the gap between the end of Phase II and the marketplace, or Phase III. The two agencies that use the standard model (DoE and NASA) also do not provide any Phase II+ program.

Various Phase II+ efforts at the other agencies are described elsewhere in this chapter in more detail (see below). The point here is to note that the gap-reducing model extends beyond the Phase I-Phase II gap, into the gap after Phase II.

5.5.3 Conclusions

While the standard selection model has the virtues of simplicity and of close adherence to SBA guidelines, there are solid reasons for supporting efforts to close the funding gaps that recipient companies confront, especially where this can be done through program redesign with relatively minimal additional costs.

Best-practice analysis across agencies should enable significant improvements in this area if agencies commit to the gap-reduction model, though such a commitment may require the provision of additional administrative funding.

5.6 AWARD SIZE AND BEYOND

Both the relevant legislation and SBA guidelines make it appear that SBIR award sizes are tightly delineated. However, this is a misconception.

The "standard" awards from the SBIR program, as defined in the SBA Guidelines, are:

- Phase I: $100,000 for 6 months, to cover feasibility studies; and
- Phase II: up to $750,000 for 2 years to fund research leading toward a product.

No agency follows these guidelines precisely. Differences emerge in five areas:

- Size/duration of Phase I.
- Size/duration of Phase II.
- Funding to cover the Phase I-Phase II gap.
- Supplementary funding to cover unexpected costs.
- Follow-on funding to help completed Phase II projects move closer to market.

The recent GAO report, focused specifically on DoD and NIH, noted that a majority of awards at NIH, and almost 20 percent of awards at DoD, were larger than the guidelines permitted.

5.6.1 Size/Duration of Phase I

As Table 5-4 shows, the formal size and duration of Phase Is vary considerably at the five agencies and, in the case of DoD, between its components.

Most of the deviations from the $100,000/6-month norm are accounted for by DoD components that reserve $30,000 of the award for an option to cover the gap between the end of Phase I funding and the start of Phase II.

5.6.1.1 Larger Awards at NIH

NIH uses Phase I awards differently than the other agencies:

- It has begun to make much larger awards in some cases.
- It regularly extends Phase I awards to one year (at no additional cost), and has begun to offer a second year of Phase I support in some cases, compared to strict six or nine month limits at most other agencies.
- It provides administrative supplements that boost Phase I awards, when additional resources are needed to complete the proposed research.

TABLE 5-4 Phase I Awards

Agency Component	Award Size ($)	Duration (months)
NIH	150,000	12
NSF	100,000	6
DoE	100,000	6
NASA	70,000	6
DoD		
Army	70,000	6
Navy	70,000	6
AF	100,000	6
DARPA	99,000	6
DTRA	100,000	6
MDA	100,000	6
SOCOM	100,000	6
CBD	70,000	9
OSD	100,000	9
NIMA	100,000	6
Education	75,000	6
EPA	70,000	6
Agriculture	80,000	6
DoC		
NOAA	n/a	
NIST	75,000	6
DoT	100,000	6

SOURCE: Agency Reports and Web pages.

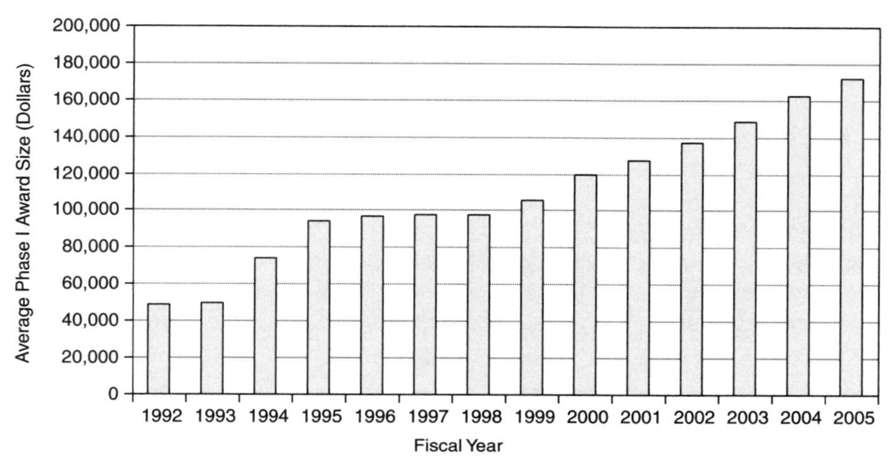

FIGURE 5-3 Average size of Phase I awards at NIH, FY1992-2005.
SOURCE: NIH Awards Database.

Figure 5-3 shows that starting in 1999, the average size of NIH Phase I awards reached $171,000 in 2005. In a growing number of cases, NIH provided Phase I funding of more than $1 million, although the median size of award has grown much more slowly than the average.

No specific policy decision appears to have been taken to make these larger awards, and the large Phase I awards have been explained in various ways by different staff at NIH. (See NIH Report: Chapter 2[32], for detailed discussion).

The median award size has remained close to the guideline, although it continues to trend up, while the mean award size has increased quite rapidly. This implies that extra large awards are (relatively) rare and (relatively) large.

5.6.1.2 Second Year Awards at NIH

Just as the size of awards has changed, NIH has extended the period of support as well. In FY2002 and FY2003, more than 5 percent of all Phase I awards were receiving a second year of support, with a median value of about $200,000 for the second year alone (see Figure 5-4).

NIH staff and recipients noted in interviews that six months is often too short to complete many biomedical Phase I projects, and that at NIH it is standard practice to grant a "no-cost" extension to one year or even more. This extends the

[32]National Research Council, *An Assessment of the SBIR Program at the National Institutes of Health*, op. cit.

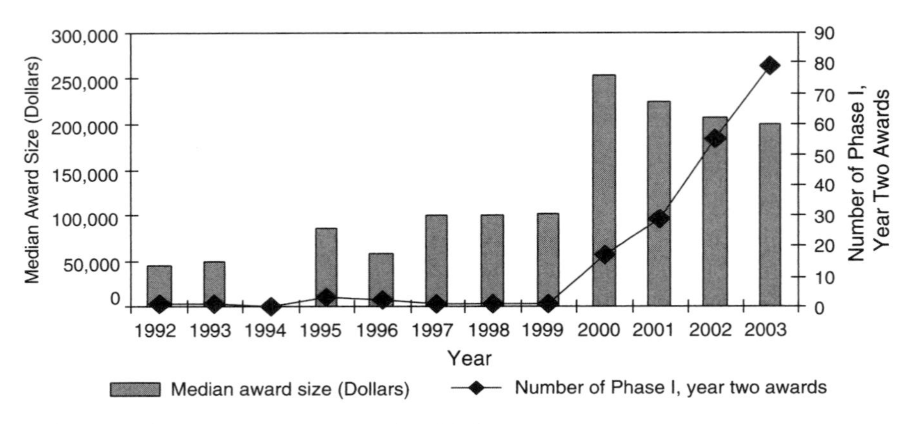

FIGURE 5-4 Phase I, year two awards at NIH.
SOURCE: NIH Awards Database.

term of the award without providing additional funding. No other agency offers such a liberal extension program.

5.6.1.3 Supplementary Funding

NIH offers a further form of funding flexibility. In principle, program officers can add limited additional funds to an award in order to help a recipient pay for unexpected costs. While practices vary at individual ICs, it appears that awards of up to 25 percent of the annual funding awarded can be made by the program manager, without further IC or NIH review. More substantial supplements are possible, but these are reviewed more extensively.

For Phase I, supplementary funding remains relatively rare, averaging less than 20 cases per year in recent years.

5.6.2 Size/Duration of Phase II

While most agencies follow the $750,000 limit mandated by SBA, others such as NSF offer less, while NIH offers more. In each case, this is an agency decision, although a waiver from SBA is required for funding beyond $750,000. NIH has received a "blanket" SBA waiver for all extra sized awards.

Beyond the formal limits, NIH once again is an important exception, as it has begun to offer some awards that are much bigger than the norm, and it also extends support far beyond 24 months.[33] GAO's 2006 report indicates that using

[33]The design of the NIH database makes it impossible to comprehensively and systematically track

TABLE 5-5 Phase II Awards

Agency Component	Award Size ($)	Duration (months)
NIH	860,000	n/a
NSF	500,000	24
DoE	750,000	24
NASA	600,000	24
DoD		
Army	730,000	24
Navy	600,000	24
AF	750,000	24
DARPA	750,000	24
DTRA	750,000	24
MDA	750,000	24
SOCOM	750,000	24
CBD	750,000	24
OSD	750,000	24
NIMA	250,000	24
Education	500,000	24
EPA	225,000	24
Agriculture	325,000	24
DoC		
NOAA		
NIST	300,000	24
DoT	750,000	15

SOURCE: Data from agency Web sites.

a slightly different methodology than NRC analysis, it found that more than half of FY2001-2004 NIH awards were above the guidelines, as were about 12 percent of DoD awards (see Table 5-6).[34]

Table 5-6 indicates that oversized awards altogether accounted for 69.3 percent of all SBIR funding at NIH (2001-2004) and 22.6 percent of funding at DoD. The larger NIH awards may reflect the rapidly rising cost of medical research (exceeding the inflation rate) and the lack of any change in the guidance for award size by the SBA. The larger awards noted above, reflecting the needs of those agencies, are mirrored by the smaller awards made, for example, by NSF and NASA. In both cases, the award amounts reflect agency judgments on how to best adapt the program to meet their mission needs.

awards across Phases. Thus, these NIH data are derived from reflect the sum of the average awards for Phase II year 1 support in FY2002 and year 2 or more support in FY2003. The correct average for total phase II award is now somewhat higher, as awards sizes have continued to grow since FY2002. Data for other agencies reflect formal limits on award sizes.

[34]U.S. General Accountability Office, *Small Business Innovation Research: Information on Awards made by NIH and DoD in Fiscal Years 2001 through 2004*, GAO-06-565, Washington, DC: U.S. General Accountability Office, 2006, p.21.

TABLE 5-6 Awards Beyond Guidelines at NIH and DoD, FY2001-2004

	NIH		DoD	
	Within the Guidelines—Number and Percentage	Above the Guidelines—Number and Percentage	Within the Guidelines—Number and Percentage	Above the Guidelines—Number and Percentage
Phase I				
Number of Awards	1,738 (49%)	1,842 (51%)	6,826 (90%)	740 (10%)
Dollar Value of Awards (in Millions)	171 (29%)	411 (71%)	587 (87%)	91 (13%)
Phase II				
Number of Awards	549 (43%)	734 (57%)	2,830 (83%)	562 (17%)
Dollar Value of Awards (in Millions)	376 (31%)	823 (69%)	1,964 (75%)	653 (25%)
Fast Track				
Number of Awards	100 (51%)	98 (49%)	[a]	[a]
Dollar Value of Awards (in Millions)	59 (28%)	152 (72%)	[a]	[a]
Total				
Number of Awards	2,387 (47%)	2,674 (53%)	9,656 (88%)	1,302 (12%)
Dollar Value of Awards (in Millions)	606 (30%)	1,386 (70%)	2,550 (77%)	743 (23%)

NOTE:
[a]All DoD Fast Track awards are included with DoD's Phase I and Phase II awards.
Almost all of DoD's Phase I awards above the guidelines are attributable to the Army's effort to provide funding for the transition from Phase I to Phase II. However, the amount above the guidelines generally reduces subsequent Phase II award amounts.
SOURCE: U.S. General Accountability Office, *Small Business Innovation Research: Information on Awards made by NIH and DoD in Fiscal Years 2001 through 2004*, GAO-06-565, Washington, DC: U.S. General Accountability Office, 2006.

5.6.3 Extra-large Phase II Awards at NIH

The data recording mechanisms at NIH make it extremely difficult to compute overall award sizes with any degree of certainty, as awards are recorded on an annual basis and the tools for linking the different years of a given award are not always accurate.

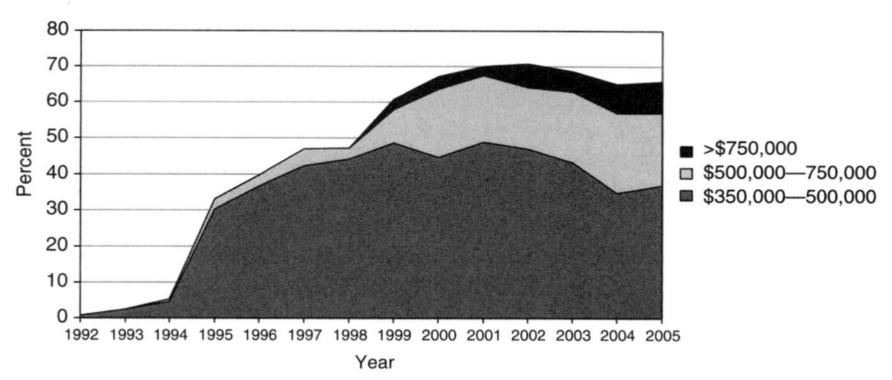

FIGURE 5-5 Phase II, year one award size at NIH.
SOURCE: National Institutes of Health.

The data show that by 2000, more than two-thirds of Phase II awards at NIH received more than $350,000, and 20 percent received more than $500,000 during the first year of the award. While the average size of the first year of an NIH Phase II award has increased to approximately $500,000, this increase has been driven by some large awards—such as the $3.5 million award in FY2003. Median award sizes have changed less.

Beyond larger awards, NIH also offers longer awards, as Phase II is sometimes continued beyond 2 years (see Figure 5-6).

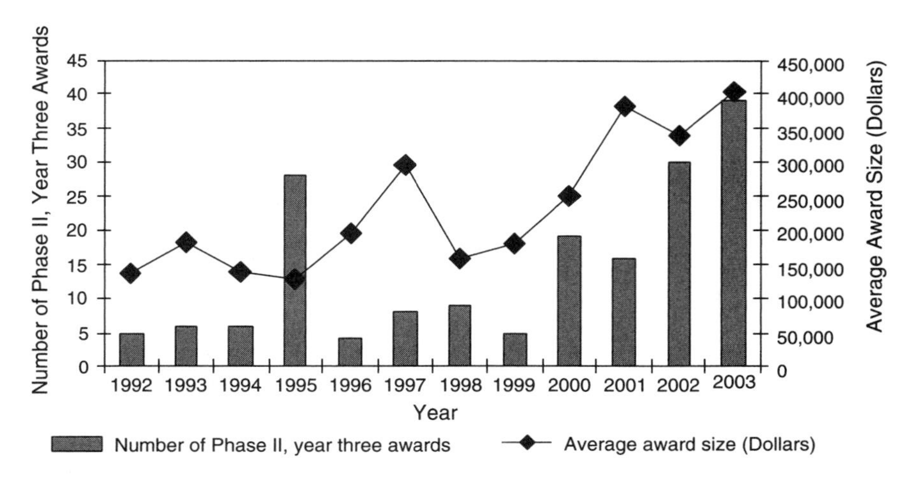

FIGURE 5-6 Third year of support for Phase II awards at NIH.
SOURCE: NIH Awards Database.

The steadily rising numbers of third-year support awards in recent years suggest that they are becoming an important component of NIH SBIR activity. In FY2002 and FY2003, more than 10 percent of awards received a third year of support.

In a few cases, NIH goes further. Ten awards have received a 5th overall year of SBIR support, and a few have received support beyond the fifth year.

5.6.3.1 Conclusions

These trends indicate that the median and average size of NIH SBIR awards are both rising, with the average award size now comfortably exceeding the statutory limit. Large awards have been made in recent years and multi-year awards have also become increasingly common.

As discussed in the NIH volume, a shift away from standardized award sizes carries important implications for fairness and efficiency that NIH has not yet confronted. Varied award size implies a trade-off between a few larger awards and more smaller ones. Yet the award selection process at NIH does not seek guidance on this from the peer reviewers who make the key funding recommendations, and there is no standard process within the Institutes and Centers at NIH to ensure that this tradeoff is fully assessed.

Absent some balancing mechanism, some agency staff agreed that this could lead to a "race to upsize," as larger projects tend—all other things being equal—to generate higher technical merit scores because they are more ambitious.

5.6.4 Supplementary Funding

NIH is also a leader in finding flexibility within the guidelines for supplementary funding. No other agency provides supplementary funding to cover expected costs.

At NIH, there are two kinds of supplementary funding:

- Administrative supplements.
- Competing supplements.

Each IC has its own guideline for supplements. Some will not permit them at all, and others allow them for only limited amounts. Administrative supplements of up to $50,000 can be authorized by an IC without going to the IC's governing council.

Only a few ICs allow competing supplements, mainly because these must be re-reviewed by peer reviewers. They reflect large adjustments for change in scope from the original project.

All supplemental requests are considered formal and require documentation. A full application is required for competing supplements, and a budget page and letter justification is required for administrative supplements. However, discus-

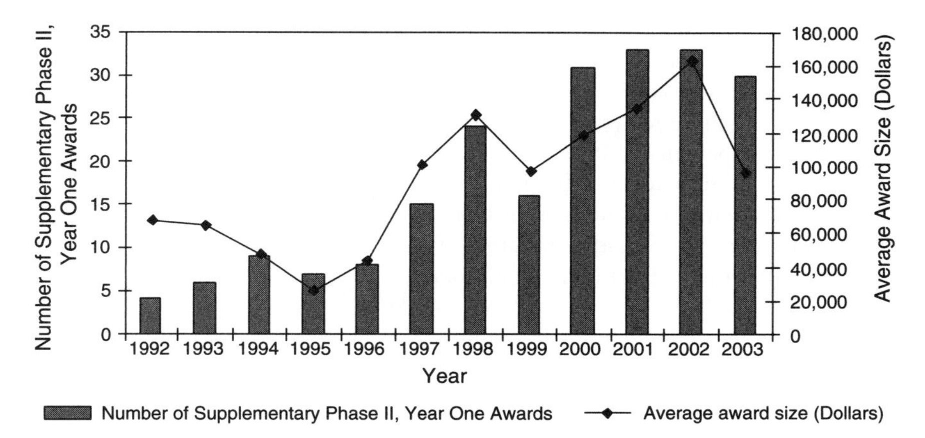

FIGURE 5-7 Supplementary Phase II, year one awards at NIH.
SOURCE: NIH Awards Database.

sions with awardees who have received administrative supplements suggest that agency staff have considerable discretion.

The data indicates that the trend for both the number and size of supplementary awards are growing at NIH (see Figure 5-7).

5.6.5 Bridge Funding to Phase III

Traversing the "Valley of Death" between the end of Phase II research funding and the commercial marketplace is the single most important challenge confronting SBIR companies, as they seek to convert research into viable projects and services. As a result, success in helping awardees traverse this crucial period should be a major consideration in assessing each agency's program.

All the agencies are now using mechanisms for addressing the Phase III/commercialization problem, such as the training and support services described in the section below on commercialization.

In addition, though, some agencies are developing funding mechanism to bridge transition to the market. Two models can be identified:

- The NSF matching funds model (also used in parts of DoD, where it known as Phase II+).
- The NIH focus on support for addressing regulatory requirements.

5.6.5.1 The NSF Matching Funds Model

NSF has taken perhaps the most aggressive approach to bridge funding. NSF Phase II awards are limited to $500,000. This leaves NSF with $250,000 under the SBA guidelines maximum, which the agency uses to provide Phase IIB awards which are made once companies can show matching third party funding. In some cases, NSF is prepared to offer Phase IIB+ awards—larger than $250,000—but these require a 2:1 match rather than 1:1.

About 30 percent of Phase II grantees have applied for Phase IIB grants. Approximately 80 percent of these applicants have been successful in receiving a Phase IIB grant. It does not appear that any applicants who met the third-party investment requirement had been turned down.

Phase IIB funding is highly conditional:

- It requires a matching contribution from bona fide third parties;
- It matches at a maximum rate of 50 percent, and provides a maximum of $250,000 in new funding, up to a total Phase II-IIB maximum of $750,000; and
- It is not automatic—companies have to apply.

Other agencies are also using variants on the matching model.

The Army's Phase II+ program is similar to the NSF model. Matching funds of up to $250,000 are made available on a competitive basis to companies with third-party funding, which can come either from the private sector or from an acquisition program at DoD.[35] This funding must be generated after the Phase II contract is signed, and must be related to that contract. Applications must be submitted at least 90 days before the end of the Phase II contract.

All other DoD components use similar but not identical approaches, based on the provision of matching funds for outside investment. Some components offer advantageous matches for funds from within DoD, which is taken as a signal that the technology is finding interest within the acquisitions community.

Although NASA does not have a Phase II+ program, agency staff have indicated that Phase II award decisions themselves can be influenced by the existence of matching non-SBIR funds from a NASA research center.

5.6.5.2 The NIH Regulatory Support Model

NIH has noticed that some of its SBIR awardees have difficulty in dealing with the financial requirements for addressing the FDA regulatory approval process for medical products and devices. The process of regulatory approval is time-consuming and, in some cases, extremely expensive. Deep-pocket investors from the private sector often do not wish to invest in high-risk products before at least some of the regulatory hurdles have been overcome.

Figure 5-8 shows that of the 768 respondents to the 2003 NIH Phase II recipient survey, about 40 percent required FDA approval, but only about 15 per-

---

[35]Accessed at <*http://www.aro.army.mil/arowash/rt/sbir/sbir_phaseii.htm*>.

BOX 5-4
Characteristics of the Army Phase II Plus Awards

- 126 total Phase II plus awards
- 8 awards greater than the stated maximum
- 51 awards at $250,000 or more
- Median award size: $160,000
- Average award size: $165,670
- Average match: $198,416
- Private investors—35
- Other government agencies—6

cent had received it. It should be understood that clinical trials involve a whole series of activities, testing both the efficacy and the safety of proposed drugs and devices, using first animal trials in many cases and then human subjects. This is generally not a short or inexpensive process.

What remains is a substantial funding gap, between the end of Phase II—

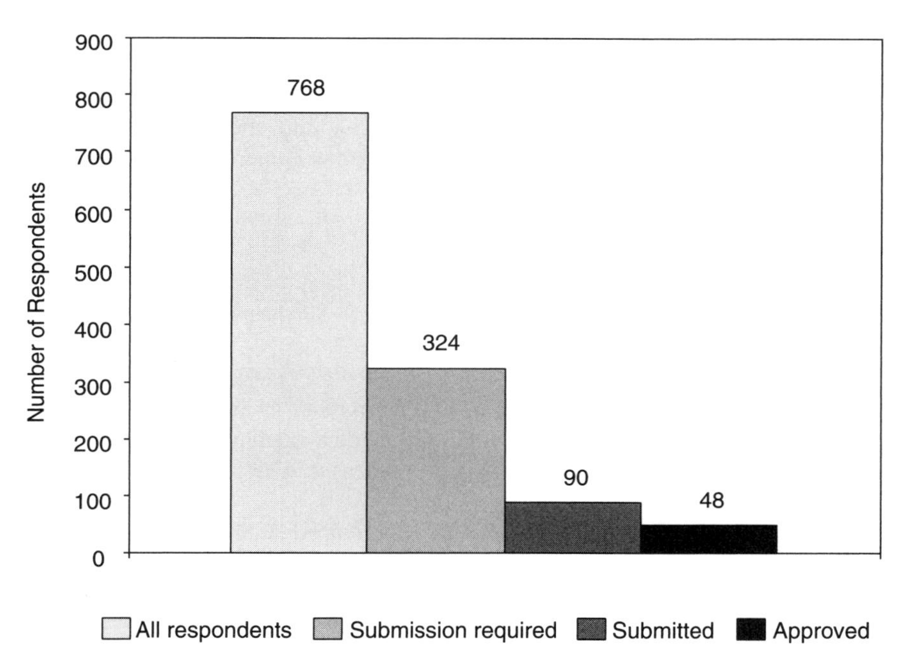

FIGURE 5-8 FDA requirements and NIH survey respondents.
SOURCE: National Institutes of Health, *National Survey to Evaluate the NIH SBIR Program: Final Report*, July 2003.

which at NIH may not even create a product that is ready for first phase clinical trials—and the possible adoption of the product by subsequent investors.

NIH has addressed this gap, in part, through a new program called Competing Continuation Awards (CCAs). Competing Continuation Awards offer up to $1 million annually for up to 3 years to companies that complete Phase II, but require additional support as they enter the regulatory approval process.

The introduction of these awards is still too recent to develop a clear picture of their eventual success or utilization level at NIH. However, they have certainly been greeted with enthusiasm by the awardee community, although recent data from NIH show that only three were awarded in FY2004.[36]

5.7 REPORTING REQUIREMENTS AND OTHER TECHNICAL ISSUES

In general, reporting requirements for SBIR awards appear to be comparable or better than those for other federal government R&D programs.

Award winners at all agencies must provide the following:

- A Phase I application, nominally limited to 25 pages, in reality may be closer to 40 pages. Includes detailed worksheet of proposed expenses.
- A Phase I contract (if a contracting agency).
- Phase I final report, showing outcomes of the research.
- Phase II application may be closer to 60-70, as letters of support etc. become more important. This phase also includes the commercialization plan—about 15 pages—outlining plans for commercializing the output from Phase II.
- Phase II final report.
- Phase II+ application (if applicable).
- Phase II+ final report (if applicable).
- Longer-term reports on downstream results from the award (varies strongly by agency).

Each agency has its own mechanisms for paying awardees. In each case, however, it is important to note that funds are made available in the course of the research, not as reimbursement for specific expenses. The internal funding gap for these projects is therefore much smaller.

Discussions with awardees and with experienced program managers suggest that there is little dissatisfaction with the level of paperwork required by the various agencies.

[36]This raises the question of whether this is an appropriate number, given the low rate at which successful laboratory discoveries get through clinical trials. Awards data from NIH, available online at <*http://grants.nih.gov/grants/funding/award_data.htm*>.

5.8 COMMERCIALIZATION SUPPORT

In the course of the most recent congressional reauthorization of SBIR, considerable emphasis was placed on the need to support commercialization efforts by awardees. Agencies have taken these comments as a direction to create or enhance quite a wide range of efforts to help companies commercialize their research.

The following major initiatives are under way at some or all of the agencies:

5.8.1 Commercialization Planning

All agencies now require applicants to submit a commercialization plan with their Phase II application. The extent to which this plan actually affects selection decisions varies widely. At DoE, it accounts for one-sixth of the final score.[37] At NIH, the effect varies depending on the make-up and interests of the study section. Several program managers at NIH institutes and centers noted that their study sections paid little heed to commercialization in the selection process. However, they also observed that having to develop a commercialization plan was a valuable exercise for the companies, even if it had no impact on selection.

5.8.2 Business Training

As agency interest in Phase III outcomes has increased, the agencies have begun to see that the technically sophisticated winners of Phase II awards are in many cases inexperienced business people with only a limited understanding of how to transition their work into the marketplace. Most agencies have developed initiatives to address this area, although these efforts are constrained by limited funds. While the transfer of best practices does occur between agencies, the extent to which agencies invent their own solutions to what are, in general, common problems is striking.

5.8.2.1 DoD

The various components at DoD operate their own business training programs; the Navy is generally believed to have put the most effort and resources into this area.

The Navy's Transition Assistance Program (TAP) has several distinct and important components. Possibly more important than any of the specific elements is the clear commitment to supporting the transition of SBIR technologies into acquisition streams at the Navy.

The TAP involves a number of components.

---

[37] DoE "Overview" PowerPoint presentation.

- All Phase II companies must attend the TAP orientation, which kicks off the program and provides detailed information about the program and its possible benefits.
- The program itself is optional—a service provided by a contractor (Dawnbreaker, Inc.) on behalf of the Navy.

The DoD components do not offer a comparable commercialization assistance program, although change is clearly under way in some areas. Defense Advanced Research Projects Agency (DARPA) has teamed with Larta Institute, a technology commercialization assistance organization, to develop and deliver a pilot program designed to assist DARPA's SBIR Phase II awardees in the commercialization of their technologies. The program began in May 2006 and will comprise individual mentoring, coaching sessions, and a showcase event where SBIR awardees will present their technologies to the investment community, to potential strategic partners and licensees—an approach similar to the Navy's. The program is available to current DARPA Defense Sciences Office SBIR Phase II Awardees.[38] DARPA is also funding technical assistance through the Foundation for Enterprise Development (FED).[39]

In FY2005, following the NRC meeting on this subject, Congress established a new commercialization pilot program (CPP), which permits DoD to use 1 percent of SBIR funding (about $11.3 million in FY2006) to pilot various approaches that will further support commercialization of successful Phase II research.[40]

- The CPP was established in the 2006 Defense Bill and is still being defined by DoD.
- The goal is to provide additional resources and emphasis on SBIR insertion into Acquisition Programs.
- New incentives are to be established and CPP projects will receive additional assistance.

Under guidance from the Under Secretary of Defense for Acquisition, Technology & Logistics, USD(AT&L), the Army, Navy and Air Force began establishing commercialization pilot programs in FY2006. Subsequent USD(AT&L) guidance in FY2007 encouraged all remaining DoD components to establish CPP activities.[41]

[38]Larta Institute Web site, accessed at <*http://www.larta.org/darpa/*>.

[39]This is a correction of the text in the prepublication version released on July 27, 2007.

[40]For a summary of the key points raised on this topic at the conference, see National Research Council, *SBIR and the Phase III Challenge of Commercialization*, op. cit.

[41]This is a correction of the text in the prepublication version released on July 27, 2007.

5.8.2.2 NSF

NSF's program emphasizes commercialization and begins support for commercialization during Phase I. All grantees are required to attend a business development and training workshop during Phase I. Phase II grantees meet annually, where they are briefed on Phase IIB opportunities and requirements and are provided with workshops intended to assist with commercialization. The Phase IIB program provides additional funding predicated on the companies obtaining third-party funding.

NSF's SBIR program operates a "MatchMaker" program, which seeks to bring together SBIR recipient companies and potential third-party funders such as venture capitalists. Just recently, NSF has begun to support participation of its grantees in DoE's SBIR Opportunity Forum, which brings grantees together with potential investors and partners.

5.8.2.3 NASA

Like other agencies, NASA does not formally market its SBIR-derived technologies. However, because the agency understands the difficulties facing companies as they approach Phase III, it has developed a range of mechanisms through which to publicize the technologies developed using SBIR (and other NASA R&D activities). These include:

- *Spin-off* (an annual publication).
- *Technology Innovation* (quarterly).
- *Technology Briefs* (monthly).
- *Success Stories* publications.
- *TechFinder*, NASA's heavily used electronic portal and database.
- Other materials such as a 2003 DVD portfolio of then-current NASA SBIR projects.

NASA also sponsors nine small business incubators in different regions of the country. Their purpose is to provide assistance in creating new businesses based on NASA technology.

5.8.2.4 NIH

NIH has managed a successful annual conference for some years, with growing attendance and positive attendee feedback.

NIH has also completed two pilot programs, one at the National Cancer Institute and one agency wide. The positive response to these programs has led the agency to roll out a larger Commercialization Assistance Program (CAP).

Larta Institute (Larta) of Los Angeles, California[42] was selected by a competitive process to be the contractor for this program.[43] The Larta contract began in July 2004, and will run for five years. During the first three years, three cohorts of SBIR Phase II winners will receive assistance. Years four and five will cover follow-up work, as each cohort is tracked for 18 months after completion of the assistance effort.

CAP Program Details. The assistance process for each group typically includes:

- Provision of consultant time for business planning and development.
- Business presentation training.
- Development of presentation materials.
- Participation in a public investment event organized by Larta.
- 18 months for follow-up and tracking.

It is still too soon to draw conclusions about the effectiveness of this effort, although NIH reports that of the 114 companies participating in the CAP program in 2004-2005, 22 have received investments totaling $22.4 million.[44]

5.8.2.5 DoE

To aid Phase II awardees that seek to speed the commercialization of their SBIR technology, the DoE has sponsored a Commercialization Assistance Program (CAP). Formally launched in 1989 by Program Manager Samuel Barish with a $50,000 contribution from 11 DoE research departments, the CAP is now the oldest commercialization focused program among all federal SBIR programs.[45] The CAP provides, on a voluntary basis, individual assistance in developing business plans and in preparing presentations to potential investment sponsors.

The CAP is operated by a contractor, Dawnbreaker, Inc., a private firm based in Rochester, New York. In order to make the forum a more attractive event for the potential partners/investors (i.e., more business opportunities), the DoE SBIR program often partners with other agencies or with other DoE offices. For the FY2006 Forum DoE SBIR has agreed to partner with the National Science Foundation's SBIR Program and with DoE's Office of Industrial Technologies.

---

[42]Larta Institute Web site, accessed at <*http://www.larta.org*>.

[43]Larta was founded by Rohit Shukla who remains as its chief executive officer. It assists technology oriented companies by bringing together management, technologies, and capital to accelerate the transition of technologies to the marketplace.

[44]Jo Anne Goodnight, NIH SBIR Coordinator, Personal Communication, March 16, 2007.

[45]Interview with Sam Barish, DoE Director of Technology Research Division, May 6, 2005.

BOX 5-5
NASA's Sponsored Business Incubators

- Business Technology Development Center
- Emerging Technology Center
- Florida/NASA Business Incubation Center
- Hampton Roads Technology Incubator
- Lewis Incubator for Technology
- Mississippi Enterprise for Technology
- NASA Commercialization Center/California State Polytechnic University
- University of Houston/NASA Technology Commercialization Incubator
- NASA Illinois Commercialization Center

5.9 CONCLUSIONS

The wide variety of organizational arrangements used to implement the SBIR program at the various agencies—and the differences even within programs at the two largest programs, at DoD and NIH—mean that generalized conclusions about the program should be treated with some caution,. Nonetheless, the NRC research permits us to draw the following conclusions, some which form the basis for recommendations.

5.9.1 Differentiation and Flexibility

It is difficult to overstress the importance of diversity within the SBIR program. In fact, given the differences, it is a mistake to talk about "the" SBIR program at all: There are agency SBIR programs, and they are—and should be—different enough that they must be considered separately, as the NRC has done through the companion agency volumes.

These differences—outlined above and emerging through a comparison of the agency volumes—stem of course partly from the differing agency missions and histories, and partly from the differing cultures within them. It is not our mandate to disentangle the sources of these complexities. The point here is to assert that the diversity of the program—which reflects the flexibility permitted within the SBIR structure, one largely enhanced by SBA interpretations of its guidelines—is fundamental to the program's success. The fact is that DoD programs are managed to generate technologies for use within DoD. It would make little sense to manage them like NIH programs, which aim to provide technologies that will eventually find their way largely to private sector providers in health care. The different objectives impose different needs and constraints and hence different strategies.

The inescapable conclusion from this is that efforts to standardize the operation of different aspects of the program are likely to prove counterproductive.

A further corollary is that managers need substantial flexibility to adapt the program to the context that they confront within their own agency and to their own technical domains. Nothing in the following points should therefore be taken as asserting the need for a single one-size-fits-all approach. We believe that every agency will find value in considering the suggestions and recommendations in this report—but also that each agency must judge those recommendations in light of its own needs, objectives, and programs.

5.9.2 Fairness, Openness, and Selection Procedures

The evidence suggests that the programs are by and large operated in a fair and open manner. Opportunities for funding are widely publicized, and the process for applying and for selection is largely transparent at all agencies.

It is also true that at some agencies—in particular NIH, where peer reviewers from outside the agency have a predominant role in selection—concerns about possible conflict of interest were widely reported by interviewees. Conflicts of interest may occur in some areas, but the different selection procedures used by the agencies require that this issue be addressed at the agency level, as it is recommended, for example, in the NIH volume. It is by no means a general problem.

While open, the agency selection procedures do differ, yet none of the agencies has made any effort to assess the success of their particular selection methodologies, or to determine ways of piloting improvements in them. Linking outcomes to selection has been done only at DoD, through the use of the Commercialization Index, but efforts to integrate this into selection procedures have not been systematically analyzed.

5.9.3 Topics

In agencies where topics define the proposal boundaries, the topic development process is also an important part of outsider perceptions of the program, and here there is less transparency: in none of the agencies is the process of topic development spelled out clearly on the Web site, which at least at DoD results in perceptions that some topics are designed—wired—for specific companies. As noted above, there is no available evidence to support this claim, not least because some companies have developed expertise, giving them a competitive advantage. Increased transparency in this area might enhance perceptions of the program's fairness more generally.

At DoD in particular, the agency began making efforts to reduce cycle times and to increase the relevancy of topics, starting in 1997 if not before. These efforts have had an important impact on relevancy, where a growing number of topics—now more than 50 percent DoD-wide—are sponsored by acquisition offices. At NASA, DoE, and NSF, topics are all drawn from technical experts with area responsibilities in the agencies.

However, two other topic issues have emerged:

- **Cycle Time.** The DoD process does reduce duplication and enhance relevance, but at the cost of timeliness. Recent initiates to develop quick response topics are therefore a promising way of balancing the needs for effectiveness and speed.
- **Breadth.** At least two factors seem to drive agencies toward narrow definitions of topics, including definitions that mandate the technical solution to be used as well as the problem to be solved. At DoD, the effort to gain interest from acquisition and other non-research communities could lead to the use of SBIR as a form of off-budget contract research. At NSF and DoE, topics have been narrowed in an effort to reduce the flow of applications to what the agency staff believes to be manageable proportions. Neither of these motives seems justified, and we would suggest that agencies seek to keep their topics as broadly defined as possible, at least in terms of the technical solutions that might be acceptable.

5.9.4 Cycle Time

In interviews with awardees and other observers of the program, concerns about cycle time surfaced regularly, at every agency. The fact is that it takes time to develop topics, publish a solicitation, assess applications, make awards, and in the end sign a contract. And that is just for Phase I.

Despite efforts at some agencies—including DoD, and in other ways DoE—the cycle time issue has not been fully addressed. Agencies have not in general fully committed to a "gap reduction" model where every effort is made to squeeze days out of the award cycle. For small businesses, especially those with few awards and few other resources, cycle time is a huge disincentive to applying for SBIR, and constitutes a significant problem for companies in the program.

Every aspect of cycle time—from lags between Phases to the number of annual solicitations, the breadth of topics and the permeability of topic borders, to contracting procedures—should be monitored, assessed, and, where possible, improved. Currently, agencies are in general insufficiently focused on this issue, as detailed in the agency volumes.

The additional funding recommended below should, in part, be used to address this problem.

5.9.5 Award Size

The fact that award size has not been officially increased since the 1992 reauthorization, and has therefore not kept pace with inflation, in itself raises the question of whether the size of awards should be increased. But this is not a simple question. Related questions include: by what amount? One-time only or possibly tied to inflation? Both Phases? At all agencies?

The primary justification for raising the nominal limits is that the cost of research has increased with inflation, and hence that these limits do not buy the agencies the same amount of research results that the Congress intended when this guidance on award size was first introduced.

Agency staff have offered a range of additional justifications for larger awards. At NIH, which has been most active in experimenting with larger awards, justifications include the need to focus on the highest quality research, the likelihood that more funding will lead to more commercial success, the impact of inflation, the need to support companies through regulatory hurdles, and the possibility that higher funding levels will expand the applicant pool by attracting, in particular, high-quality applicants who currently believe SBIR is too small to justify the effort to apply. Finally, and not insignificantly, there is a relatively higher overhead cost of administering more, smaller awards.

The latter is tied to the minimal administrative funding for SBIR, discussed below. The other NIH-specific points are discussed in the NIH volume.

Awardees in interviews have also favored larger awards—until they are asked to make an explicit trade-off between the size of awards and the number of awards. At that point, awardees often become less supportive of larger awards.

There are also arguments against larger (and fewer) awards. Because it is extremely difficult to predict which awards will generate large returns (commercial or otherwise), it may be wise to spread the awards as widely as possible. SBIR awards also play a critical role (described in Chapter 4: SBIR Program Outputs) in supporting the transition of research from the academy to the market place. This kind of motivation may not require more than the existing level of support. There is as yet also no evidence to support the assertion that larger awards generate larger returns, although further analysis at NIH might test that connection.

If we conclude—as we do—that the SBIR programs at the agencies do work as intended by Congress, and do generate significant benefits, we should recommend change only with caution. It therefore seems that while there is a case to increase award size, there are risks involved, and it would be prudent for agencies taking this step to increase the awards incrementally over, perhaps, three years to avoid a sharp contraction of the program and to allow hope for increases in R&D funding to mitigate the impact on applicant success rates of increasing award sizes.

5.9.6 Multiple-Award Winners

Multiple-award winners do not appear to constitute a problem for the SBIR program at any agency. At all agencies except DoD, only a limited number of companies win a sufficiently large number of awards to meet even the loosest definition of a "mill."

Even at DoD, we find arguments aimed at limiting a company's participation in SBIR to be unconvincing, for a number of reasons:

(1) **Successful Commercialization.** Aggregate data from the DoD commercialization database indicates that the basic charge against "mills," i.e., no commercialization, is simply incorrect. Companies winning the most awards are on average more successful commercializers than those winning fewer awards.
  - While data from this source are not comprehensive, they do cover the vast majority of MAWs—and the data indicate that on average, firms with the largest number of awards commercialize as much or more than all other groups of awardees; that in the aggregate, there is no MAW problem of companies living off SBIR awards.

(2) For some multiple winners, at least, even though they continue to win a considerable number of awards, the contribution of SBIR to overall revenues has declined.[46]

(3) Case studies show that some of the most prolific award winners have successfully commercialized, and have also in other ways met the needs of sponsoring agencies.

(4) **Graduation.** Some of the biggest Phase II winners have graduated from the program either by growing beyond the 500-employee limit or by being acquired—in the case of Foster-Miller, for example, by a foreign-owned firm. Legislating to solve a problem with companies that are in any event no longer eligible seems inappropriate.

(5) **Contract Research.** This can be valuable in and of itself. Agency staff indicate that SBIR fills multiple needs, many of which do not show up in sales data. For example, efficient probes of the technological frontier, conducted on time, on budget, to effectively test technical hypotheses, may save extensive time and resources later, according to agency staff.

(6) **Spin-offs** Some MAWs spin off companies—like Optical Sciences, Creare, and Luna. Creating new firms can be a valuable contribution.

(7) **Valuable Outputs.** Some MAWs have provided the highly efficient and flexible capabilities needed to solve pressing problems rapidly.

(8) **Compared to What?** Agency programs do not impose limits. It is hard to see why small businesses should be subjected to limits on the number of awards annually when successful universities and prime contractors are not subject to such limits.

---

[46]At Radiation Monitoring, for example, SBIR has fallen steadily and is now only 16 percent of total firm revenues.

All these points suggest that while there have been companies that depend on SBIR as their primary source of revenue for a considerable period to time, and there are some who fail to develop commercial results, the evidence strongly supports the conclusion that there is no multiple winner problem. Moreover, those who advocate a limit on the annual number of awards to a given company should explain how this limit is to be addressed across multiple agencies, and why technologies that may be important and unique to a given company should be excluded on this basis.

Given that SBIR awards meet multiple agency needs and multiple congressional objectives, it is difficult to see how the program might be enhanced by the imposition of an arbitrary limit on the number of applications per year, as is currently the case at NSF. However, if agencies continue to see issues in this area, they should consider adopting some version of the DoD "enhanced surveillance" model, in which multiple winners are subject to enhanced scrutiny in the context of the award process.

5.9.7 Information Flows

The shift toward Web-based information delivery has occurred unevenly at the different agencies. DoD has perhaps moved farthest; along with NASA, it was the first agency to require electronic submission of applications, and the online support for applicants is strong. It is, moreover, well integrated with non-electronic information sources, with two innovations in particular being well-received by awardees:

- **The Pre-release Period**, during which topics are released on the Web along with contact information for the topic authors. This enables potential applicants to directly determine how well their proposed research will fit with the agency's needs, and provides opportunities for tuning applications, so they are a better fit. This innovation also connects applicants specifically to the technical officers running a particular topic, who will, in the end, also make finding decisions.
- **The Help Desk**, which is staffed by contractors and is designed to divert non-technical questions (for example about contracts and contracting) to staff with relevant experience in the SBIR program, who may well know these materials better than topic authors for example. Program managers at NIH and NASA have complained in interviews that SBIR applicants require much more help than academic applicants for other kinds of awards. Some of that burden might be alleviated with a better resourced help desk function.[47]

[47]This is a correction of the text in the prepublication version released on July 27, 2007.

Overall, the growing size of the program, the clear interest exhibited by state economic development staff, and the increasingly positive view of SBIR at many universities, suggest that knowledge about the program is increasingly being diffused to potential applicants. The rise of the Internet—and the high quality Web sites developed by the agencies—mean that general interest can be translated into specifics quickly and inexpensively for both applicants and agencies. This conclusion is buttressed by the continuing flow of new companies into the program; these new companies account for more than 30 percent of all Phase I awards at every agency every year.

Thus it appears that the general outreach function historically fulfilled by the SBIR Agency Coordinators/Program Managers may now be changing toward a more nuanced and targeted role, focused on enhancing opportunities for underserved groups and underserved states, or on specific aspects of the SBIR program (e.g., the July 2005 DoD Phase III meeting in San Diego)—while relying on the Web and other mechanisms to meet the general demand for information. This seems entirely appropriate.

5.9.8 Commercialization Support

While some agencies have been working to support the commercialization activities of their companies for a number of years, this has clearly become a higher priority at most agencies in the recent past. Congress has always permitted agencies to spend a small amount per award ($4,000) on commercialization support, and most agencies have done so.

Commercialization support appears likely to have a significant pay-off for the agencies, partly because many SBIR firms have limited commercial experience. They are often founded by scientists and engineers who are focused on the technology, and interviews with awardees, agency staff, and commercialization contractors all indicate that the business side of commercial activities is often where companies experience the most difficulty.

Important recent initiatives include the extensive set of services provided at Navy through the TAP, and the NIH commitment to roll out the CAP program. These add to the long-running program at DoE.

It is important to understand that the character of commercialization differs quite fundamentally between DoD, NASA, and the remaining nonprocurement agencies respectively (DoE is partly a procurement agency, but for our purposes here, it purchases such a small amount of SBIR outputs for internal consumption, that it is best grouped with the nonprocurement agencies).

At DoD, where the agency provides a substantial market *if* companies can find a connection to the acquisitions programs, the critical focus of commercialization is on bridging the gaps to adequate Technology Readiness Levels and on finding ways to align companies with potential downstream acquisition programs.

At NASA, where the market within NASA for technologies, though important, may not be large enough to sustain long term development and profitability, the focus is increasingly on the spinout of technologies into the private sector.

At NIH, NSF, and DoE, commercialization means finding markets in the private sector.

These differences mean that while in general all commercialization assistance programs provide help in formulating business plans, in developing strategic business objectives, and in tuning pitches for more funding, there are important differences. In particular, DoD, which accounts for about half of the entire SBIR program, has commercialization programs that are largely (though not exclusively) focused on markets internal to DoD and on the particularly complex process of finding a way into the acquisition stream. This requires different training, different analysis, and different benchmarks than do other commercialization programs.

All of these suggest that it is important to find appropriate benchmarks against which to measure success.

Appendixes

Appendix A

NRC Phase II Survey and NRC Firm Survey

The first section of this appendix describes the methodology used to survey Phase II SBIR awards (or contracts.) The second part presents the results—first of the awards (NRC Phase II Survey) and then of the NRC Firm Survey. (Appendix B presents the NRC Phase I Survey.)

ABOUT THE SURVEYS

Starting Date and Coverage

The survey of SBIR Phase II awards was administered in 2005, and included awards made through 2001. This allowed most of the Phase II awarded projects (nominally two years) to be completed, and provided some time for commercialization. The selection of the end date of 2001 was consistent with a GAO study, which in 1991, surveyed awards made through 1987.

A start date of 1992 was selected. The year 1992 for the earliest Phase II project was considered a realistic starting date for the coverage, allowing inclusion of the same (1992) projects as the Department of Defense (DoD) 1996 survey, and of the 1992, and 1993 projects surveyed in 1998 for the Small Business Administration (SBA). This adds to the longitudinal capacities of the study. The 10 years of Phase II coverage spanned the period of increased funding set-asides and the impact of the 1992 reauthorization. This time frame allowed for extended periods of commercialization and for a robust spectrum of economic conditions. Establishing 1992 as the cut-off date for starting the survey helped to avoid the problems that older awards suffer from, including meager early data collection as well as potentially irredeemable data loss; the fact that some firms and principal investigators are no longer in place; and fading memories.

Award Numbers

While adding the annual awards numbers of the five agencies would seem to define the larger sample, the process was more complicated. Agency reports usually involve some estimating and anticipation of successful negotiation of selected proposals. Agencies rarely correct reports after the fact. Setting limitations on the number of projects to be surveyed from each firm required knowing how many awards each firm had received from all five agencies. Thus, the first step was to obtain all of the award databases from each agency and combine them into a single database. Defining the database was further complicated by variations in firm identification, location, phone numbers, and points of contact within individual agency databases. Ultimately, we determined that 4,085 firms had been awarded 11,214 Phase II awards (an average of 2.7 Phase II awards per firm) by the five agencies during the 1992-2001 time frame. Using the most recent awards, the firm information was updated to the most current contact information for each firm.

Sampling Approaches and Issues

The Phase II survey used an array of sampling techniques, to ensure adequate coverage of projects, to address a wide range of both outcomes and potential explanatory variables, and also to address the problem of skew—(that is, a relatively small percentage of funded projects typically account for a large percentage of commercial impact in the field of advanced, high-risk technologies).

- **Random Samples.** After integrating the 11,214 awards into a single database, a random sample of approximately 20 percent was sampled. Then a random sample of 20 percent was ensured for each year; e.g., 20 percent of the 1992 awards, of the 1993 awards, etc. Verifying the total sample one year at a time allowed improved ability to adapt to changes in the program over time, as otherwise the increased number of awards made in recent years might dominate the sample.

- **Random Sample by Agency.** Surveyed awards were grouped by agency; additional respondents were randomly selected as required to ensure that at least 20 percent of each agency's awards were included in the sample.

- **Firm Surveys.** After the random selection, 100 percent of the Phase IIs that went to firms with only one or two awards were polled. These are the hardest firms to find for older awards. Address information is highly perishable, particularly for earlier award years. For firms that had more than two awards, 20 percent were selected, but no less than two.

- **Top Performers.** The problem of skew was dealt with by ensuring that all Phase IIs known to meet a specific commercialization threshold (total of $10 million in the sum of sales plus additional investment) were surveyed (derived from the DoD commercialization database). Since 56 percent of all awards were in the random and firm samples described above, only 95 Phase IIs were added in this fashion.

- **Coding.** The project database tracks the survey sample, which corresponds with each response. For example, it is possible for a randomly sampled project from a firm that had only two awards to be a top performer. Thus, the response could be analyzed as a random sample for the program, a random sample for the awarding agency, a top performer, and as part of the sample of single or double winners. In addition, the database allows examination of the responses for the array of potential explanatory or demographic variables.

- **Total Number of Surveys.** The approach described above generated a sample of 6,410 projects and 4,085 firm surveys—an average of 1.6 award surveys per firm. Each firm receiving at least one project survey also received a firm survey. Although this approach sampled more than 57 percent of the awards, multiple award winners, on average, were asked to respond to surveys covering about 20 percent of their projects.

Administration of the Survey

The questionnaire drew extensively from the one used in the 1999 National Research Council assessment of the SBIR program at the Department of Defense, *The Small Business Innovation Research Program: An Assessment of the Department of Defense Fast Track Initiative.*[1] That questionnaire, in turn, built upon the questionnaire for the 1991 GAO SBIR study. Twenty-four of the twenty-nine questions on the earlier NRC study were incorporated. The researchers added twenty-four new questions to attempt to understand both commercial and non-commercial aspects, including knowledge base impacts, of SBIR, and to gain insight into impacts of program management. Potential questions were discussed with each agency, and their input was considered. In determining questions that should be in the survey, the research team also considered which issues and questions were best examined in the case studies and other research methodologies.

[1]National Research Council, *The Small Business Innovation Research Program: An Assessment of the Department of Defense Fast Track Initiative*, Charles W. Wessner, ed., Washington, D.C.: National Academy Press, 2000.

Many of the resultant thirty-two NRC Phase II Survey questions and fifteen NRC Firm Survey questions had multiple parts.

The surveys were administered online, using a Web server. The formatting, encoding and administration of the survey was subcontracted to BRTRC, Inc., of Fairfax, Virginia.

There are many advantages to online surveys (including cost, speed, and possibly response rates). Response rates become clear fairly quickly, and can rapidly indicate needed follow up for non-respondents. Hyperlinks provide amplifying information, and built-in quality checks control the internal consistency of the responses. Finally, online surveys allow dynamic branching of question sets, with some respondents answering selected sub-sets of questions but not others, depending on prior responses.

Prior to the survey, we recognized two significant advantages of a paper survey over an online one. For every firm (and thus every award), the agencies had provided a mailing address. Thus, surveys could be addressed to the firm's president or CEO at that address. That senior official could then forward the survey to the correct official within the firm for completion. For an on-line survey we needed to know the email address of the correct official. Also, each firm needed a password to protect its answers. We had an SBIR point of contact (POC) and an email address and a password for every firm which had submitted for a DoD SBIR 1999 survey. However, we had only limited email addresses and no passwords for the remainder of the firms. For many, the email addresses that we did have were those of principal investigators rather than an official of the firm. The decision to use an on-line survey meant that the first step of survey distribution was an outreach effort to establish contact with the firms.

Outreach by Mail

This outreach phase began with establishing a National Academy of Sciences (NAS) registration Web site, which allowed each firm to establish a POC, an email address, and a password. Next, the Study Director, Dr. Charles Wessner, sent a letter to those firms for which email contacts were not available. Ultimately, only 150 of the 2,080[2] firms provided POC/email after receipt of this letter. The U.S. Postal Service returned 650 of those letters as invalid addresses. Each returned letter required thorough research by calling the agency-provided phone number for the firm, then using the Central Contractor Registration database, Business.com (powered by Google), and Switchboard.com to try to find correct address information. When an apparent match was found, the firm was called to verify that it was, in fact, the firm, which had completed the SBIR. Two

[2]The letter was also erroneously sent to an additional 43 firms that had received only STTR awards.

hundred thirty-seven of the 650 missing firms were so located. Another ten firms that had gone out of business and had no POC were located.

Two months after the first mailing, a second letter from the study director was mailed to firms whose first letter had not been returned, but which had not yet registered a POC. This letter also went to 176 firms for whom we had a POC email address, but no password, and to the 237 newly corrected addresses. The large number of letters (277) from this second mailing that were returned by the U.S. Postal Service indicated that there were more bad addresses in the first mailing than were indicated by its returned mail. (If the initial letter was inadvertently delivered, it may have been thrown away.) Of the 277 returned second letters, 58 firms were located using the search methodology described above. These firms were asked on the phone to go to the registration Web site to enter POC/email/password. A total of 93 firms provided POC/email/password on the registration site subsequent to the second mailing. Three additional firms were identified as out of business.

The final mailing, a week before survey, was sent to those firms that had not received either of the first two letters. It announced the study/survey and requested support of the 1,888 CEOs for which we had assumed good POC/email information from the DoD SBIR submission site. That letter asked the recipients to provide new contact information at the DoD submission site if the firm information had changed since their last submission. One hundred seventy-three of these letters were returned. We were able to find new addresses for 53 of these, and to ask those firms to update their information. One hundred fifteen firms could not be found, and 5 more were identified as out of business.

The three mailings had demonstrated that at least 1,100 (27 percent) of the mailing addresses were in error, 734 of which firms could not be found, and 18 were reported to be out of business.

Outreach by Email

We began Internet contact by emailing the 1,888 DoD POCs to verify their email and to give them an opportunity to identify a new POC. Four hundred ninety-four of those emails bounced. The next email went to 788 email addresses that we had received from agencies as PI emails. We asked that the PI have the correct company POC identify themselves at the NAS Update registration site. One hundred eighty-eight of these emails bounced. After a more detailed search of the list used by the National Institutes of Health (NIH) to send out their survey, we identified 83 additional PIs and sent them the PI email discussed above. Email to the POCs not on the DoD Submission site resulted in 110 more POCs/emails/passwords being registered on the NAS registration site.

We began the survey at the end of February with an email to 100 POCs as a beta test and followed that with another email to 2,041 POCs (total of 2,141) a week later.

SURVEY RESPONSES

By August 5, 2005, five months after release of the survey, 1,239 firms had begun and 1,149 firms had completed at least fourteen of fifteen questions on the Firm Survey. Project surveys were begun on 1,916 Phase II awards. Of the 4,085 firms that received Phase II SBIR awards from DoD, NIH, NASA, NSF, or DoE from 1992 to 2001, an additional 7 firms were identified as out of business (total of 25), and no email addresses could be found for 893. For an additional 500 firms, the best email addresses that were found were also undeliverable. These 1,418 firms could not be contacted and thus had no opportunity to complete the surveys. Of these firms, 585 had mailing addresses known to be bad. The 1,418 firms that could not be contacted were responsible for 1,885 of the individual awards in the sample.

Using the same methodology as the GAO had used in the 1992 report of their 1991 survey of SBIR, undeliverables and out-of-business firms were eliminated prior to determining the response rate. Although 4,085 firms were surveyed, 1,418 firms were eliminated as described. This left 2,667 firms, of which 1,239 responded, representing a 46 percent response rate by firms,[3] which could respond. Similarly, when the awards, which were won by firms in the undeliverable category, were eliminated (6,408 minus 1,885), this left 4,523 projects, of which 1.916 responded, representing a 42 percent response rate. Table App-A-1 displays by agency the number of Phase II awards in the sample, the number of those awards, which by having good email addresses had the opportunity to respond, and the number that responded.[4] Percentages displayed are the percentage of awards with good addresses, the percentage of the sample that responded, and the responses as a percentage of awards with the opportunity to respond.

The NRC Methodology report had assumed a response rate of about 20 percent. Considering the length of the survey and its voluntary nature, the rate achieved was relatively high and reflects both the interest of the participants in the SBIR program and the extensive follow-up efforts. At the same time, the possibility of response biases that could significantly affect the survey results must be recognized. For example, it may be possible that some of the firms that could not be found have been unsuccessful and folded. It may also be possible that unsuccessful firms were less likely to respond to the survey.

[3]Firm information and response percentages are not displayed in Table App-A-1, which displays by agency, since many firms received awards from multiple agencies.

[4]The average firm size for awards, which responded, was 37 employees. Nonresponding awards came firms that averaged 38 employees. Since responding Phase II were more generally more recent than nonresponding, and awards have gradually grown in size, the difference in average award size ($655,525 for responding and $649,715 for nonresponding) seems minor.

TABLE App-A-1 NRC Phase II Survey Responses by Agency as of August 4, 2005

Agency	Phase II Sample Size	Awards with Good Email Addresses	Percent of Sample Awards with Good Email Addresses	Answered Survey as of 8/4/2005	Surveys as a Percentage of Sample	Surveys as a Percentage of Awards Contacted
DoD	3,055	2,191	72	920	30	42
NIH	1,680	1,127	67	496	30	44
NASA	779	534	69	181	23	34
NSF	457	336	74	162	35	48
DoE	439	335	76	157	36	47
Total	6,408	4,523	70	1,916	30	42

NRC Phase II Survey Results

NOTE: SURVEY RESPONSES APPEAR IN BOLD, AND EXPLANATORY NOTES ARE IN TYPEWRITER FONT.

Project Information 1,916 respondents answered the first question. Since respondents are directed to skip certain questions based on prior answers, the number that responded varies by question. Also some respondents did not complete their surveys. 1,736 completed all applicable questions. For computation of averages, such as average sales, the denominator used was 1,916, the number of respondents who answered the first question. Where appropriate, the basis for calculations is provided after the question.

PROPOSAL TITLE:
AGENCY:
TOPIC NUMBER:
PHASE II CONTRACT/GRANT NUMBER:

Part I. Current status of the Project.

1. What is the current status of the project funded by the referenced SBIR award? *Select the one best answer.* Percentages are based on the 1,916 respondents who answered this question.
   a. **5%** Project has not yet completed Phase II. *Go to question 21.*
   b. **22%** Efforts at this company have been discontinued. No sales or additional funding resulted from this project. *Go to question 2.*
   c. **10%** Efforts at this company have been discontinued. The project did result in sales, licensing of technology, or additional funding. *Go to question 2.*
   d. **24%** Project is continuing post-Phase II technology development. *Go to question 3.*
   e. **15%** Commercialization is underway. *Go to question 3.*
   f. **23%** Products/Processes/ Services are in use by target population/customer/consumers. *Go to question 3.*
2. Did the reasons for discontinuing this project include any of the following? ***PLEASE SELECT YES OR NO FOR EACH REASON AND NOTE THE ONE PRIMARY REASON.***

This question answered only by those who answered a or b to first question. 628 projects were discontinued. The percentages below are the percent of the discontinued projects that responded with the indicated response.

	Yes	No	Primary Reason
a. Technical failure or difficulties	29%	71%	14%
b. Market demand too small	54%	46%	25%
c. Level of technical risk too high	20%	80%	2%
d. Not enough funding	52%	48%	15%
e. Company shifted priorities	37%	63%	9%
f. Principal investigator left	13%	87%	3%
g. Project goal was achieved (e.g., prototype delivered for federal agency use)	51%	49%	8%
h. Licensed to another company	8%	92%	4%
i. Product, process, or service not competitive	25%	75%	4%
j. Inadequate sales capability	20%	80%	3%
k. Other (please specify): ______________	19%	81%	13%

The next question to be answered depends on the answer to question 1. If c, go to question 3. If b, skip to question 16.

Part II. Commercialization activities and planning.

Questions 3-7 concern actual sales to date resulting from the technology developed during this project. Sales includes all sales of a product, process, or service, to federal or private sector customers resulting from the technology developed during this Phase II project. A sale also includes licensing, the sale of technology or rights, etc.

3. Has your company and/or licensee had any actual sales of products, processes, services or other sales incorporating the technology developed during this project? *Select all that apply.* This question was not answered for those projects still in Phase II (5%) or for projects, which were discontinued without sales or additional funding (22%). The denominator for the percentages below is all projects that answered the survey. Only 73% of all projects, which answered the survey, could respond to this question. Responses a and b are exclusive; however c through f are not. A respondent could have checked all four or any combination of those responses.

    a. **18%** No sales to date, but sales are expected. *Skip to Question 8.*
    b. **7%** No sales to date nor are sales expected. *Skip to Question 11.*
    c. **37%** Sales of product(s)
    d. **6%** Sales of process(es)
    e. **18%** Sales of services(s)
    f. **8%** Other sales (e.g., rights to technology, licensing, etc.)

From the combination of responses 1b, 3a, and 3b, we can conclude that 29% had no sales and expect none, and that 18% had no sales but expect sales.

4. For your company and/or your licensee(s), when did the first sale occur, and what is the approximate amount of total sales resulting from the technology developed during this project? If multiple SBIR awards contributed to the ultimate commercial outcome, report only the share of total sales appropriate to this SBIR project. *Enter the requested information for your company in the first column and, if applicable and if known, for your licensee(s) in the second column. Enter approximate dollars. If none, enter 0 (zero).*

	Your Company	Licensee(s)
a. Year when first sale occurred.	☐☐☐☐	☐☐☐☐

45% reported a year of first sale. 60% of these first sales occurred in 2000 or later. 17% reported a licensee year of first sale. 55% of these first sales occurred in 2001 or later.

	Your Company	Licensee(s)
b. Total Sales Dollars of Product(s), Process(es), or Service(s) to date. (Average of 1,916 survey respondents)	**$925,020**	**$329,137**

Although 857 reported a year of first sale, only 790 reported sales >0. The average sales for these 790 were $2,403,255. Over half of the total sales dollars were due to 26 projects, each of which had $15,000,000 or more in sales. The highest reporting project had over $129,000,000 in sales. Similarly, of the 326 projects that reported a year of first licensee sale, only 108 reported actual licensee sales >0. Their average sales were $6,484,685. Over half of the licensee total sales dollars were due to 3 projects, each of which had $70,000,000 or more in licensee sales. The highest reporting project had $200,000,000 in licensee sales.

	Your Company	Licensee(s)
c. Other Total Sales Dollars (e.g., Rights to technology, Sale of spin-off company, etc.) to date. (Average of 1,916 survey respondents)	**$65,884**	**$36,388**

Combining the responses for b and c, the average for each of the 1,916 projects that responded to the survey is thus sales of almost one million dollars by the SBIR company and over three hundred sixty-five thousand in sales by licensees.

Display this box for Q 4 & 5 if project commercialization is known.

Your company reported sales information to DoD as a part of an SBIR proposal or to NAS as a result of an earlier NAS request. This information may be useful in answering the prior question or the next question. You reported as of *(date)*: DoD sales *($ amount)*, Other Federal Sales *($ amount)*, Export Sales *($ amount)*, Private Sector sales *($ amount)*, and other sales *($ amount)*.

5. To date, approximately what percent of total sales from the technology developed during this project have gone to the following customers? *If none enter 0 (zero). Round percentages. Answers should add to about 100%.*[5]
851 firms responded to this question as to what percent of their sales went to each agency or sector.

Domestic private sector	**35%**	
Department of Defense (DoD)	**32%**	
Prime contractors for *DoD or NASA*	**10%**	
NASA	**2%**	
Other federal agencies *(Pull down)*	**1%**	
State or local governments	**4%**	
Export Markets	**14%**	
Other (Specify)____________	**2%**	**(Specified Answers included many of the above choices as well as over 1% listing Universities or other educational institutions**

The following questions identify the product, process, or service resulting from the project supported by the referenced SBIR award, including its use in a fielded federal system or a federal acquisition program.

6. Is a federal system or acquisition program using the technology from this Phase II?
If yes, please provide the name of the Federal system or acquisition program that is using the technology. **7% reported use in a federal system or acquisition program.**

7. Did a commercial product result from this Phase II project? **32% reported a commercial product.**

[5]Please note: If a NASA SBIR award, the prime contractor's line will state "Prime contractors for NASA." The "Agency that awarded the Phase II" will only appear if it is not DoD or NASA. The name of the actual awarding agency will appear.

8. If you have had no sales to date resulting from the technology developed during this project, what year do you expect the first sales for your company or its licensee? Only firms that had no sales but answered that they expected sales got this question.

   18% expected sales. The year of expected first sale is ☐☐☐☐
   79% of those expecting sales expected sales to occur before 2008.

9. For your company and/or your licensee, what is the approximate amount of total sales expected between now and the end of 2006 resulting from the technology developed during this project? *If none, enter 0 (zero).* This question was seen by those who already had sales and by those w/o sales who reported expecting sales; however, averages are computed for all who took the survey since all could have expected sales.

   a. Total sales dollars of product(s), process(es) or services(s) expected between now and the end of 2006. (Average of 1,916 projects) **$706,090**

   b. Other Total Sales Dollars (e.g., rights to technology, sale of spin-off company, etc.) expected between now and the end of 2006. (Average of 1,916 projects) **$114,308**

   c. Basis of expected sales estimate. *Select all that apply.*
      a. **19%** Market research
      b. **27%** Ongoing negotiations
      c. **37%** Projection from current sales
      d. **4%** Consultant estimate
      e. **32%** Past experience
      f. **36%** Educated guess

10. How did you (or do you expect to) commercialize your SBIR award?
    a. **3%** No commercial product, process, or service was/is planned.
    b. **31%** As software
    c. **54%** As hardware (final product, component, or intermediate hardware product)
    d. **22%** As process technology
    e. **18%** As new or improved service capability
    f. **1%** As a drug
    g. **2%** As a biologic
    h. **22%** As a research tool
    i. **8%** As educational materials
    j. **9%** Other, please explain. **2% listed licensing.**

11. Which of the following, if any, describes the type and status of marketing activities by your company and/or your licensee for this project? *Select one for each marketing activity.* This question answered by 1,313 respondents, which completed Phase II and have not discontinued the project, w/o sales or additional funding.

	Marketing activity	Planned	Need Assistance	Underway	Completed	Not Needed
a.	Preparation of marketing plan	**10%**	**7%**	**20%**	**32%**	**31%**
b.	Hiring of marketing staff	**8%**	**6%**	**9%**	**22%**	**55%**
c.	Publicity/advertising	**13%**	**8%**	**24%**	**22%**	**34%**
d.	Test marketing	**10%**	**6%**	**15%**	**19%**	**50%**
e.	Market Research	**8%**	**9%**	**22%**	**27%**	**35%**
f.	Other *(Specify)*	**1%**	**1%**	**3%**	**1%**	**33%**

Part III. Other outcomes.

12. As a result of the technology developed during this project, which of the following describes your company's activities with other companies and investors? *Select all that apply.* Percentage of the 1,310 that answered this question.

		U.S. Companies/ Investors		Foreign Companies/ Investors	
	Activities	Finalized Agreements	Ongoing Negotiations	Finalized Agreements	Ongoing Negotiations
a.	Licensing Agreement(s)	**16%**	**16%**	**6%**	**6%**
b.	Sale of Company	**1%**	**4%**	**0%**	**1%**
c.	Partial sale of Company	**2%**	**4%**	**0%**	**1%**
d.	Sale of technology rights	**5%**	**9%**	**1%**	**3%**
e.	Company merger	**0%**	**3%**	**0%**	**1%**
f.	Joint Venture agreement	**3%**	**8%**	**1%**	**2%**
g.	Marketing/distribution agreement(s)	**14%**	**9%**	**8%**	**4%**
h.	Manufacturing agreement(s)	**5%**	**7%**	**2%**	**2%**
i.	R&D agreement(s)	**14%**	**13%**	**3%**	**4%**
j.	Customer alliance(s)	**11%**	**13%**	**4%**	**2%**
k.	Other *Specify*__________	**2%**	**2%**	**0%**	**1%**

13. In your opinion, in the absence of this SBIR award, would your company have undertaken this project?
Select one. Percentage of the 1,308 that answered this question.
a. **3%** Definitely yes
b. **10%** Probably yes *If selected a or b, Go to question 14.*
c. **16%** Uncertain

d. **33%** Probably not
e. **38%** Definitely not *If c, d or e, skip to question 16.*

14. If you had undertaken this project in the absence of SBIR, this project would have been Questions 14 and 15 were answered only by the 13% who responded that they definitely or probably would have undertaken this project in the absence of SBIR.
    a. **7%** Broader in scope
    b. **39%** Similar in scope
    c. **54%** Narrower in scope

15. In the absence of SBIR funding, *(Please provide your best estimate of the impact)*
    a. The start of this project would have been delayed about **an average of 12** months.
    62% of the 170 firms expected the project would have been delayed. 47% (80 firms) expected the delay would be at least 12 months. 26% anticipated a delay of at least 24 months.
    b. The expected duration/time to completion would have been
        1) **71%** Longer
        2) **16%** The same
        3) **1%** Shorter
        12% No response
    c. In achieving similar goals and milestones, the project would be
        1) **1%** Ahead
        2) **16%** The same place
        3) **64%** Behind
        19% No response

16. Employee information. Enter number of employees. You may enter fractions of full time effort (e.g., 1.2 employees). Please include both part time and full time employees, and consultants, in your calculation.

Number of employees (if known) when Phase II proposal was submitted.	**Ave = 31** **4% report 0** **36% report 1–5** **31% report 6–20** **12% report 21–50** **8% report >100**

Current number of employees.	**Ave = 58** **2% report 0** **21% report 1–5** **33% report 6–20** **19% report 21–50** **15% report >100**
Number of current employees who were hired as a result of the technology developed during this Phase II project.	**Ave = 2** **47% report 0** **46% report 1–5** **5% report 6–20** **2% report >20**
Number of current employees who were retained as a result of the technology developed during this Phase II project.	**Ave = 2** **44% report 0** **49% report 1–5** **5% report 6–20** **1% report >20**

17. The Principal Investigator for this Phase II Award was a *(check all that apply)*
    a. **10%** Woman
    b. **10%** Minority
    c. **82%** Neither a woman or minority

18. Please give the number of patents, copyrights, trademarks and/or scientific publications for the technology developed as a result of this project. *Enter numbers. If none, enter 0 (zero).* Results are for 1,701 respondents.

	Number Applied For/Submitted	Number Received/Published
Patents	**1,600**	**937**
Copyrights	**412**	**389**
Trademarks	**491**	**418**
Scientific Publications	**3,013**	**2,882**

Part IV. Other SBIR funding.

19. How many SBIR awards did your company receive prior to the Phase I that led to this Phase II?
    a. Number of previous Phase I awards. **Average of 18. 32% had no prior Phase I and another 42% had 5 or less prior Phase I.**

b. Number of previous Phase II awards. **Average of 7. 48% had no prior Phase II and another 36% had 5 or less prior Phase II.**

20. How many SBIR awards has your company received that are related to the project/technology supported by this Phase II award ?
    a. Number of related Phase I awards. **Average of two awards. 47% had no prior related Phase I and another 47% had 5 or less prior related Phase I.**
    b. Number of related Phase II awards. **Average of one award. 59% had no prior related Phase II and another 39% had 5 or less prior related Phase II.**

Part V. Funding and other assistance.

21. Prior to this SBIR Phase II award, did your company receive funds for research or development of the technology in this project from any of the following sources? Of 1,777 respondents.
    a. **21%** Prior SBIR *Excluding the Phase I, which proceeded this Phase II.*
    b. **10%** Prior non-SBIR federal R&D
    c. **3%** Venture capital
    d. **9%** Other private company
    e. **8%** Private investor
    f. **30%** Internal company investment (including borrowed money)
    g. **3%** State or local government
    h. **2%** College or University
    i. **5%** Other *Specify* _________

Commercialization of the results of an SBIR project normally requires additional developmental funding. Questions 22 and 23 address additional funding. Additional Developmental Funds include non-SBIR funds from federal or private sector sources, or from your own company, used for further development and/or commercialization of the technology developed during this Phase II project.

---

22. Have you received or invested any additional developmental funding in this project?
    a. **56%** Yes *Continue.*
    b. **44%** No *Skip to Question 24.*

23. To date, what has been the **total additional developmental funding for the technology** developed during this project? Any entries in the **Reported** column are based on information previously reported by your firm to DoD or NAS. They are provided to assist you in completing the **Developmental**

funding column. Previously reported information did not include investment by your company or personal investment. *Please update this information to include breaking out Private investment and Other investment by subcategory. Enter dollars provided by each of the listed sources.* *If none, enter 0 (zero).* The dollars shown are determined by dividing the total funding in that category by the 1,916 respondents who started the survey to determine an average funding. Only 989 of these respondents reported any additional funding.

Source	Reported	Developmental Funding
a. Non-SBIR federal funds	$_ _, _ _ _, _ _ _	**$259,683**
b. Private Investment	$_ _, _ _ _, _ _ _	
(1) U.S. venture capital		**$164,060**
(2) Foreign investment		**$ 40,682**
(3) Other private equity		**$125,690**
(4) Other domestic private company		**$ 64,304**
c. Other sources	$_ _, _ _ _, _ _ _	
(1) State or local governments		**$ 9,329**
(2) College or Universities		**$ 1,202**
d. Not previously reported		
(1) Your own company (Including money you have borrowed)		**$113,454**
(2) Personal funds		**$ 15,706**
Total average additional developmental funding, all sources, per award		$794,110

24. Did this award identify matching funds or other types of cost sharing in the Phase II Proposal?[6]
    a. **82%** No matching funds/co-investment/cost sharing were identified in the proposal. *If a, skip to question 26.*
    b. **17%** Although not a DoD Fast Track, matching funds/co-investment/cost sharing were identified in the proposal.
    c. **2%** Yes. This was a DoD Fast Track proposal.

25. Regarding sources of matching or co-investment funding that were proposed for Phase II, check all that apply. The percentages below are

[6]The words underlined appear only for DoD awards.

computed for those 322 projects, which reported matching funds.

a. **50%** Our own company provided funding (includes borrowed funds).
b. **14%** A federal agency provided non-SBIR funds.
c. **43%** Another company provided funding.
d. **12%** An angel or other private investment source provided funding.
e. **10%** Venture capital provided funding.

26. Did you experience a gap between the end of Phase I and the start of Phase II? 1,761 respondents.
   a. **73%** Yes *Continue.*
   b. **27%** No *Skip to question 29.*
   The average gap reported by 1,271 respondents was 9 months. 5% of the respondents reported a gap of two or more years.

27. Project history. Please fill in for all dates that have occurred. This information is meaningless in aggregate. It has to be examined project by project in conjunction with the date of the Phase I end and the date of the Phase II award to calculate the gaps.

   Date Phase I ended *Month/year* ☐☐☐☐

   Date Phase II proposal submitted *Month/year* ☐☐☐☐

28. If you experienced funding gap between Phase I and Phase II for this award, (select all answers that apply)
   a. **52%** Stopped work on this project during funding gap.
   b. **40%** Continued work at reduced pace during funding gap.
   c. **6%** Continued work at pace equal to or greater than Phase I pace during funding gap.
   d. **5%** Received bridge funding between Phase I and II.
   e. **2%** Company ceased all operations during funding gap.

29. Did you receive assistance in Phase I or Phase II proposal preparation for this award? Of 1,630 respondents.
   a. **2%** State agency provided assistance.
   b. **2%** Mentor company provided assistance.
   c. **0%** Regional association provided assistance.
   d. **5%** University provided assistance.
   e. **91%** We received no assistance in proposal preparation.

   Was this assistance useful?
   a. **64%** Very Useful
   b. **35%** Somewhat Useful
   c. **1%** Not Useful

30. In executing this award, was there any involvement by university faculty, graduate students, and/or university developed technologies? Of 1,736 respondents.
    36% Yes
    64% No

31. This question addresses any relationships between your firm's efforts on this Phase II project and any University(ies) or College(s). The percentages are computed against the 1,736, which answered question 30, not just those who answered yes to question 30.
    Select all that apply.
    a. **2%** The Principal Investigator (PI) for this Phase II project was at the time of the project a faculty member.
    b. **3%** The Principal Investigator (PI) for this Phase II project was at the time of the project an adjunct faculty member.
    c. **22%** Faculty member(s) or adjunct faculty member(s) work on this Phase II project in a role other than PI, e.g., consultant.
    d. **15%** Graduate students worked on this Phase II project.
    e. **13%** University/College facilities and/or equipment were used on this Phase II project.
    f. **3%** The technology for this project was licensed from a University or College.
    g. **5%** The technology for this project was originally developed at a University or College by one of the participants in this Phase II project.
    h. **17%** A University or College was a subcontractor on this Phase II project.

In remarks enter the name of the University or College that is referred to in any blocks that are checked above. If more than one institution is referred to, briefly indicate the name and role of each.

32. Did commercialization of the results of your SBIR award require FDA approval? Yes **7%**

    In what stage of the approval process are you for commercializing this SBIR award?
    a. **0.3%** Applied for approval
    b. **0.2%** Review ongoing
    c. **2.5%** Approved
    d. **0.3%** Not Approved
    e. **1.1%** IND: Clinical trials
    f. Other **Of the 2.4%, which indicated other, half indicated that they were preparing or intending to apply.**

NRC Firm Survey Results

NOTE: ALL RESULTS APPEAR IN BOLD. RESULTS ARE REPORTED FOR ALL 5 AGENCIES (DoD, NIH, NSF, DoE, AND NASA).

1,239 firms began the survey. 1,149 completed through question 14. 1,108 completed all questions.

If your firm is registered in the DoD SBIR/STTR Submission Web site, the information filled in below is based on your latest update as of September 2004 on that site. Since you may have entered this information many months ago, you may edit this information to make it correct. In conjunction with that information, the following additional information will help us understand how the SBIR program is contributing to the formation of new small businesses active in federal R&D and how they impact the economy. Questions A-G are autofilled from Firm database, when available.

A. Company Name: ______________________________
B. Street Address: ______________________________
C. City: ____________________ State: ____ Zip: ________
D. Company Point of Contact: ______________________________
E. Company Point of Contact Email: ______________________________
F. Company Point of Contact Phone: (___) ___ - ____ Ext: ________
G. The year your company was founded: __________

1. Was your company founded because of the SBIR Program?
   a. **79%** No
   b. **8%** Yes
   c. **13%** Yes, In part

2. Information on company founders. *Please enter zeros or the correct number in each pair of blocks.*

   a. Number of founders. ☐☐
      5% unknown
      40% 1
      30% 2
      13% 3
      8% 4
      2% 5
      2% >5
      Average = 2 founders/firm

b. Number of other companies started by one or more of the founders.

5% unknown
46% started no other firms
23% started 1 other firm
13% started 2 other firms
7% started 3 other firms
3% started 4 other firms
3% started 5 or more other firms
Average number of other firms founded is one.

c. Number of founders who have a business background.

5% Unknown
50% No founder known to have business background
30% One founder with business background
14% More than one founder with business background

d. Number of founders who have an academic background

5% Unknown
29% No founder known to have academic background
38% One founder with academic background
28% More than one founder with academic background

3. What was the most recent employment of the company founders prior to founding this company? *Select all that apply.* **Total >100% since many companies had more than one founder.**

a. **65%** Other private company
b. **36%** College or University
c. **9%** Government
d. **10%** Other

4. How many SBIR and/or STTR awards has your firm received from the federal government?

a. Phase I: ________ **Average number of Phase I reported was 14.**

13% 1 Phase I
34% 2 to 5 Phase I
24% 6 to 10 Phase I
14% 11 to 20 Phase I
11% 21 to 50 Phase I
3% 51 to 100 Phase I
2% >100 Phase I Five firms reported >300 Phase I

What year did you receive your first Phase I Award? _______

3% **reported 1983 or sooner.**
33% **reported 1984 to 1992.**
40% **reported 1993 to 1997.**
24% **reported 1998 or later.**

b. Phase II: _______ **Average number of Phase II reported was 7**

27% **1 Phase II**
44% **2 to 5 Phase II**
15% **6 to 10 Phase II**
8% **11 to 20 Phase II**
5% **21 to 50 Phase II**
1% **>50 Phase II Four firms reported >100 Phase II**

What year did you receive your first Phase II Award? _______

3% **reported 1983 or sooner.**
22% **reported 1984 to 1992.**
35% **reported 1993 to 1997.**
41% **reported 1998 or later.**

5. What percentage of your company's growth would you attribute to the SBIR program after receiving its first SBIR award?
   a. **31%** Less than 25%
   b. **25%** 25% to 50%
   c. **20%** 51% to 75%
   d. **24%** More than 75%

6. Number of company employees (including all affiliates):
   a. At the time of your company's first Phase II Award: ____

   56% **5 or less**
   28% **6 to 20**
   9% **21 to 50**
   8% **> 50 Fourteen firms (1.3%(had greater than 200 employees at time of first Phase II.**

   b. Currently: ______

   29% **5 or less**
   37% **6 to 20**
   17% **21 to 50**
   13% **51 to 200**
   5% **> 200 Eleven firms report over 500 current employees.**

7. What Percentage of your Total R&D Effort (Man-hours of Scientists and Engineers) was devoted to SBIR activities during the most recent fiscal year?___%

   22% **0% of R&D was SBIR during most recent fiscal year.**
   16% **1% to 10% of R&D was SBIR during most recent fiscal year.**
   11% **11% to 25% of R&D was SBIR during most recent fiscal year.**
   18% **26% to 50% of R&D was SBIR during most recent fiscal year.**
   14% **51% to 75% of R&D was SBIR during most recent fiscal year.**
   19% **>75% of R&D was SBIR during most recent fiscal year.**

8. What was your company's total revenue for the last fiscal year?

   a. **10%** <$100,000
   b. **18%** $100,000 - $499,999
   c. **16%** $500,000 - $999,999
   d. **33%** $1,000,000 - $4,999,999
   e. **14%** $5,000,000 - $19,999,999
   f. **6%** $20,000,000 - $99,999,999
   g. **1%** $100,000,000 +
   h. **0.4%** Proprietary information

9. What percentage of your company's revenues during its last fiscal year is federal SBIR and/or STTR funding (Phase I and/or Phase II)? ____________

   30% **0% of revenue was SBIR (Phase I or II) during most recent fiscal year.**
   17% **1% to 10% of revenue was SBIR (Phase I or II) during most recent fiscal year.**
   11% **11% to 25% of revenue was SBIR (Phase I or II) during most recent fiscal year.**
   13% **26% to 50% of revenue was SBIR (Phase I or II) during most recent fiscal year.**
   13% **51% to 75% of revenue was SBIR (Phase I or II) during most recent fiscal year.**
   13% **76% to 99% of revenue was SBIR (Phase I or II) during most recent fiscal year.**
   4% **100% of revenue was SBIR (Phase I or II) during most recent fiscal year.**

10. **This question eliminated from the survey as redundant.**

11. Which, if any, of the following has your company experienced as a result of the SBIR Program? *Select all that apply.*

    a. **Fifteen** firms made an initial public stock offering in calendar year
    Seven reported prior to 2000; two in 2000; four in 2004; and one in both 2006 and 2007

    b. **Six** planned an initial public stock offering for 2005/2006.

    c. **14%** Established one or more spin-off companies.

    How many spin-off companies?
    242 Spin-off companies were formed.

    d. **84%** reported None of the above.

12. How many patents have resulted, at least in part, from your company's SBIR and/or STTR awards?
    43% reported no patents resulting from SBIR/STTR.
    16% reported one patent resulting from SBIR/STTR.
    27% reported 2 to 5 patents resulting from SBIR/STTR.
    13% reported 6 to 25 patents resulting from SBIR/STTR.
    1% reported >25 patents resulting from SBIR/STTR.

A total of over 3,350 patents were reported; an average of almost 3 per firm

---

The remaining questions address how market analysis and sales of the commercial results of SBIR are accomplished at your company.

---

13. This company normally first determines the potential commercial market for an SBIR product, process or service
    a. **66%** Prior to submitting the Phase I proposal
    b. **21%** Prior to submitting the Phase II proposal
    c. **9%** During Phase II
    d. **3%** After Phase II

14. Market research/analysis at this company is accomplished by: *(Select all that apply.)*
    a. **28%** The Director of Marketing or similar corporate position

b. **7%** One or more employees as their primary job
c. **41%** One or more employees as an additional duty
d. **23%** Consultants
e. **53%** The Principal Investigator
f. **67%** The company President or CEO
g. **1%** None of the Above

15. Sales of the product(s), process(es) or service(s) that result from commercialising an SBIR award at this company are accomplished by: *Select all that apply.*
   a. **35%** An in-house sales force
   b. **52%** Corporate officers
   c. **30%** Other employees
   d. **30%** Independent distributors or other company(ies) with which we have marketing alliances
   e. **26%** Other company(ies), which incorporate our product into their own
   f. **9%** Spin-off company(ies)
   g. **26%** Licensing to another company
   h. **11%** None of the Above

Appendix B

NRC Phase I Survey

SURVEY DESCRIPTION

This section describes a survey of Phase I SBIR awards over the period 1992-2001. The intent of the survey was to obtain information on those which did not proceed to Phase II, although most of the firms that did receive a Phase II were also surveyed.

Over that period the five agencies (DoD, DoE, NIH, NASA, and NSF) made 27,978 Phase I awards. Of the total number for the five agencies, 7,940 Phase I awards could be linked to one of the 11,214 Phase II awards made from 1992-2001. To avoid putting an unreasonable burden on the firms that had many awards, we identified all firms that had over 10 Phase I awards that apparently had not received a Phase II. For those firms, we did not survey any Phase I awards that also received a Phase II. This meant that 1,679 Phase Is were not surveyed.

We chose to survey the Principal Investigator (PI) rather than the firm to reduce the number of surveys that any one person would have to complete. In addition, if the Phase I did not result in a Phase II, the PI was more likely to have a better memory of it than firm officials. There were no PI email addresses for 5,030 Phase I awardees. This reduced the number of surveys sent since the survey was conducted by email.

Thus there were 21,269 surveys (27,978 – 1,679 – 5,030 = 21,269) emailed to 9,184 PIs). Many PIs had received multiple Phase I awards. Of these surveys, 6,770 were undeliverable. This left possible responses of 14,499. Of these, there were 2,746 responses received. The responses received represented 9.8 percent of all Phase I awards for the five agencies, or 12.9 percent of all surveys emailed, and 18.9 percent of all possible responses.

The agency breakdown, including Phase I survey results, is given in Table App-B-1.

TABLE App-B-1 Agency Breakdown for NRC Phase I Survey

Phase I Project Surveys by Agency	Phase I Awards, 1992-2001	Answered Survey (Number)	Answered Survey (%)
DoD	13,103	1,198	9
DoE	2,005	281	14
NASA	3,363	303	9
NIH	7,049	716	10
NSF	2,458	248	10
TOTAL	27,978	2,746	10

SURVEY PREFACE

This survey is an important part of a major study commissioned by the U.S. Congress to review the SBIR program as it is operated at various federal agencies. The assessment, by the National Research Council (NRC), seeks to determine both the extent to which the SBIR programs meet their mandated objectives, and to investigate ways in which the programs could be improved. Over 1,200 firms have participated earlier this year in extensive survey efforts related to firm dynamics and Phase II awards. This survey attempts to determine the impact of Phase I awards that do not go on to Phase II. We need your help in this assessment. We believe that you were the PI on the listed Phase I.

We anticipate that the survey will take about 5-10 minutes of your time. If this Phase I resulted in a Phase II, this survey has only 3 questions; if there was not a Phase II, there are 14 questions. Where $ figures are requested (sales or funding,) please give your best estimate. Responses will be aggregated for statistical analysis and not attributed to the responding firm/PI, without the subsequent explicit permission of the firm.

Since you have been the PI on more than one Phase I from 1992 to 2001, you will receive additional surveys. These are not duplicates. Please complete as many surveys for those Phase Is that did not result in a Phase II as you deem to be reasonable.

Further information on the study can be found at <*http://www7.nationalacademies.org/sbir*>. BRTRC, Inc., is administering this survey for the NRC. If you need assistance in completing the survey, call 877-270-5392. If you have questions about the assessment more broadly, please contact Dr. Charles Wessner, Study Director, NRC.

Project Information
Proposal Title:
Agency:
Firm Name:
Phase I Contract / Grant Number:

NRC Phase I Survey Results

NOTE: RESULTS APPEAR IN BOLD. RESULTS ARE REPORTED FOR ALL 5 AGENCIES (DoD, NIH, NSF, DoE, AND NASA). EXPLANATORY NOTES ARE IN TYPEWRITER FONT.

2,746 responded to the survey. Of these 1,380 received the follow on Phase II. 1,366 received only a Phase I.

1. Did you receive assistance in preparation for this Phase I proposal?

Phase I only			Received Phase II	
95%	No	*Skip to Question 3.*	**93%**	No
5%	Yes	*Go to Question 2.*	**7%**	Yes

2. If you received assistance in preparation for this Phase I proposal, put an X in the first column for any sources that assisted and in the second column for the most useful source of assistance. Check all that apply. Answered by 74 Phase I only and 91 Phase II who received assistance.

	Phase I only Assisted/Most Useful	**Received Phase II** Assisted/Most Useful
State agency provided assistance	**10/3**	**11/10**
Mentor company provided assistance	**15/9**	**21/15**
University provided assistance	**31/17**	**34/22**
Federal agency SBIR program managers or technical representatives provided assistance	**16/8**	**25/19**

3. Did you receive a Phase II award as a sequential direct follow-on to this Phase I award? *If yes, please check yes. Your survey would have been automatically submitted with the HTML format. Using this Word format, you are done after answering this question. Please email this as an attachment to jcahill@brtrc.com, or fax to Joe Cahill 703 204 9447. Thank you for you participation.* 2,746 responses

50% No. We did not receive a follow-on Phase II after this Phase I.
50% Yes. We did receive the follow-on Phase II after this Phase I.

4. **Which statement correctly describes why you did not receive the Phase II award after completion of your Phase I effort. *Select best answer.*** All questions which follow were answered by those 1,366 who did not receive the follow-on Phase II. % based on 1,366 responses.

33% The company did not apply for a Phase II. *Go to question 5.*
63% The company applied, but was not selected for a Phase II. *Skip to question 6.*
1% The company was selected for a Phase II, but negotiations with the government failed to result in a grant or contract. *Skip to question 6.*
3% Did not respond to question 4.

5. **The company did not apply for a Phase II because: Select all that apply.** % based on 446 who answered "The company did not apply for a Phase II" in question 4.

38% Phase I did not demonstrate sufficient technical promise.
11% Phase II was not expected to have sufficient commercial promise.
6% The research goals were met by Phase I. No Phase II was required.
34% The agency did not invite a Phase II proposal.
3% Preparation of a Phase II proposal was considered too difficult to be cost effective.
1% The company did not want to undergo the audit process.
8% The company shifted priorities.
5% The PI was no longer available.
6% The government indicated it was not interested in a Phase II.
13% Other—explain:

6. **Did this Phase I produce a noncommercial benefit? Check all responses that apply.** % based on 1,366.

59% The awarding agency obtained useful information.
83% The firm improved its knowledge of this technology.
27% The firm hired or retained one or more valuable employees.
17% The public directly benefited or will benefit from the results of this Phase I. *Briefly explain benefit.*
13% This Phase I was essential to founding the firm or to keeping the firm in business.
8% No

7. Although no Phase II was awarded, did your company continue to pursue the technology examined in this Phase I? *Select all that apply.* % based on 1,366.

46% The company did not pursue this effort further.
22% The company received at least one subsequent Phase I SBIR award in this technology.
14% Although the company did not receive the direct follow-on Phase II to the this Phase I, the company did receive at least one other subsequent Phase II SBIR award in this technology.
12% The company received subsequent federal non-SBIR contracts or grants in this technology.
9% The company commercialized the technology from this Phase I.
2% The company licensed or sold its rights in the technology developed in this Phase I.
16% The company pursued the technology after Phase I, but it did not result in subsequent grants, contracts, licensing or sales.

Part II. Commercialization

8. How did you, or do you, expect to commercialize your SBIR award? *Select all that apply.* % based on 1,366.

33% No commercial product, process, or service was/is planned.
16% As software
32% As hardware (final product component or intermediate hardware product)
20% As process technology
11% As new or improved service capability
15% As a research tool
4% As a drug or biologic
3% As educational materials

9. Has your company had any actual sales of products, processes, services or other sales incorporating the technology developed during this Phase I? *Select all that apply.* % based on 1,366.

5% Although there are no sales to date, the outcome of this Phase I is in use by the intended target population.
65% No sales to date, nor are sales expected. *Go to question 11.*
15% No sales to date, but sales are expected. *Go to question 11.*
9% Sales of product(s)
1% Sales of process(es)

6% Sales of services(s)
2% Other sales (e.g., rights to technology, sale of spin-off company, etc.)
2% Licensing fees

10. For you company and/or your licensee(s), when did the first sale occur, and what is the approximate amount of total sales resulting from the technology developed during this Phase I? If other SBIR awards contributed to the ultimate commercial outcome, estimate only the share of total sales appropriate to this Phase I project. (Enter the requested information for your company in the first column and, if applicable and if known, for your licensee(s) in the second column. Enter dollars. If none, enter 0 (zero); leave blank if unknown.)

	Your Company	Licensee(s)
a. Year when first sale occurred	**89 of 147 after 1999**	**11 of 13 after 1999**
b. Total Sales Dollars of Product(s), Process(es), or Service(s) to date		
(Sale Averages)	**$84,735**	**$3,947**
Top 5 Sales Accounts for 43% of all sales	1. **$20,000,000** 2. **$15,000,000** 3. **$5,600,000** 4. **$5,000,000** 5. **$4,200,000**	
c. Other Total Sales Dollars (e.g., Rights to technology, Sale of spin-off company, etc.) to date		
(Sale Averages)	**$1,878**	**$0**

Sale averages determined by dividing totals by 1,366 responders.

11. If applicable, please give the number of patents, copyrights, trademarks and/ or scientific publications for the technology developed as a result of Phase I. (Enter numbers. If none, enter 0 [zero]; leave blank if unknown.)

Applied For or Submitted / # Received/Published

319 / 251 Patent(s)
50 / 42 Copyright(s)
52 / 47 Trademark(s)
521 / 472 Scientific Publication(s)

12. In your opinion, in the absence of this Phase I award, would your company have undertaken this Phase I research? Select only one lettered response. If you select c, and the research, absent the SBIR award, would have been different in scope or duration, check all appopriate boxes. Unless otherwise stated, % are based on 1,366.

5% Definitely yes
7% Probably yes, similiar scope and duration
16% Probably yes, but the research would have been different in the following way
% based on 218 who responded probably yes, but research would have . . .
75% Reduced scope
4% Increased scope
21% No Response to scope
5% Faster completion
51% Slower completion
44% No Response to completion rate
14% Uncertain
40% Probably not
16% Definitely not
4% No Response to question 12

Part III. Funding and other assistance

Commercialization of the results of an SBIR project normally requires additional developmental funding. Questions 13 and 14 address additional funding. Additional Developmental Funds include non-SBIR funds from federal or private sector sources, or from your own company, used for further development and/or commercialization of the technology developed during this Phase I project.

13. Have you received or invested any additional developmental funding in this Phase I? % based on 1,366.

25% Yes. Go to question 14.
72% No. Skip question 14 and submit the survey.
3% No response to question 13.

14. To date, what has been the approximate total additional developmental funding for the technology developed during this Phase I? (Enter numbers. If none, enter 0 [zero]; leave blank if unknown).

Source	# Reporting that source	Developmental Funding (Average Funding)
a. Non-SBIR federal funds	**79**	**$72,697**
b. Private Investment		
(1) U.S. Venture Capital	**13**	**$4,114**
(2) Foreign investment	**8**	**$4,288**
(3) Other private equity	**20**	**$7,605**
(4) Other domestic private company	**39**	**$8,522**
c. Other sources		
(1) State or local governments	**20**	**$1,672**
(2) College or Universities	**6**	**$293**
d. Your own company (Including money you have borrowed)	**149**	**$21,548**
e. Personal funds of company owners	**54**	**$4,955**

Average funding determined by dividing totals by 1,366 responders.

Appendix C

Case Studies

CASE STUDY COMPANIES

1. Advanced Ceramics Research
2. Creare, Inc.
3. Faraday Technology, Inc.
4. Immersion Corporation
5. ISCA Technology, Inc.
6. Language Weaver
7. MicroStrain, Inc.
8. National Recovery Technology, Inc.
9. NVE Corporation
10. Physical Sciences, Inc.
11. SAM Technologies
12. Savi Technology
13. Sociometrics Corporation

Advanced Ceramics Research[1]

Irwin Feller
American Association for the Advancement of Science

Advanced Ceramics Research (ACR) was originally incorporated as a start-up, self-financed firm in 1989 by Anthony Mulligan who had recently graduated in mechanical engineering from the University of Arizona, and Mark Angier who was still a student in mechanical engineering, also at the University of Arizona. Shortly after, they were joined by Dr. Donald Uhlmann, a professor at the University of Arizona, and Kevin Stuffle, a chemical engineer previously employed at Ceramatec Corporation, Salt Lake City, Utah. In late 1996 Dr. Daniel Albrecht, retired CEO of Buehler Corporation, joined as a shareholder and officer until 2000. Since 2000, Angier and Mulligan have remained as the only shareholders and are active in the management of the company.

From its inception, ACR sought to become a product development company, capable of manufacturing products for a diverse set of industries based on its technological developments. Although its competitive advantage has been in its advanced technology, it has sought to avoid being limited to being a contract R&D house. Over its history, the relative emphasis on R&D, product development, and manufacturing has varied, being primarily shaped by market demand conditions for its end-user products. The firm has both an extended set of collaborative, network relationships with university researchers, who conduct basic research on materials, and "downstream" customers, for its products.

Also, from its early inception, the firm knew about the SBIR program, but viewed its profit ceiling margins, placed at 5-7 percent, as too low to warrant much attention. Only commercial products were seen as yielding an adequate profit margin. Over time though, it has participated in the SBIR program of several federal agencies, including DoD, NASA, Department of Energy, and the National Science Foundation.

ACR's initial 2 products were PVA-SIC grinding stones and Polyurethane friction drive belts for the aluminum memory disk manufacturing industry. These two products were a direct result of a NASA Phase I SBIR program entitled "Laser Induced Thermal Micro-cracking for Ductile Regime Grinding of Large Optical Surfaces." While the program did not go on to Phase II, the commercial sales generated from the first two products was significant for the growth of the company.

The firm also saw market potential in developing products from advanced ceramics. The attractiveness of the SBIR program was that it would underwrite concept development. Firm representatives had several discussions with DoD

[1]Based on interview with Dr. Ranji Vaidyanthan, May 3, 2005, at the Navy Opportunity Forum and publicly available information

ADVANCED CERAMICS RESEARCH:
COMPANY FACTS AT A GLANCE

- **Address:** 3292 E. Hemisphere Loop
  Tucson, Arizona 85706
- **Phone:** 520-573-6300
- **Web site:** *<http://www.acrtucson.com>*
- **Year Started:** 1989
- **Ownership:** Private
- **Annual Sales:**
  FY2002: $5 million
  FY2003: $8.3 million
  FY2004: $11.5 million
  FY2005: $20+ million
- **Number of Employees:** 83
- **3-Year Sales Growth Rate:** 250 percent
- **4-Year Sales Growth Rate:** 400 percent
- **SIC:**
- **Technology Focus:** Advanced composite materials; rapid prototyping, UAV's, sensors
- **Number of SBIR Awards—Phase I**
  (DoD Phase I)—75
- **Number of SBIR Awards—Phase II**
  (DoD Phase II)—18
- **Number of Patents:**
- **Number of Publications:**
- **Number of Presentations:**
- **Awards:** 2002 R&D 100 Awards (Fibrous monolith wear-resistant components that increased the wear life of mining drill bits), 2001 R&D 100 Award for water-soluble composite tooling material, 2000 R&D 100 Award for water-soluble rapid prototyping support material

officials about the SBIR program, but the catalytic event was a meeting with a DARPA program officer, Bill Coblenz. Coblenz already held a patent (issued in 1988) on ceramic materials. He was interested in supporting 'far out' ideas related to the development of low-cost production processes on advanced ceramics, based on the technique of rapid prototyping. DARPA already was supporting research at the University of Michigan.

ACR was encouraged to begin work on low-cost production techniques. It did this under a series of DARPA awards and SBIR awards, although never concentrating on SBIR. Drawing in part on the advanced research being done at the University of Michigan and drawing on its expertise in both advanced ceramics and manufacturing, ACR developed a general purpose technology of being able to convert autoCAD drawings into machine readable code, then to direct generation of ceramic, composite, and metal parts.

Initially SBIR awards accounted for nearly all of ACR's revenues. By 1993 the firm had transitioned to nearly 50 percent of its revenues from the commercial sector and about 25 percent of its revenues from non-SBIR government R&D funding, with the remaining 25 percent as SBIR revenues. For 2005 the company projects about $20 million in sales with about 15-20 percent of the revenues coming from STTR/SBIR Phase I and Phase II programs. The firm's R&D also has been underwritten by revenues generated by its manufacturing operations. Its primary use of SBIR awards was to develop specific application technologies based on its core technology.

One market that it saw as having considerable potential was that of developing and manufacturing "flexible carriers for hard-disk drives" for the electronics industry. After aggressively "knocking on doors" to gain customers, it soon became a major supplier to firms such as SpeedFam Corporation, Komag, Seagate, and IBM. ACR's competitive advantage rested in its ability to make prototypes accurately, quickly, and at competitive prices. Demand for this product line grew rapidly, enabling the firm to go to a 3-shift, 7-day-a-week operation. In addition, ACR developed ancillary products related to testing and quality control tied to this product line.

The firm financed its expansion through a combination of retained earnings and license revenues, primarily from Smith Tools International, an oil and rock drilling company, and Kyocera, a Japan-based firm, which specialized in ceramics for communications applications, which licensed its Fibrous Monolith technology. ACR also reports receiving approximately $100,000 in the form of a bridge loan between a Phase I and Phase II award from a short-lived Arizona's state economic development program, funded from state lottery revenues. It reports no venture capital financing. It remains a privately held firm.

Demand for ACR's electronic products seemed to be on an upward trajectory through the 1990s. In response to demands from its primary customers for an increase in output from 5,000 to 60,000 units monthly, ACR built a new 30,000 square foot plant. The electronics market for ACR's products however declined abruptly in 1997, when two of its major customers-Seagate and Komag, two of the largest producers of hard disk drives, shifted production to Asia. This move represented both the shift from 8-inch to 5-inch and then 3.5-inch disks, and lower production costs, which drove down the price of the carrier components they produced from $16 to $1.50 per unit. The loss of its carrier business was a major reversal for the firm. Heavy layoff resulted, with employment declining to low of about 28 employees in 1998.

1998-1999 are described as years of reinvention for survival for ACR. The firm's R&D division, which formerly had been losing money, was now seen as having to become its primary source of revenue. The explicit policy was to undertake only that R&D which had discernible profit margins and the opportunity for near term commercialization. Previously, ACR had conducted a small number of Phase I SBIR awards, but had not actively pursued Phase II awards unless it could readily see the commercial product that was likely to flow from this research or it had a commercial partner.

ACR reports several outcomes from its participation in the SBIR program. As of 2005, it has received 75 Phase I and 21 Phase II awards. The larger number of awards have been from DoD, followed by NASA, with a few from the other agencies such as NSF and DoE. Products based on SBIR awards received from DARPA and NASA have had commercial sales of approximately $14 million.

ACR is now actively engaged in development and marketing of Silver Fox, a small unmanned aerial vehicle (UAV). R&D for the Silver Fox has been supported by awards under DoD's STTR program, and involves collaboration between ACR and researchers at the University of Arizona, University of California-Berkeley, the University of California-Los Angeles, and MIT.

The genesis of the project highlights the multiple uses of technological innovations. In 2000, while in Washington, DC, to discuss projects with Office of Naval Research (ONR) program managers, ACR representatives also had a chance meeting with a program manager for the Navy interested in small SWARM unmanned air vehicles (UAVs). At the time, ONR expressed an interest and eventually provided funding for developing a new low-cost small UAV as a means to engage in whale watching around Hawaii, with the objective of avoiding damage to the Navy's underwater sonic activities. Once developed however, the UAV's value as a more general purpose battlefield surveillance technology soon became apparent and ONR provided additional funding to further refine the UAV for war-fighter use in Operation Iraqi Freedom.

ACR has a bonus compensation plan that rewards employees for invention disclosures, patents, licenses, and presentations at professional meetings. These incentives are seen as fostering outcome from SBIR awards (as with all other company activities).

ACR owns a 49 percent stake in a joint venture manufacturing company called Advanced Ceramics Manufacturing, LLC, which is located on the Tohono O'Odham Reservation south of Tucson, Arizona. Fifty-one percent is owned by Tribal Land Allotees. The company, which employs about 10 people who manufacture ceramic products in a multi-million dollar facility (15,000 square feet), is expected to do about $2.5 million in sales revenues over the next 12 months.

ACR also has also recently opened 2500 square feet of laboratory and office space in Arlington, Virginia where it is basing its new Sensors Division and providing customer support to its military customers with an initial staff of eight persons.

Funding delays between Phase I and Phase II awards have been handled primarily through a process of shared decision making, leading to consensus-based reallocations of firm resources and staff assignments. ACR typically has several R&D projects occurring simultaneously. When delays occur, researchers are assembled to determine whether the firm's internal funds, including its IR&D funds, will be used to continue a specific project.

DoD's SBIR review and award procedures are seen as fair and timely. The dollar amounts of Phase I and Phase II awards and SBIR "paperwork" requirements likewise are seen as reasonable.

The Navy is seen as especially good in the speed with which it handles the selection process. It has reduced the length of time to make awards from 3-4 months to 2 months; NSF, by way of contrast, takes 6 months.

The length of the selection process across federal agencies does influence ACR's decisions. It is more likely to pursue Phase I awards from agencies such as DoD that have short selection cycles than those with long(er) ones.

The company has seen great benefit in accelerating commercialization of its SBIR/STTR programs through participation of the Navy's Technology Assistance Program (TAP). ACR first participated in the TAP program for its Water Soluble Tooling Technology, its Fibrous Monolith Technology, and its UAV technology. ACR's diligent following to what it learned in the Navy's TAP program has assisted it in receiving three separate Indefinite Deliverables, Indefinite Quantities (ID/IQ) Phase III contracts totaling $75 million. Each of the three technologies has received a $25 million ID/IQ contract to facilitate continued government use of the technology.

Creare, Inc.[2]

Philip A. Auerswald
George Mason University

OVERVIEW

Creare Inc. is a privately held engineering services company located in Hanover, New Hampshire. The company was founded in 1961 by Robert Dean, formerly a research director at Ingersoll Rand. It currently has a staff of 105 of whom 40 are engineers (27 PhDs) and 21 are technicians and machinists. A substantial percentage of the company's revenue is derived from the SBIR program. As of Fall 2004, Creare had received a total of 325 Phase I awards, 151 Phase II awards—more in the history of the program than all but two other firms.[3] While its focus is on engineering problem solving rather than the development of commercial products, since its founding, it has been New Hampshire's version of Shockley Semiconductor, spawning a dozen spin-off firms employing over 1,500 people in the immediate region, with annual revenues reportedly in excess of $250 million.[4]

Creare's initial emphasis was on fluid mechanics, thermodynamics, and heat transfer research. For its first two decades its client base concentrated in the turbo-machinery and nuclear industries. In the 1980s the company expanded to energy, aerospace, cryogenics, and materials processing. Creare expertise spans many areas of engineering. Research at Creare now bridges diverse fields such as biomedical engineering and computational fluid and thermodynamics.

At any given point in time Creare's staff is involved in approximately 50 projects. Of the 40 engineers, 10-15 are active in publishing, external relations with clients, and participation in academic conferences. The company currently employs one MBA to manage administrative matters (though the company has operated for long periods of time with no MBAs on staff). As Vice President and Principal Engineer Robert Kline Schoder states, "Those of us who are leading

---

[2]This case is based primarily on primary material collected by Philip Auerswald during an interview at Creare, Inc., in Hanover, New Hampshire, on September 16, 2004, with Robert J. Kline-Schoder (Vice President, Principal Engineer), James J. Barry (Principal Engineer), Nabil A. Elkouh (Engineer). It is also based on preliminary research. A source on the early history of Creare was Philip Glouchevitch, "The Doctor of Spin-Off." *Valley News*, December 8, 1996, pp. E1 and E5. We are indebted to Creare, Inc., for their willingness to participate in the study and in offering both a wealth of information to cover the various aspects of the study and his broad experience with the SBIR program and with high technology in the context of small business. Views expressed are those of the authors, not of the National Academy of Sciences.

[3]The other two firms are Foster-Miller (recently sold, and no longer eligible for the SBIR program) and Physical Science, Inc.

[4]A list is given in annex to this case study.

business development also lead the projects, and also publish. We wear a lot of hats."

The company's facilities comprise a small research campus, encompassing over 43,000 square feet of office, laboratory, shop, and library space. In addition to multipurpose labs, Creare's facilities include a chemistry lab, a materials lab with a scanning electron microscope, a clean-room, an electronics lab, cryogenic test facilities, and outdoor test pads. On-site machine shops and computer facilities offer support services.

FIRM DEVELOPMENT: FOUNDING AND GROWTH

Creare's founder, Robert (Bob) Dean, earned his PhD in engineering (fluid/thermal dynamics) from MIT. He joined Ingersoll Rand as a director of research. Not finding the research work in a large corporation to his liking, he took an academic position at Dartmouth's Thayer School. Soon thereafter, he and two partners founded Creare. One of the two left soon after the company's founding; the other continued with the company. But for its first decade, Robert Dean was the motive force at Creare.

Engineer Nabil Elkouh relates that the company was originally established to "invent things, license the inventions, and make a lot of money that way." Technologies that would yield lucrative licensing deals proved to be difficult to find. The need to cover payroll led to a search for contract R&D work to cover expenses until the proverbial "golden eggs" started to hatch.

The culture of the company was strongly influenced by the personality of the founder, who was highly engaged in solving research and engineering problems, but not interesting in building a commercial company—indeed, it was precisely to avoid a "bottom line" preoccupation that he had left Ingersoll Rand. Thus, even the "golden eggs" that Bob Dean was focused on discovering were innovations to be licensed to other firms, not innovations for development at Creare.

As Elkouh observes "the philosophy was—even back then—that what a product business needs isn't what an R&D business needs. You're not going to be as creative as you can be if you're doing this to support the mother ship. . . . Products go through ebbs and flows and sometimes they need a lot of resources." Furthermore, Dean was a "small organization person," much more comfortable only in companies with a few dozen people than in a large corporation. A case in point: In 1968, Hypertherm was established as a subsidiary within Creare to develop and manufacture plasma-arc metal-cutting equipment. A year later Creare spun off Hypertherm. Today, with 500 employees, it is the world leader in this field.

By 1975, an internal division had developed within Creare. Where Dean, the founder, continued to be focused on the search for ideas with significant commercial potential, others at Creare preferred to maintain the scale and focus consistent with a contract research firm. The firm split, with Dean and some engineers leaving to start Creare Innovations. Creare Innovations endured for a decade, during which time it served as an incubator to three successful companies: Spectra, Verax, Creonics.

The partners who remained at Creare Inc. instituted "policies of stability" that would deemphasize the search for "golden eggs"—ultimately including policies, described below, to make it easy for staff members to leave and start companies based upon Creare technologies.

The nuclear power industry became the major source of support for Creare. That changed quickly following the accident at Three Mile Island. At about the same time, the procurement situation with the federal government changed. Procurement reform made contracting with the federal government a far more elaborate and onerous process than it had been previously. As research funds from the nuclear industry disappeared and federal procurement contracts became less accessible to a firm of Creare's size, the company was suddenly pressured to seek new customers for its services.

In the wake of these changes came the SBIR program. The company's president at the time, Jim Block, had worked with New Hampshire Senator Warren Rudman, a key congressional supporter of the original SBIR legislation. As a consequence, the company knew that SBIR was on its way. Creare was among the first firms to apply for, and to receive, an SBIR award.

Elkouh notes that "early in the program, small companies hadn't figured out how to use it. Departments hadn't figured out how to run the program." The management of the project was ad hoc. The award process was far less competitive than it is today." Emphasis on commercialization was minimal. Program managers defined topics according to whether or not they would represent an interesting technical challenge. There was little intention on the part of the agency to use the information "other than just as a report on the shelf."

IMPACTS

From the earliest stages of its involvement in the SBIR program, Creare has specialized in solving agency-initiated problems. Many of these problems required multiple SBIR projects, and many years, to reach resolution. In most instances, the output of the project was simply knowledge gained—both by Creare employees directly, and as conveyed to the funding agency in a report. Impacts of the work were direct and indirect. As Elkouh states: "You're a piece in the government's bigger program. The Technical Program Officer learns about what you're doing. Other people in the community learn about what you're doing—both successes and failures. That can influence development of new programs."

Notwithstanding the general emphasis within the company on engineering problem solving without an eye to the market, the company has over thirty years generated a range of innovative outputs. The firm has 21 patents resulting from SBIR funded work.[5] Staff members have published dozens of papers. The firm has licensed technologies including high-torque threaded fasteners, a breast cancer surgery aid, corrosion preventative coverings, an electronic regulator for firefighters, and mass vaccination devices (pending). Products and services developed at Creare include thermal-fluid modeling and testing, miniature vacuum pumps, fluid dynamics simulation software, network software for data exchange, and the NCS Cryocooler used on the Hubble Space Telescope to restore the operation of the telescope's near-infrared imaging device.

In some cases, the company has developed technical capabilities that have remained latent for years until a problem arose for which those capabilities were required. The cryogenic cooler for the Hubble telescope is an example. The technologies that were required to build that cryogenic refrigerator started being developed in the early 1980s as one of Creare's first SBIR projects. Over 20 years, Creare received over a dozen SBIR projects to develop the technologies that ultimately were used in the cryogenic cooler. Additionally, Creare has been awarded "Phase III" development funds from programmatic areas that were 10 times the magnitude of all of the cumulative total of SBIR funds received for fundamental cryogenic refrigerator technology development. However, until the infrared imaging device on the Hubble telescope failed due to the unexpectedly rapid depletion of the solid nitrogen used to cool it, there had been no near-term application of the technologies that Creare had developed. The company has built five cryogenic cooler prototypes, and has been contacted by DoD primes and other large corporations seeking to have Creare custom build cryogenic coolers for their needs.[6]

Cooling systems for computers provide another example. The company worked intensively for a number of years in two-phase flow for the nuclear industry. This work branched into studies of two-phase flow in space—that is, a liquid-gas flow transferring heat under microgravity conditions. In the course of this work, the company developed a design manual for cooling systems based on this technology. The manual sold fifteen copies. As Elkouh observes, "there aren't that many people interested in two-phase flow in space." A Creare-developed computer modeling program for two-phase flows under variable gravity had a similar limited market. Ten years later, Creare received a call from a large semiconductor manufacturing company seeking new approaches to cooling its equipment because fans and air simply were not working any more. This led to a sequence of large industrial projects doing feasibility studies and design work to assist the client in evaluating different possible cooling systems, including two-phase approaches. The work covered the spectrum from putting together

[5]Numbers as of Fall 2004.

[6]See National Aeronautics and Space Administration, "Small Business/SBIR: NICMOS Cryocooler—Reactivating a Hubble Instrument," *Aerospace Technology Innovation*, 10(4):19-21, 2002 (July/August), accessed at *<ipp.nasa.gov/innovation/innovation104/6-smallbiz1.html>*. See also *<http://www.nasatech.com/spinoff/spinoff2002/goddard.html>*.

complete design methods—based on work performed under SBIR awards—to building experimental hardware. Most recently, NASA has contacted Creare with a renewed interest in the technology. From the agency standpoint, there is a benefit to Creare's relative stability as a small firm: They don't have to go back to square one to develop the technologies, if a need disappears and then arises again years later.

As academic research in the 1990s demonstrated the power of small firms as machines of job creation, the perception of the program changed. In the process, the relationship of perennial SBIR recipient firms such as Creare changed as well. These new modes of relationship, and some recommendations for the future, are described below.

Spin-off Companies

The success of the numerous companies that have spun off from Creare naturally leads to the question: Is fostering spin-offs an explicit part of the company's business model?

The answer is no to the extent that the company does not normally seek an equity stake in companies that it spins off. The primary reason has to do with the culture of Creare. Elkouh states that, as a rule, Creare has sought to inhibit firms as little as possible. "If you encumber them very much, they're going to fail. They are going to have a hard enough row to hoe to get themselves going. So, generally, we've tried to institute fairly minimal encumbrances on them. We've even licensed technology to companies who've spun off on relatively generous terms for them."

Does the intermittent drain of talent and technology from Creare due to the creation of spin-off firms create a challenge to the firm's partners? According to Kline-Schoder, no: "It has not happened all that often and when it has, opportunities for people who stay just expand. It's not cheap [to build a company] starting from scratch. So there's a barrier to people leaving and doing that. The other thing—in some sense, is that Creare is a lifestyle firm. Engineers are given a lot of freedom—a lot of autonomy in terms of things to work on. We think that Creare is a rather attractive place to work. So there's that barrier too."

ROLE OF THE SBIR PROGRAM

The founding of Creare pre-dated the start of the SBIR program by 20 years. However, SBIR came into being at an extremely opportune moment for the firm. It is very difficult to say whether or not the firm would have continued to exist without the program, but it is plain that the streamlined government procurement process for small business contracting ushered in by the SBIR program facilitated its sustainability and growth. In the intervening years, the SBIR program and technologies developed under the program have become the primary sources of revenue for the firm.

What accounts for the company's consistent success in winning SBIR awards? Kline-Schoder relates that "I've come across companies that have spun-out of a university or a larger organization. I routinely receive calls—five years or more after I met these start-ups—calling us and asking 'We were wondering, how you guys have been so successful? Can you tell us how do you do it?'"

As reported by the firm's staff members, Creare's rate of success in competitions where it has no prior experience with the technology or no prior relationship with the sponsor—"cold" proposals—is about the same as the overall average for the program. However, in domains where it has done prior work, the company's success rate is higher than that of the program overall. In some of these cases the author of the technical topic familiar with Creare's work may contact the firm to make them aware of the topic (this phenomenon is not unique to Creare).

Where the company has success with "cold proposals," it is often because the company successfully bridges disciplinary boundaries. In these instances, as Elkouh states, "We may have done something in one field. Someone in a different field needs something that's related to our previous work and we carry that experience over."

IMPROVING THE ADMINISTRATION OF THE SBIR PROGRAM

According to Creare's current staff members, the single most significant determinant of the Phase III potential of a project is the engagement of the author of the technical topic. Kline-Schoder states: "If your goal is to, at the end, have something that transitions (either commercially or to the government) having well written topics with authors who are energetic enough and know how to make that process happen. Oftentimes we see that you develop something, it works—it's great—and then the person on the other side doesn't know what to do. Even if you sat it on a table, the government wouldn't know how to buy it. There's no mechanism for them to actually buy it."

It is something of an irony that today, forty years after its founding, Creare is increasingly fulfilling the original ambitions of its founder: earning an increasing share of its revenue from the licensing of its technologies. Here, also, the active engagement of the topic author is critical. In one instance Elkouh worked with a Navy technical topic manager who saw the potential in a covering that had been developed at Creare with SBIR funds. This individual introduced him to over 300 people, and helped set up 100 presentations. That process led to Creare making a connection with a champion within a program area in the Navy who had the funds and was willing to seek a mechanism to buy the technology from Creare for the Navy's use.

However, even in this instance, concluding the license was not a simple matter. The appropriation made it into the budget—but that funding was still two years away. Elkouh: "The government funded the development of the technology because there was a need. Corrosion is the most pervasive thing that the Navy actually fights—a ship is a piece of metal sitting in salt water. There were reports from the fleet of people saying 'We want to cover our whole ship in this.' So now you have the people who use it say they want it, but who buys it? There is this vacuum right there—*who buys it*?"

With regard to contracting challenges, the SBIR program has largely solved the problem of a small business receiving R&D funds. From the standpoint of the staff interviewed at Creare, the contracting process directly related to the award is straightforward. What the SBIR program has not solved is the challenge of taking a technology developed under the SBIR program and finding the place within the agency, or the government, that could potentially purchase the technology.

Large corporations are no more willing to fund technology development than are government agencies. Kline-Schoder reports being approached by a large multinational interested in a technology that had been developed at Creare. The company offered to assist Creare with marketing and distribution once the technology had been fully developed into a product. However, the company was unwilling to offer any of the development funds required to get from a prototype to production.

Further obstacles to the commercial development of SBIR-funded technology are clauses within the enabling legislation pertaining to technology transfer. Kline-Schoder: "FAR clauses were in existence before the SBIR program. They were inherited by the SBIR program, but they don't fit. For instance, they state that the government is entitled to a royalty-free license to any technology developed under SBIR. But there has never been a clear definition of what that means." In one instance Creare developed a coating of interest to a private company for use in a specific product. The federal government was perceived ultimately to be the major potential market for the product in question. The issue arose: Could the company pay a royalty to Creare for its technology, given that it would be prohibited from passing on the cost to the federal buyer? Contracting challenges related to the FAR clauses created a significant obstacle to the commercialization of the technology, even when two private entities were in agreement on its potential value. "We could potentially be sitting here now looking at fairly substantial licensing revenues from that product as would [the corporate partner] and it's not happening because of that IP issue."

A second issue pertaining to the intellectual property pertains to timing. As the clause is written, a company that invents something under an SBIR is obliged to disclose the invention to the government. Two years from the day that the company discloses, it must state whether or not it will seek a patent for the invention. However, the gap between the start of Phase I and the end of Phase II is most often longer than two years. So the SBIR-funded company is placed in the awkward position of being compelled to state whether or not it intends to seek a patent on a technology essentially before it is clear if the technology works. Pressure to disclose inventions has increased over time, as the commercial focus of the program has intensified. The time pressure is even more severe when Creare seeks to find the specific corporate partner who wants to use the technology in a product. The requirement also, importantly, precludes the SBIR-funded company from employing trade secrets as an approach to protecting its intellectual property—in certain contexts, a significant constraint. Kline-Schoder: "Patenting is not the only way to protect intellectual property. The way things are structured now, you don't have that choice. No matter what invention you disclose, you have to decide within two years whether or not to patent. If you don't patent, then the rights revert to the government." In this context, Creare has a much longer time horizon than most small companies.

The view expressed by the Creare staff members interviewed was that the size of awards is adequate for the scope of tasks expected. The variation in program administration among agencies is a strength of the program—although creating uniform reporting requirements for SBIR Phase III and commercialization data would significantly reduce the burdens on the company.

Finally, from an institutional standpoint, no substitutes exist for the SBIR program. Private firms often will not pay for the kind of development work funded by SBIR. Once the scale of a proposed project grows over $100K, a private company will question the value of outsourcing the project. Lack of control is also a concern.

CONCLUSION

Creare appears to occupy a singular niche among SBIR funded companies. The company's forty year history as a small research firm is one characteristic that sets it apart from other SBIR-funded firms. The many spin-offs it has produced is a second. However, from the standpoint of its ongoing success in the SBIR program and in providing corporate consulting services, Creare's most significant differentiating characteristic may be its range of expertise. The scope of the SBIR-funded work at Creare is very broad. The reports of staff members suggest that the firm's competitive advantage relative to other small research firms is based to a significant extent on that breadth. "A lot of companies compartmentalize people," as Elkouh observes. "Everybody here is free to work on a variety of projects. At the end of the day, the companies I work with think that is where we bring the value." The same factor may account for the longevity of the firm. "We diversified internally by hiring people in different areas. That is when the cross-pollination happened." Areas come and go. Small product companies or small start-up companies focused in one area will struggle when the money disappears for whatever reason. Having evolved into a diversified research firm, Creare has endured.

CREARE—ANNEX: SAMPLE OF INDEPENDENT COMPANIES WITH ORIGINS LINKED TO CREARE

- Hypertherm, now the world's largest manufacturer of plasma cutting tools, was founded in 1968 to advance and market technology first developed at Creare. Hypertherm is consistently recognized as one of the most innovative and employee-friendly companies in New Hampshire.
- Creonics, founded in 1982, is now part of the Allen-Bradley division of Rockwell International. It develops and manufactures motion control systems for a wide variety of industrial processes.
- Spectra, a manufacturer of high-speed ink jet print heads and ink deposition systems (now a subsidiary of Markem Corporation) was formed in 1984 using sophisticated deposition technology originally developed at Creare.
- Creare's longstanding expertise in computational fluid dynamics (CFD) gave birth to a uniquely comprehensive suite of CFD software that is now marketed by Fluent (a subsidiary of Aavid Thermal Technologies, Inc.), a Creare spin-off company that was started in 1988.
- Mikros, founded in 1991, is a provider of precision micromachining services using advanced electric discharge machining technology initially developed at Creare.

Faraday Technology, Inc.[7]

Rosalie Ruegg
TIA Consulting

THE COMPANY

After a stint in a large company research lab where few of the research ideas actually became products, Dr. E. Jennings Taylor was eager to test the waters in a small company environment. He subsequently worked at first one, then another small research company in the Boston area. During this period the entrepreneurial bug bit and he added an MS in technology strategy and policy at Boston University to his PhD in material science from the University of Virginia. Shortly afterwards, he left Boston for Ohio where he launched his own company, Faraday Technology.

He chose Ohio for two reasons: It was his home state, and, while at Boston University, he had heard about the Ohio Thomas Edison Program, which offered an incubator system for business start-ups. The incubator turned out to be an old school building in Springfield. Basement space was provided at the rate of about \$2.00 per ft^2, plus telephone answering and part-time use of a conference facility. It was modest assistance, but it gave the company inexpensive space to get started. Two years later, the company was able to move into a research park near Dayton, and, two years after this, into a custom-built facility, which has since been expanded. The custom facility provides space for the development of pilot-scale prototypes of electrochemical-based processes.

The staff of approximately 10 full-time and 9 part-time employees includes researchers and experienced manufacturing engineers. Dr. Taylor, who is a registered patent agent, serves not only as CTO, but also as IP Director. The company has developed core business competencies in patent analysis. The staff also includes a full-time marketing director who oversees implementation of the company's strategic marketing plan for developing new implementation areas and customers.

The company has collaborative arrangements with a number of universities, including Columbia University, Case Western Reserve University, University of South Carolina, University of Dayton Research Institute, University of Cincinnati, Ohio State University, Wright State University, University of Nebraska, University of California-San Diego, United States Naval Academy, University

[7]The following informational sources informed the case study: interview at the company with company founder, CTO, and IP Director, Dr. E. Jennings Taylor; telephone discussion with company marketing director, Mr. Phillip Miller; company Web site, <*http://www.faradaytechnology.com*>; company brochures and other company documents; news articles; Dun & Bradstreet Company Profile Report; and earlier interview results compiled by Ritchie Coryell, NSF (retired).

**FARADAY TECHNOLOGY, INC.:
COMPANY FACTS AT A GLANCE**

- **Address:** 315 Huls, Clayton, OH 45315
- **Telephone:** 937 836-7749
- **Year Started:** 1991 (incorporated in 1992)
- **Ownership:** Private; majority woman-owned
- **Revenue:** Approx. $2 million annually
  Approx. $6.6 in direct cumulative commercial sales
  Approx. $22.9 in cumulative licensee sales
  —Revenue share from SBIR/STTR grants & contracts: 48 percent
  —Revenue share from sales, licensing, & retained earnings: 52 percent
- **Number of Employees:** 10 full-time, 9 part-time
- **Issued Patent Portfolio:** 23 U.S., 3 foreign
- **Issued Patents per Employee:** 1.4
- **3 Year Issued Patent Growth:** 130 percent
- **SIC:** Primary SIC: 8731, Commercial Physical Research
  87310300, Natural Resource Research
  Secondary SIC: 8732, Commercial Nonphysical Research
  87320108, Research Services, Except Laboratory
- **Technology Focus:** Electrochemical technologies
- **Application Areas:** Electronics, edge and surface finishing, industrial coatings, corrosion countermeasures, environmental systems, and emerging areas, e.g., fuel cell catalysis and MEMS manufacturing.
- **Funding Sources:** State and federal government grants & contracts, government sales, commercial sales, licensing fees, reinvestment of retained earnings, and private investment.
- **Number of SBIR grants:** 47
  —From NSF: 10
  —From other agencies: 37

of Virginia, and others. It often employs students, professors, and post-docs in a research capacity. The company also has collaborated with national laboratories, including Los Alamos National Laboratory.

Asked what drives the company, Dr. Taylor responded, "What drives us is we are technologists and we want to see our stuff implemented. . . . A company like Faraday is an innovation house for a number of companies that are not well positioned to innovate themselves."

THE TECHNOLOGY AND ITS USE

The company's mission, which has not changed over time, is to develop and commercialize novel electrochemical technology. Called the Faradayic™ Process, the company's platform technology is an electrically mediated manufacturing process that offers advantages of robust control, enhanced performance, cost effectiveness, and reduced hazards to the environment and to workers as compared with using chemical controls. Electrical mediation entails the sophisticated manipulation of non-steady-state electric fields as a process control method for inventing and innovating electrochemical processes, which add to or remove material from targeted devices and other media.

The technology's advantages of cleaner, faster, more precise, and cost-effective save money for the company's customers and support value-added manufacturing for them. It will allow, for example, electronics manufacturers to make smaller circuit boards with 20 or more layers stacked on a single board, with each layer connected by tiny holes uniformly plated with copper.

Faraday is applying its technology platform in multiple applications. Developing each new application area entails a new set of technical problems and is research intensive. Over the company's first decade, it has developed more than six application areas. About 25 percent of the business is currently in the electronics sectors, and about 28 percent is in edge and surface finishing. Environmental applications account for 15 percent of the business and include effluent recycling and monitoring. Industrial coatings account for about 6 percent of the company's business. Countermeasures to corrosion account for another 20 percent. Emerging technologies, including nanocoatings, 3-D MEMs manufacturing, and fuel cell catalysis make up the remaining 6 percent of the business. The company is always looking for the next manufacturing process for which it can solve a problem using its Faradayic™ process, and attract new customers.

THE ROLE OF SBIR IN COMPANY FUNDING

Dr. Taylor became aware of the SBIR program from working in two SBIR-funded companies during the early part of his career. Were it not for this experience, he is doubtful that he would have become aware of the SBIR. With his knowledge of the program, he applied for an SBIR grant early in 1993, soon after the company was incorporated. The first SBIR grant was from DoE to harness electrical mediation for monitoring contaminants in soils and groundwaters. A follow-on Phase II application was not successful. Next the company received an SBIR grant from the Navy for a sensor application, followed by non-SBIR funding of more than $1 million from DARPA for developing process technology to clean up circuit board waste. At that point the company received several EPA SBIR grants to address additional environmental problems. A Phase I SBIR grant from the Air Force followed, and still later the company received SBIR grants from other agencies including NSF. In total, the company has received 47 SBIR grants, 28 of them Phase I, 16 Phase II, and 3 Phase IIB or Phase II enhancements. From NSF, it has received five Phase I grants, four Phase II grants, and one Phase IIB grant. Table App-C-1 summarizes the company's SBIR/STTR grants in number and amount.

SBIR funding has been an essential component of the company's funding, particularly in the early years when nearly all the funding came from SBIR grants. In fact, according to Dr. Taylor, SBIR grants are involved in all areas of application pursued by the company. There are concentrations of SBIR funding in certain areas. Some things started out under SBIR, but later other sources of funding supported further research. Some things started out under other sources of funding, but later entailed an SBIR funding component for further development.

Taking into account all funding sources, the company obtained financial support from state and federal government grants and contracts, government sales, commercial sales, licensing, retained earnings, and private investment. Historically, SBIR/STTR grants and federal research contracts have comprised approximately 48 percent of total revenue. The next largest share at 28 percent

TABLE App-C-1 Faraday Technology, Inc.: SBIR/STTR Grants from NSF and Other Agencies

NSF Awards	Number	Amount ($)	Other Agency Awards	Number	Amount ($)	Total Number	Total Amount ($)
SBIR Phase I	5	470,000	SBIR Phase I	23	1,875,000	28	2,345,000
SBIR Phase II[a]	4	1,986,000	SBIR Phase II[a]	12	6,624,315	16	8,610,315
SBIR Phase IIB/ Enhancements	1	350,000	SBIR Phase IIB/ Enhancements	2	749,900	3	1,099,900
STTR Phase I	2	200,000	STTR Phase I	0	0	2	200,000
STTR Phase II	1	532,000	STTR Phase II	0	0	1	532,000
STTR Phase IIB/ Enhancements	0	0	STTR Phase IIB/ Enhancements	0	0	0	
Totals	13	3,538,000		37	9,249,215	50	12,787,215

[a]Exludes Phase IIB/Enhancement awards which are listed separately.
SOURCE: Faraday Technology, Inc.

has come from commercial sales. Sales to the government have comprised about 15 percent. Licensing has provided approximately 3 percent. Reinvestment of retained earnings and facilities reinvestment has comprised another 9 percent and 2 percent, respectively.

According to company sources, the SBIR program has enabled the company in a variety of ways to do what it otherwise would not have done. Reportedly, it has allowed the company to undertake research that otherwise would not have been undertaken. It has sped the development of proof of concepts and pilot-scale prototypes, opened new market opportunities for new applications, and led to the formation of new business units in the company. It has enabled the company to increase licensing agreements for intellectual property. It has led to key strategic alliances with other firms. It has also enabled the hiring of key professional and technical staff.

SBIR conveys more than dollars to the grantee, according to Dr. Taylor. "It is well structured to allow taking on higher risk, and it is highly competitive. The larger government programs tend to have specific deliverables instead of looking at the feasibility of high-risk activities. So to me, it [the SBIR program] is very unique. It is understood that the program is highly competitive, therefore, there is prestige associated with gaining an SBIR grant."

BUSINESS STRATEGY, COMMERCIALIZATION, AND BENEFITS

Historically the company's business strategy has been to determine a market need that can potentially be met by an adaptation of its Faradayic™ Process. Then it has pursued an SBIR grant or other sources of research funding to support the necessary research and to develop a pilot-scale prototype of the process. The company actively files patents to protect intellectual property as it develops new technical capabilities. It also investigates who is citing Faraday's patents in different application areas, obtaining a patent file wrapper to see the documentations that occur in the prosecution of each citing patent. This allows Faraday to see how other companies have claimed around Faraday's patents, and gives Faraday background knowledge about potential customers in different areas of interest. Patents and the fees they generate are the central focus of Faraday's business strategy. Thus far, the company has 23 U.S issued patents and three foreign issued patents, which, historically, amounts to 1.4 issued patents per employee.

A major route to commercialization has been to license "fields of use" to interested customers. Company staff members regularly participate in conferences and trade shows to help inform potential customers of Faraday's existing and newly emerging capabilities. The company's strategic marketing plan identifies potential customers for further contact. Once engaged, potential customers issue a purchase order to Faraday to adapt its technology for the customers' needs. In addition to the purchase order, the customer typically pays additional consideration to Faraday contractually to encumber the technology into a "no-shop/stand-still" position, effectively taking the product off the market during the period of adaptation and evaluation. The potential customer has an option to acquire exclusive rights to the technology by paying a negotiated up-front fee and a license fee in the range of 3 percent to 5 percent of the user's generated revenue for the life of the patent. The company also performs contract research for hire, and engages in product design and vending for equipment manufacturers. In the future, Faraday hopes more frequently to form strategic partnerships at the outset of a research program both for the purpose of securing research funding, but also to have a better defined path to market.

Although the company's annual revenue of roughly $2 million is relatively modest, Dr. Taylor makes the point that its license fees signal on the order of 20 to 30 times as much revenue generated by customers who are using Faraday's manufacturing processes. In Dr. Taylor's words, "Based on calculations we have done using our royalty and licensee revenue and associated multipliers, we believe that we have created on the order of $30 million in market value. Of course, Faraday only reaped a small part of this." Over its history, Faraday has generated direct commercial sales of approximately $6.6 million. Customers realize value from Faraday's processes in several main ways: lower cost manufacturing processes, higher quality output, and a combination of the two.

Beyond developing technical capabilities that lead to revenue for Faraday and value-added for its customers, Dr. Taylor pointed to a whole "under current" of effects that the SBIR is having that nobody is really able to capture. "For example, one of our customers likely was going to move off-shore if it could not find a cost-reducing solution to a manufacturing problem it had. Faraday was able to meet the need through innovative research. Of course, I can't prove it, but I think the technology solution figured importantly in their decision not to move. So what's the value of having a company remain in the United States? That's an example of the benefit of innovation funded by SBIR that is not usually factored into the value of the SBIR program. I can't quantify the value; yet I feel strongly that it is true based on what I know."

As an example of another difficult-to-capture type of benefit, the technology also offers the potential of environmental effects in several ways. For one thing, by using electricity to achieve results, the Faradayic™ Process reduces the need for polluting chemical catalysts. For another, the process enables the capture of materials from industrial process waste streams. Yet another emerging application is to control the flow of contaminants through soil for more cost-effective capture and clean up.

Educational benefits also result from the company's activities. Largely as a result of participating in NSF's annual conference, the company became active in encouraging young people to pursue careers in science. It has provided internships to three junior high students; it has employed several high school teachers during the summer; and it annually hosts a high school science day. Moreover, as a result of the company's many collaborative relationships with universities, it has employed about 20 undergraduate and graduate students, one of whom did a PhD dissertation and another, a master's thesis under the Dr. Taylor's supervision.

The innovation process and the multifaceted roles played by the SBIR are complex and nonlinear, noted Dr. Taylor. He recalled several Phase I grants that did not go on to Phase II. By one standard, these would be considered "failed grants." Yet, he explained, after some twists and turns, the concepts explored in these earlier Phase II grants eventually came to fruition and became important application areas. For example, a "failed" Phase I grant provided the seed for later electronics work that now provides 35 percent of the company's business and accounts for eight of its patents. "It is not a tidy path; it is a cumulative process."

VIEWS ON THE SBIR PROGRAM AND PROCESSES

Dr. Taylor made a number of observations about the SBIR program and its processes that may serve to improve the program. These are summarized as follows:

Need to Recognize Multiple Paths to Commercialization

Dr. Taylor expressed the hope that it will be recognized that there are multiple paths to commercialization that have merit. He pointed out that it is particularly important for agencies "to understand the various ways to get to the commercial end game—which could involve venture capitalists, could involve strategic partners, could involve an ongoing company trying to augment its business. Agencies need to be flexible. It would be a myopic view if we were to conclude that SBIR funding should only go to companies that are going to do no more than, say, four years of SBIR work and then go public. That is a model, but another model is that innovation is an ongoing thing. . . . The idea of some limit to the number of grants a company can receive cannot be addressed well in absolute terms. Rather, it is important to look at a company's history and see if it is accomplishing something in the longer run—helping to meet R&D needs of an agency or seeding work that eventually turns into something useful."

He went on to raise the issue of a company that receives many DoD and NASA SBIR grants, posing the question: "Would that make it a mill? Well, I would expect that they are providing a research service that DoE and NASA want. . . . If the grants process is modeled correctly, with effective criteria and review, and if it is functioning well, there should be no mills without value added, because nobody would keep funding them unless they do have value." In short, the existence of a mill implies a program breakdown, where SBIR is not taken seriously, where inadequate attention is given to proposal review and project selection.

Mr. Phillip Miller, company marketing director, noted that most SBIR grantees are not OEM suppliers of product; that most grantees develop technology and intellectual property and in turn sell the innovations to customers through a variety of means—not just through products shipped. Yet, the agencies who collect information about the SBIRs impact, typically ask only about products.

Recommendation for Simpler Accounting

Dr. Taylor's opinion is that there are a lot of misconceptions among prospective applicants about accounting requirements, particularly in terms of the indirect rate and what is allowable and what is not. A company's overhead rate may look higher than others because it puts items in it that others put in the direct rate. It is important to look at the overall rate in comparing costs across companies. Furthermore, Dr. Taylor mused, "If the program is geared toward commercialization, and patenting is an important component of this, why wouldn't they allow you to charge patent costs? After all, the government has so called 'march-in' rights,' Furthermore, what is a patent cost? Clearly filing fees and maintenance fees are patent costs. But, are patent attorney's fees associated with evaluating and assessing the technical and patent literature also patent costs? We often use consultants and professors to do the same work and allocate these costs to professional services." He also questioned why the DoD SBIR forms allow you to charge fees and to pay royalties, but the other programs he is familiar with do not have these features. Noting that it is the financial side that is the most daunting to technical people, he urged the SBIR program to give more attention to education on financial issues. He noted that some small companies have very poor accounting systems, and they could benefit from learning how to set up an appropriate system.

SBIR Application Process

According to Dr. Taylor, other agencies' online submission application processes are easier to manage than the system implemented at NSF. His opinion is that NSF's application seems a bit strange from a business standpoint because the form is geared to universities, which comprise the majority of NSF's customers. It is not a dedicated, customized form for SBIR.

Commercialization Issues

Dr. Taylor had heard that a commercialization index is being used to rate SBIR companies, but he did not know how it is computed. He expressed the view that license revenue should be treated differently in computing the index than product sales and other revenue, because a license fee represents on the order of 3 percent to 5 percent of the revenue generated by the licensee. This means that a multiplier of 20 to 30 would need to be applied to license fees to put them on a comparable basis with product sales.

He also emphasized the importance that should be placed on matching funds as a way to indicate commercial potential. "Bringing in cash matching funds is a more powerful signal of commercialization than any review panel's opinion."

Misconceptions about the SBIR and Other Government R&D Partnerships

Dr. Taylor noted that many of the other companies he has worked with have had no awareness of the SBIR program. He recalled a strategic partner who had misunderstandings and misconceptions about the SBIR that interfered with negotiations on a strategic alliance. Another partner, he noted, was afraid of march-in rights, and this was a major encumbrance to making a deal. "More public education might help, or even the elimination of the march-in rights clause."

Turning to the NSF's SBIR program, Dr. Taylor commented on some of its features as follows:

Support of Manufacturing Innovations

According to Dr. Taylor the NSF SBIR program is unique in its strong support of manufacturing innovations. In his words: "NSF seems more supportive of manufacturing-type innovations. . . . NSF seems to actively appreciate the importance of innovating in manufacturing." He contrasted this interest with a lack of interest in manufacturing on the part of most other SBIR programs.

Specification of Topics

Dr. Taylor saw NSF's SBIR as having more topic flexibility than the other programs. He indicated that this flexibility is helpful for a business like his that wishes to pursue various application areas. Further, he noted that the NSF SBIR is very responsive to national priorities and needs in crafting its topics.

Portfolio or Program Managers

"NSF's special strength is in what they call their portfolio managers. They have people who manage the different technology sectors. NSF is more proactive in helping grantees through the commercial stuff. . . . To me, it is a very, very good, solid interactive group. . . . They hold you to task, but I like that. . . . I am impressed with the NSF group because they themselves are an innovative entity—they are looking for ways to continually improve the program; improve their grantee's conference. . . . For example, NSF is trying to expand its matchmaker program—which was geared towards the venture capital community—to include strategic industrial partners . . . and to me it just makes sense. They had about 12 potential strategic industrial partners in Phoenix" [location of the last NSF annual grantee conference].

Commercialization Assistance

A three-day patenting workshop offered at the NSF annual conferences won special praise from Dr. Taylor. At the same time, he found it interesting that many grantees said they couldn't afford the time to attend. To him, this indicated not only an excellent opportunity missed, but a lack of seriousness about patenting, and, therefore, a possible lack of effective commercialization plans.

Lack of Travel Funds

Noting that "NSF hasn't had one person out here," Dr. Taylor expressed his view that it is unfortunate that NSF staff does not have the funding to travel. At the same time, he acknowledged that NSF requires grantees to attend the annual conference, and the conference provides a forum for interacting with NSF program managers and other people "who are at a similar stage of business as you." He added that NSF does not appear to be unique in a lack of travel funding.

Proposal Review Process

Commenting on NSF's review process, Dr. Taylor noted that he has served on panels, and "it is an extensive process." He explained that the technical and business reviewers sat together on the panels in which he participated, and commented that this combining of the reviewers, which may be unique to NSF, is a valuable approach.

SUMMARY

This case study shows how SBIR grants enabled a start-up company in Ohio, Faraday Technology, Inc., to develop an underlying electrochemical technology platform and, through continuing innovation, to leverage it into multiple lines of business. The company's main focus is on developing cleaner, faster, more precise, and more cost-effective processes to add to or remove target materials from many different kinds of media, ranging from metal coatings to fabricated parts, to electronic components, to contaminants in soil. By offering innovative processes that reduce costs for customers and support value-added manufacturing, the company serves as an innovation house for a number of manufacturing companies that are not well positioned to innovate themselves.

The case illustrates that a basic technology platform can be leveraged through additional research into many novel applications to solve specific problems. Because the application areas are so different, they each have required advances in scientific and technical knowledge for success. The challenges of devising a process to uniformly coat tiny holes through 20 or more layers of circuits stacked on a printed wiring board, for example, are quite different from the challenges of developing a process to produce super smooth surface finishing for titanium jet engine components.

The case illustrates a business model that relies primarily on an aggressive patenting strategy and licensing in multiple fields of use to generate business revenue. Essential to leveraging the technology platform is the alignment of intellectual property and marketing strategies. The company continually assesses market drivers to identify needs that may be addressed by Faraday's platform technology. The company works closely with patent firms, has a patent professional on staff and has its engineers and scientists trained in patent drafting. Furthermore, the company has long employed a full-time Marketing Director.

The case also demonstrates how relatively modest licensing fees rest on a much larger revenue stream realized by the innovating firm's customers. Further, by leveraging advances from one application area into the next, customers in different industry sectors benefit from the company's past advances in other industry sectors. Economists studying the rationale for government support of scientific and technical research have identified the licensing of technology as one of the factors conducive to generating higher than average spillover benefits.

Finally, this case compares and contrasts aspects of different agency SBIR programs. It suggests ways for improving the SIBR program.

Immersion Corporation[8]

Rosalie Ruegg
TIA Consulting

THE COMPANY

As a Stanford graduate student in mechanical engineering, Louis Rosenberg, investigated computer-based and physical simulations of remote space environments to provide a bridge across the sensory time gap created when an action is performed remotely and the resulting effect is known only after a time delay. For example, a satellite robot tightens a screw and scientists on the ground find out with a delay if the screw was stripped. As earlier described by Dr. Rosenberg, "I was trying to understand conceptually how people decompose tactile feeling. How do they sense a hard surface? Crispness? Sponginess? . . . Vision and sound alone do not convey all the information a person needs to understand his environment. Feel is an important information channel."[9]

From aerospace researcher, Dr. Rosenberg turned entrepreneur with a focus on the less-studied sensory problem of feel, which was also closely attuned to his specialization in mechanical engineering. He took as his first business challenge to convert a $100,000, dishwasher-sized NASA test flight simulator into a $99 gaming joystick. To take advantage of his breakthroughs, he founded Immersion Corporation in 1993 in San Jose, initially drawing heavily on other Stanford graduates to staff the company. Reflecting the early NASA-inspired challenge, Immersion's first products were computer games with joysticks and steering wheels that move in synch with video displays. Other application areas followed.

The company has now grown to 141 employees. Growth over the first seven years reflected internal gains mainly in the entertainment area. Then, in 2000, Immersion grew mainly by acquiring two companies: Haptic Technologies, located in Montreal, Canada, and Virtual Technologies, Inc.[10], located in Palo Alto, California, both acquisitions now an integral part of Immersion Corporation.

---

[8]The following informational sources informed the case study: interview at the company with Mr. Chris Ullrich, Director of Applied Research; company Web site, <*http://www.immersion.com*>; company 2004 Annual Report; company product brochures; Stanford University School of Engineering alumni profile of Dr. Rosenberg in its 1997-1998 Annual Report; Dun & Bradstreet Company Profile Report; and ownership information obtained from Charles Schwab Investment Service.

[9]Quote is taken from the Stanford University School of Engineering alumni profile of Dr. Rosenberg contained in its 1997-1998 Annual Report.

[10]Mr. Ullrich, who was interviewed for this case, joined the company in 2000, in conjunction with the acquisition of Virtual Technologies.

IMMERSION CORPORATION:
COMPANY FACTS AT A GLANCE

- **Address:** 801 Fox Lane, San Jose, CA 95131
- **Telephone:** 408-467-1900
- **Year Started:** 1993
- **Ownership:** Publicly traded on NASDAQ: IMMR
- **Revenue:** Approx. $23.8 million in 2004
  —Revenue share from SBIR/STTR grants & contracts: approx. 4 percent
  —Revenue share from sales, licensing, & retained earnings: 96 percent
- **Number of Employees:** 141
- **Patent Portfolio:** Over 550 issued or pending patents, U.S. and foreign
- **SIC:** Primary SIC: 3577, Computer Peripheral Equipment
  35779907, Manufacture Input/output Equipment, Computer
  Secondary SIC: 7374, Data Processing and Preparation
  73740000, Data Processing and Preparation, Computer
- **Technology Focus:** Touch-feedback technologies
- **Application Areas:** Computer peripherals, medical training systems, video and arcade games, touch-screens, automotive controls, 3-D modeling, and other
- **Funding Sources:** Licensing fees, product sales, contracts, stock issue, commercial loans, federal government grants, and reinvestment of retained earnings
- **Number of SBIR grants:**
  —From NSF: 10 (4 Phase I, 3 Phase II, and 3 Phase IIB)
  —From other agencies: 33 (20 Phase I and 13 Phase II)

And, in 2001, Immersion acquired HT Medical Systems, located in Gaithersburg, Maryland, renamed it Immersion Medical, and made it a subsidiary of Immersion Corporation. In the case of Haptic Technology and HT Medical Systems, the acquisitions brought into the company competitors' technologies in application areas new for Immersion. In the case of Virtual Technologies, the acquisition brought in a complementary technology.

THE TECHNOLOGY AND ITS USE

Of our five senses, the sense of touch differs from the others in that "it requires action to trigger perception." Development of a technology to sense touch draws on the disciplines of mechanical and electrical engineering, computer science, modeling of anatomy and physiology, and haptic content design. The technology uses extensive computer power to bring the sense of touch to many kinds of computer-based applications, making them more compelling or more informative processes. As a company publication puts it, "At last, the world inside your computer can take on the physical characteristics of the world around you. . . . Tactile feedback makes software programs more intuitive."

The technology was brought to life for the interviewer by a series of demonstrations. The first demonstration was of a medical training simulator that teaches and reinforces the skills doctors need to perform a colonoscopy. Low grunts from "the patient" informed the performer that a small correction in technique was needed for patient comfort. "Stop, you are really hurting me!" informed the performer in no uncertain terms that her technique was in need of substantial improvement.

Immersion has developed five main AccuTouch® platforms for helping to teach medical professionals. The five platforms teach skills needed for endoscopy, endovascular, hysteroscopy, laparoscopy, and vascular access—all minimally invasive procedures.

The next demonstration was of a gaming application. The weight of a ball on the end of a string was "felt" to swing in different directions in response to manipulating a joystick. The technology is used also to enhance the computer feedback experience when using a mouse or other peripheral computer controllers for PC gaming systems, arcade games, and theme park attractions, as well as for other PC uses.

A third demonstration was of a "haptic interface control knob" to provide human-machine touch interface on an automobile dash to help manage the growing number of feedbacks from navigational, safety, convenience, and other systems. The purpose is to lessen the risk of overloading the driver.

A fourth demonstration was of Immersion's "Vibe-Tonz" system for mobile phones. The system expands the touch sensations for wireless communications by providing vibrotactile accompaniment to ringtones, silent caller ID, mobile gaming haptics and many other tactile features.

THE ROLE OF SBIR IN COMPANY FUNDING

Though the initial funding of Immersion Corporation was through private equity, the company applied for and received its first SBIR grant in its second year, 1994. In addition, the acquired companies, HT Medical and Virtual Technologies, had received SBIR grants prior to their acquisition by Immersion, and HT Medical had also received a grant from the Advanced Technology Program

(ATP) that was nearing completion at the time Immersion acquired the company. All totaled, Immersion and its acquired companies have received 24 Phase I SBIR grants and 19 Phase II (including 3 Phase IIB) grants, summing to approximately $10.6 million. SBIR funding agencies include NIH, DoE, DoD, Navy, Army, and NSF. Table App-C-2 summarizes the company's SBIR and STTR grants in number and amount.

According to Mr. Ullrich, SBIR grants gave the company the ability to further develop its intellectual property and to help to grow its intellectual property portfolio, which is the very core of the company's commercial success. The company has leveraged its government funding by investment funding from private sources in the amount of $12.7 million. The company attributes approximately $33 million in revenue to products directly derived from Phase II SBIR research projects, including licensing, direct sales of product, and product sales due to licensees. However, due to the company's licensing model, third-party revenues and tertiary economic activity, which are very significant, are not tracked directly by Immersion.

The company now receives only a small fraction of its annual revenue from SBIR/STTR funding, with the percentage ranging variously between 4 percent and 9 percent from 2001 to 2004. Its objectives for rapid commercialization growth are expected to reduce this percentage to an even lower level in the near future.

TABLE App-C-2 Immersion Corporation: SBIR/STTR Grants from NSF and Other Agencies

NSF Grants	Number	Amount ($)	Other Agency Grants	Number	Amount ($)	Total Number	Total Amount ($)
SBIR Phase I	4	400,000	SBIR Phase I	20	~1,400,000	24	~1,800,000
SBIR Phase II[a]	3	1,300,000	SBIR Phase II[a]	13	6,600,000	16	7,900,000
SBIR Phase IIB/ Enhancements	3	850,000	SBIR Phase IIB/ Enhancements	0	0	3	850,000
STTR Phase I	0	0	STTR Phase I	3	~200,000	3	~200,000
STTR Phase II	0	0	STTR Phase II	2	1,250,000	2	1,250,000
STTR Phase IIB/ Enhancements	0	0	STTR Phase IIB/ Enhancements	0	0	0	
Totals	10	2,550,000		38	9,450,000	48	12,000,000

[a]Excludes Phase IIB/Enhancement grants which are listed separately.

SOURCE: Immersion Corporation.

BUSINESS STRATEGY, COMMERCIALIZATION, AND BENEFITS

From its beginning, Immersion's prime business strategy has been to develop intellectual property in the field of touch sense and to license it. In addition, the company performs limited manufacturing operations in its 47,000 sq. foot facility in San Jose and in Gaithersburg, and arranges for some contract manufacturing. But far and away, the company's wealth generation depends on its ever-growing portfolio of patents which it licenses to others. At the time of this interview, the company had more than 270 patents issued in the United States and another 280 pending in the United States and abroad.

Important to identifying and developing relationships with new licensing partners is the company's participation in trade shows and conferences, and its ongoing interactions with industry associations and teaching universities. The company employs a business development specialist in each of its core business areas to cultivate these contacts.

Because direct sales for Immersion's technologies are derived from the much larger markets into which its licensees typically sell, estimating ultimate market size is considered "complicated" for Immersion, and it takes a more narrow view. For example, Immersion markets its cell phone vibration technology to a limited number of cell phone OEMs, and those OEMs in turn market to millions of customers. Estimating the larger consumer markets is not Immersion's focus.

Potential benefits of the technology include boosting the productivity of software use; enhanced online shopping experiences; enhanced entertainment from computer-based games; improved skills of medical professionals resulting, in turn, in improved outcomes for patients; increased automotive safety due to reduced visual distractions to drivers; and savings to industry through the ability to experience prototypes "first hand," but virtually, before building costly physical prototypes, and the ability to capture 3-D measurements from physical objects. In addition, visually impaired computer users may benefit from the tactile feedback of the mouse, keyboard, or touch-screen.

VIEWS ON THE SBIR PROGRAM AND PROCESSES

Mr. Ullrich made several observations about the SBIR program and its processes that may serve to improve the program. These are summarized as follows:

Difference in Agency Program Intent Helpful to Companies

Mr. Ullrich thought it was clear that there is "a difference in intent" among the various SBIR programs. In particular, DoD is focused on solutions to well-specified problems, while NSF and NIH are more interested in basic technology development that has commercial potential. This distinction is helpful to companies who may wish to develop technologies under both sets of condition. Given the need to respond to fast developing commercial markets, Mr. Ullrich finds the openness and flexibility of a program to accommodate where a company needs to go to find market acceptance to be a big advantage.

SBIR Application Process

According to Mr. Ullrich, there are only minor differences among the agencies in their proposal application processes, and these differences do not pose a major concern in terms of proposal logistics. At the same time, he noted that the last time the company proposed to NIH, there was no electronic submission process, and he expressed the hope that this lack has been remedied.

SBIR Proposal Review Process

Mr. Ullrich has found the review process in support of the various agencies' SBIR grant selection to be "tough but fair." He has found the NSF review to be "much more academic" than the others. Overall, he sees no need for change in the review process.

Turning more exclusively to the NSF's SBIR program, Mr. Ullrich offered the following comments.

Timing Issue—Funding Cycle Too Long for Software Providers

According to Mr. Ullrich, the biggest drawback in NSF's SBIR program is the two deadlines per year, with six months between application and grant and 18 months to Phase II grants. This can be too slow for a software developer.

Timing Issue—Funding Gap

Mr. Ullrich pointed to an associated gap in funding that arises in the NSF program, which he thought would be a real hardship for start-up companies that had not yet developed any sales to sustain them in the interval. He pointed to the Fast Track program at NIH and DoD as being very good ideas. At the some time, he noted that having to develop both Phase I and Phase II proposals at once entails a huge investment of time for an all or nothing outcome. He suggested that providing a supplement—as he recalled some parts of DoD do—to close the funding gap would likely be a preferable approach from the company's perspective.

Phase IIB Matching Funds Requirement

For Immersion, NSF's Phase IIB matching requirement of "cash in the bank" was an easy test to meet—once the company had partners. At the same time, he found the associated review awkward in one respect: The company was required to take its business partner (the investor) to a panel review at NSF. The problem was that the company was required to discuss certain financial issues in front of its investor that it would have preferred to have discussed with NSF in private. Furthermore, it found the need to insist that the investor attend the meeting to be cumbersome and, in its opinion, unnecessary.

Commercialization Assistance

The company participated earlier in the Dawnbreaker Commercialization Assistance Program, and found that "it made sense." However, given the company's current level of business experience, Mr. Ullrich does not think the company would wish to participate again, and is glad participation is optional. Currently, the company is participating in the Foresight Commercialization Assistance Program for the first time and is "seeing if it will help."

SUMMARY

This case study describes how SBIR grants helped a young company develop a large intellectual property portfolio centered on adding the sense of touch to diverse computer applications, and how the company grew the business over its first decade to approximately 141 employees and $24 million in annual revenue. It illustrates how government funding can be used by a university spin-off to leverage funding from private sources to achieve faster growth, eventually essentially eliminating the need for government R&D support. The case also illustrates how a basic idea—adding the sense of touch to computer applications—can be used to enhance entertainment experiences, increase productivity of computer use, train doctors, and more. Immersion's technology was inspired by a NASA system, but its growth centers on its embodiment in consumer products. The case provides a number of suggestions for improving the SBIR program.

ISCA Technology, Inc[11]

Rosalie Ruegg
TIA Consulting

THE COMPANY

ISCA was founded in 1996 by Dr. Agenor Mafra-Neto, an entomologist performing basic research at the University of California-Riverside on pheromones, chemical substances produced by, in this case, insects that stimulate behavioral responses in other insects of the same species. From his background in basic research, Dr. Mafra-Neto took on the challenge of applying this knowledge to real-world applications. He contacted growers back in his native Brazil who were very supportive of putting his ideas for pest control into practice. His contacts wired up-front financing to cover a contract for pest control traps, and ISCA was born.

For the next two years, the company's principal business was export of pest control traps to Brazil. Then in January 1999, the company was caught up in a financial crisis that was a result both of the devaluation of the *Real*, the Brazilian currency, and the default by a customer on a large order of ISCA product. During the months that followed, the company was in severe financial distress, and Dr. Mafra-Neto was unsure of his company's ability to survive. It was with the help of an SBIR grant that he was able to restructure and reshape the company to provide more advanced product lines targeted at new domestic and foreign markets.

From its new start, the company has grown to 12 employees and annual revenue of $2 million. The company's offices and facilities are located in an industrial park in Riverside, California, and occupy a combined area of approximately 8,500 square feet. The staff is a multidisciplinary team of specialized researchers, including synthetic organic chemists, engineers, entomologists, and information technologists.

THE TECHNOLOGIES AND THEIR USES

ISCA synthesizes and analyses sex pheromones for a variety of insects. These pheromones are species specific, occur in nature, are environmentally friendly, and do not result in the development of insecticide resistance by the

---

[11]The following informational sources informed the case study: interviews conducted at the company's headquarters in Riverside, CA, with Dr. Agenor Mafra-Neto, President, Dr. Reginald Coler, Vice President, and Mr. Annlok Yap, Business and Finance Director; the company Web site, <*http://www.iscatech.com*>; company product brochures; and a recent Dun & Bradstreet company profile report.

ISCA TECHNOLOGY, INC.: COMPANY FACTS AT A GLANCE

- **Address:** 2060 Chicago Ave., Suite C2, Riverside, CA 92507-2347
- **Telephone:** 951-686-5008
- **Year Started:** 1996; restructured in 1999
- **Ownership:** Privately held
- **Revenue:** Approx. $2.4 million in 2004
  - —Revenue share from SBIR/STTR and other government grants: approx. 40 percent
  - —Revenue share from sale of product: approx. 60 percent
- **Number of Employees:** 12
- **SIC:** Primary SIC: 0721 Crop Planting, Cultivating, and Protecting
  2879 Insecticides and Agricultural Chemicals, NEC
  Secondary SIC: N/A
- **Technology Focus:** Pest management tools and solutions
- **Application Areas:** Insect semiochemicals (pheromones and kairomones, i.e., naturally occurring compounds that affect behavior of an organism); attractants and monitoring traps; pheromone synthesis and analysis; pheromone delivery and dispensing systems; pest management information systems; automated insect identification and field actuation devices; contract entomological R&D; insect rearing and bio-assays
- **Funding Sources:** Product sales domestic and foreign, contracts, federal government grants, and a small amount of licensing revenue
- **Number of SBIR Grants:**
  - —From NSF: 1 Phase I, 1 Phase II, and 1 Phase IIB
  - —From other agencies: 6 Phase I and 3 Phase II

pest. These properties make them an ideal alternative to insecticides for pest management.

The insect's response to pheromones and other attractants is often quantified through the use of electroantennograms (EAG), which measure the neural activity originating from the insect's antenna. In addition to EAGs, biological assays are also used to determine a variety of performance metrics, such as the optimal pheromone release method and the optimal pheromone trap design and placement. ISCA then develops pheromone delivery and dispensing systems and monitoring traps, and integrates the traps with data collection systems, including automated sensors to give pest counts and GPS/GIS analytical tools that are Internet-accessible to give pest locations.[12]

The results of monitoring provide timely information about the type, number, and location of pests captured in time to predict pest population densities, identify alarm situations, and deliver limited targeted treatments of pheromones to disrupt mating patterns. The advent of ISCA's pest information management system, equipped with smart traps and wireless communication puts an end to hand counting and the all-too-familiar-to-counters tangled balls of deteriorating insects. The resulting information enables a timely response that avoids insect proliferation throughout a field or larger area and reduces the need for blanket applications of insecticides. Although these technologies individually are not new to the world, their application in the area of integrated pest management has broken new ground.

The company has lines of pheromones, attractants, and repellents to address agricultural pests, including the boll weevil, carob moth, European corn borer, corn earworm, Mediterranean fruit fly, tomato fruit borer, olive fruit fly, peachtree borer, pecan nut casebearer, potato tuber moth, tobacco budworm, and many others. Additionally, ISCA has a product line designed for urban pests, such as the cockroach, housefly, yellow jacket wasp, and mosquito. One of ISCA's most recent lure technologies is the development of "SPLAT™" (Specialized Pheromone & Lure Application Technology), a sprayable matrix that dispenses attractants over an extended time interval substantially greater than that provided by traditional dispensing technologies.

Information technology comprises a critical component of ISCA's approach to pest management. The information technology features modular scalability and GPS/GIS capability. At its core is Moritor, an integrated and automated Internet accessible monitoring system.

In support of its R&D, ISCA operates insect rearing chambers, testing rooms, wind tunnels, and olfactometers. It uses an artificial blood membrane system to maintain its mosquito colonies. The company tests its tools and solutions through rigorous field tests as well as by feedback from user groups.

THE ROLE OF SBIR IN COMPANY FUNDING

According to the company founder and president, Dr. Mafra-Neto, the SBIR program was essential to survival of the company after it hit a major financial setback in its third year of operation. He learned about the SBIR program by re-

[12]GPS refers to Global Positioning System, which relies on a system of satellites orbiting the earth to provide precise location and navigation information. GIS is a technology that is used to view and analyze data from a geographic perspective. Looking at the distribution of features on a map can help reveal emerging patterns. The combination of GPS/GIS is an important component of the company's overall information system framework.

viewing U.S. Department of Agriculture's SBIR proposals during his days in the university. He reasoned that if the company were to get an SBIR grant, it could use the research funding to improve its approach to pest control in terms of the chemicals produced, the lure and trap design and placement, and, eventually, data collection and analysis.

NSF put out a call for sensors. The company responded with a proposal to develop an Internet accessible pest monitoring system with automated traps that would count insects. It was subsequently granted SBIR Phase I and Phase II grants, including a Phase IIB supplement. At NSF, there was interest in bringing innovation to a field not known for its use of technology. "The NSF SBIR gave us lots of prestige; it gave us credibility," said Dr. Mafra-Neto.

ISCA has received a total of seven Phase I SBIR grants, four Phase II grants, and one Phase IIB supplemental grant. It has received SBIR grants from NSF, USDA, DoD, and NIH. The amount the company has received in SBIR grants since 1999 totals a little more than $3 million. Table App-C-3 summarizes the company's SBIR/STTR grants in number and amount.

According to Dr. Mafra-Neto, the receipt of additional SBIR grants in the future is expected to be important to the company as a means of continuing the innovations necessary to maintain its technical base.

In addition to its SBIR grants, the company received a grant from the Advanced Technology Program, for the period 2002 to 2005. The grant supports

TABLE App-C-3 ISCA Technology, Inc.: SBIR/STTR Grants from NSF and Other Agencies

NSF Grants	Number	Amount ($)	Other Agency Grants	Number	Amount ($)	Total Number	Total Amount ($)
SBIR Phase I	1	100,000	SBIR Phase I	6	560,000	7	660,000
SBIR Phase II[a]	1	499,700	SBIR Phase II[a]	3	1,756,000	4	2,255,700
SBIR Phase IIB/ Enhancements	1	250,000	SBIR Phase IIB/ Enhancements	0	0	1	250,000
STTR Phase I	0	0	STTR Phase I	0	0	0	0
STTR Phase II	0	0	STTR Phase II	0	0	0	0
STTR Phase IIB/ Enhancements	0	0	STTR Phase IIB/ Enhancements	0	0	0	0
Totals	3	849,700		9	2,316,000	12	3,165,700

[a]Excludes Phase IIB/Enhancement grants which are listed separately.
SOURCE: ISCA Technology, Inc.

the integration of sensor technologies and information technology in a highly automated pest-management system.

BUSINESS STRATEGY, COMMERCIALIZATION, AND BENEFITS

The company's competitive advantage lies in its innovations to make smarter traps, which are then linked wirelessly to a centralized database located on the Internet. Automated data collection and subsequent analysis and reporting enables targeted pest control strategies. In addition, the company derives strength from its internally developed "SPLAT" technology that extends the time interval needed for effective seasonal control of pests. Sales of SPLAT products are expected to increase dramatically in the near future.

Between 60-70 percent of the company's sales now are in the domestic U.S. market. Remaining sales comprise exports to Brazil, Argentina, Chile, India, and other countries.

The company's approach to pest monitoring and control offers environmental benefits in terms of reductions in the need for and use of insecticides. These benefits result from early alerts of pest activity, targeted treatments, and use of strategies that do not involve the use of insecticides to disrupt mating patterns. Humans may benefit from reduced insecticides on products they consume, as well as from higher quality products due to less damage to fruit and vegetables from pest outbreaks.

This pest-management approach benefits growers who can avoid multiple blanket spraying of fields with insecticides that may cost as much as 10 times more for pest control than ISCA's method. Avoidance of pest outbreaks may also increase growers' yield and quality of produce.

The recent development of smart traps that automatically count mosquitoes, together with the company's pest management information system, may also offer important health benefits by providing an early alert of threats of possible outbreaks of mosquito-borne disease. The widespread use of the system could enable a near instantaneous warning of threatening trends and activities.

VIEWS ON THE SBIR PROGRAM AND PROCESSES

"The SBIR program has allowed us to get where we are today," said Dr. Mafra-Neto, emphasizing the importance he places on the program. He went on to make the following several observations about the program and its processes, some of which focused on the NSF program.

Reporting Requirements

While noting that he did not particularly like reporting requirements, Dr. Mafra-Neto acknowledged that they forced the company to stay on track. No specific need for change was noted.

Financing Gap

Dr. Mafra-Neto spoke of the difficulties posed by gaps in financing between Phase I and Phase II funding, noting that other companies have died during the gap. "The gap creates uncertainty and breaks the research cycle," he noted. A mechanism is needed to bridge this gap in those agency programs, which have not already found a solution. He pointed to an approach used by the Army's SBIR program as an example of a workable bridge.

Resubmittal of Phase II Proposals and Appeal of Funding Decisions

ISCA interviewees were of the opinion that NIH allows submittal of Phase II proposals up to three times, while NSF allows only a single submittal. Similarly, the interviewees believed that NSF does not allow appeals of its SBIR funding decisions. "Often there is a small issue that we could quickly and easily fix if only we were given the chance," said Dr. Mafra-Neto, noting that "Reviewer critiques can vary substantially."

Value of Keeping Phase I Grants as Prerequisite to Phase II

Dr. Mafra-Neto stated emphatically that Phase II grants are critically important and should be continued largely as they exist today. The Phase I grants allow companies to test ideas; they may reveal multiple solutions; they may give companies early, though typically limited, insight into markets. "Phase I grants represent a good investment of public funding," he said.

Possibly a Premature Emphasis on Venture Capital Funding

Dr. Mafra-Neto expressed a concern that NSF may be pushing companies to attempt to obtain venture capital funding too early in the innovation cycle. In the case of ISCA's approach to show matching funds needed to obtain an NSF Phase IIB SBIR grant, he said the company used ATP funding, and, alternatively, could have used sales revenue. However, had the company been without existing sales and without an ATP grant, he thought that it would have been too early for his company to have attempted to obtain venture capital funding, making it very difficult to meet the Phase IIB requirement. Thus, the comment reflected an impression and a concern for the possible plight of other companies rather than the actual experience of ISCA.

Value of Commercialization Assistance

"It is very useful to train scientists to have business points of view," said Dr. Mafra-Neto, in commenting on his company's participation in both the Dawnbreaker Commercialization Assistance Program and the Foresight Program. At the same time, he commented that he would like to see participation continue to be optional, particularly for companies that have established a degree of business acumen.

Observation about NSF's SBIR Program Manager System

In the opinion of Dr. Mafra-Neto, NSF's program manager system is good. "The program manager becomes involved with the grantee. He or she can put you in touch with other sources to help meet your special needs. For example, we were put in touch with "Iguana Robotics"

SUMMARY

This case study illustrates how SBIR grants helped a young company survive following the collapse of export sales several years after start-up. It further shows how SBIR-funded research brought needed innovation to the important but largely static field of pest monitoring and control. The development of better lures and smarter traps integrated with advanced communication tools is effectively cutting the grower's use of insecticides, and thus reducing the unwanted effects on insects (i.e., increasing their resistance to insecticides and impacting nontargeted organisms), lowering pollution, improving the quality of fruits and vegetables, and providing potential health benefits. The case also provides valuable company observations and opinions about the SBIR program and how to improve it.

Language Weaver[13]

Rosalie Ruegg
TIA Consulting

THE COMPANY

Like a newly born gazelle, Language Weaver found its legs early. In two years it has developed a fully functional commercial software product from a novel, statistics-based translation technology brought to a research prototype by the company founders, professors and researchers in the University of Southern California's Information Sciences Institute. Of course, it should not be overlooked that approximately 20 person-years of university research, heavily funded by government agencies, were critical to establishing the scientific and technical underpinnings of Language Weaver's technology. Language Weaver gained exclusive licenses to past and future patents filed by the University in the field.

With a conception date of November 27, 2001, the company's annual revenue reached several million dollars in 2004, in a market whose potential is estimated in the billions. Having a technology that rather unexpectedly turned out to be much needed at just the right time is paying off.

The company founders are still professors at the university. The company now has about 35 employees, many of them attracted from the university's Information Sciences Institute. The company is headquartered in an office building with a grand view overlooking a marina just west of downtown Los Angeles.

The company's first funding came from NSF grants. "When we were trying to start the company," related Mr. Wong, "it was a little before 9/11, and no one cared about languages. There were no Senate hearings about languages. Then 9/11 happened, and at the time we had already submitted a proposal to NSF. But we didn't hear back until November, and by then the NSF was able to bootstrap us to get us working quickly, moving code from the university to the company. . . . It was after the SBIR grant that everything happened. We started getting government interest as it became apparent that we had something interesting. But we would not have been positioned to move quickly to respond to the need if it hadn't been for that first small amount of NSF funding and the confirmation of the technology." Subsequently the company was able to obtain venture capital funding.

[13]The following informational sources informed the case study: an interview conducted at the company's headquarters in Marina del Rey, CA, with Mr. William Wong, Director of Technology Transfer; the company Web site, <*http://www.languageweaver.com*>; company brochures; two articles: "Automated Translation Using Statistical Methods—A Technology that supports communication in Hindi and other Asian languages," *Multilingual Computing & Technology*, 16(2), and "Breaking the Language Barrier," *Red Herring*, February 28, 2005; and a recent Company Commercialization Report to the Department of Defense SBIR Program.

LANGUAGEWEAVER: COMPANY FACTS AT A GLANCE

- **Address:** 4640 Admiralty Way, Suite 1210,
  Marina del Rey, CA 90292
- **Telephone:** 310-437-7300
- **Year Started:** 2002
- **Ownership:** Privately held
- **Revenue:**
  —Revenue share from government grants: approx. 60 percent
  —Revenue share from licensing fees: approx. 40 percent
- **Number of Employees:** 35
- **Technology Focus:** Statistically based automated machine language translation
- **Application Areas:** Language translation of documents, newscasts, and other source materials for defense and commercial purposes. Application languages include Arabic, Farsi, Somali, Hindi, Chinese, French, and Spanish
- **Funding Sources:** Federal government grants, venture capital, and licensing revenue
- **Number of SBIR grants:**
  —From NSF: 3
  —From other agencies: 1

"What we are trying to do," explained Mr. Wong, "is to create the best machine translation in the world, with the highest quality and readability. Our best selling system right now is Arabic to English—to the government. We have customers and we have partners."

THE TECHNOLOGY AND ITS USE

The state of the practice in commercial machine translation is rule-based, e.g., noun before verb. But 30 years of working with rules has reportedly shown that the approach does not do well in handling special cases. In contrast, Language Weaver's statistical learning approach to machine translation is designed to learn the appropriate linguistic context for distinguishing and handling words with multiple meanings. This is the kind of problem that confounds rule-based systems because it is impossible to capture all the necessary cases in rules. For example, the English word, "bank" may need to be translated differently in each of the following: "put the money in the bank;" "you can bank on it;" and "paddle the canoe near the bank."

The machine-based software uses computational algorithms and probability statistics to learn from existing translated parallel texts, analyze words and word groupings, and build translation parameters that will provide the highest statistical probability of providing a correct translation. The development of translation software for a given language entails performance of two analyses: First, a bilingual text analysis is performed using a corpus of text and statistical analysis to learn associations. For example, a bilingual corpus for Spanish/English may be found at Microsoft's Web site, which gives the same material in both Spanish and in English. From such existing one-to-one translations, the system learns. A translation model is built from the resulting analysis. Second, a great deal of monolingual text is fed into it to increase translation fluency. The process is akin to a computer chess game, whereby the computer is playing chess with itself trying to find the best move, or, in this case, the best translation. The approach requires a lot of computing, but fortunately substantial computing power is available at a reasonable cost.

"Basically, we consume this text corpus. We learn from it and develop the parameters which will be used by the runtime module we call the "decoder." What we license to the customer is the parameters and the decoder," related Mr. Wong, describing Language Weaver's approach.

What about languages for which there is not much of an existing corpus of digital translations available? Language Weaver's first two contracts involved Somali and Hindi—both of them "electronically low-density languages." In this case, the company had to employ human translators to generate a body of digital text that they could put into the learning system to create the parameters. "We don't get as high a quality taking this approach," said Mr. Wong, "but it's still readable."

"We have also developed a Chinese translator—for which a lot of existing data exists—but which is difficult because it uses characters instead of letters. And in the case of Arabic, there is a similar difficulty because we are dealing with script. Chinese and Arabic each presented special challenges that we met," noted Mr. Wong.

"In summary," explained Mr. Wong, "instead of rule based, our approach is code breaking. We just need a few weeks to create a new language set. We are getting to the point that we can deal with any language, translating it into another language by pressing buttons."

Mr. Wong brought the technology to life for the interviewer with some graphic depictions of how it can be used for defense applications: "Every day many hours of potentially important information to U.S. efforts against terrorism are broadcast in Arabic. Our technology is used to provide near simultaneous translation. Similarly, hours of taped interviews with people from different cultures, speaking in a variety of languages, may contain insights and information important for the military effort. Our technology can perform the translations and allow specific questions to be searched."

Language Weaver's translation system, for example, has been incorporated into media monitoring systems by BBN, Virage, and Z-Micro Systems. These systems can capture an Arabic satellite news broadcast. The audio track containing speech is extracted in 5-second chunks. A speech recognition system does the transcription of the chunks into Arabic text. From there, the media monitoring system sends the Arabic text to the Language Weaver translation system, which translates it into English. "An exciting part of the media monitoring systems is the way they allow an analyst or viewer to track each translated text segment in synch with the broadcast video and audio. As the speaker speaks, the English text chunk is highlighted that corresponds to what the announcer is saying."

Continuing, Mr. Wong said, "Or imagine that you are searching a house where terrorists have been and you uncover a trove of dirty or degraded documents written in Arabic, or Farsi, or some other language. These documents may contain valuable information. They can be cleaned, scanned, and transcribed using optical character recognition software, and our system can be used quickly to generate translations. These translations can be stored and searches can be performed on key words as needed."

THE ROLE OF SBIR IN COMPANY FUNDING

The company founders, Dr. Kevin Knight and Dr. Daniel Marcu, were professors at the University of Southern California's Information Sciences Institute, when they saw business potential in a new approach to machine language translation they had brought to a research prototype stage.[14] This awareness came right at the time the Internet bubble was about to burst, but at a time everyone was still excited about technology, allowing the professors to get their foot in the door with venture capitalists. But they couldn't get funding—perhaps because the venture capitalists were just then realizing that conditions were about to take an unfavorable turn.

Next the professors went to see the Tech Coast Angels, a group of investors who provide seed and early stage capital in Southern California. They were able to present to the group, but they didn't get money. However, they did get a mentor who worked with the would-be company for about nine months.

Still, during this time the professors were getting no traction from the commercial sector in terms of investment funding. It was at that point that Daniel Marcu, following something of a whim, decided to see if he could get STTR funding and start the company on that. At the university, he had worked on

[14]With backgrounds in computer science, mathematics, and linguistics, Drs. Marcu and Knight, were, and are, two of the top researchers in the field of statistical machine translation. The interviewee, William Wong, was a student of Dr. Marcu at the University of Southern California (USC), and joined the company at its inception. Prior to coming to USC, Mr. Wong had worked at Intel. At USC his intention, he said, was to go into computer animation. He took a class from Daniel Marcu on natural language processing, "fell in love with the idea," and changed his plans.

DARPA and NSF-funded research projects, and he knew from this experience about the NSF STTR grant. He submitted a proposal to NSF and got the STTR grant.

In the words of Mr. Wong, "Getting the STTR grant was a real boon to us, because we were on the verge of saying, 'This technology is not ready. No one is interested in talking to us. Let's just shelve it for a while.' But then came the grant from the National Science Foundation and with it the confirmation and redemption of the technology—an indication that it was useful, or interesting at least."

"What did that do for us?" Mr. Wong continued. "It wasn't enough to support more than one person, but it forced us to actually incorporate—that's one of the rules. The other thing was that it forced us to find a CEO—someone who could drive the formation of a whole company as you see today. And it forced us to spend some time working out the details, including those of us who were volunteering our time. It reinforced that we had something interesting. So basically November 27, 2002, was our conception date, because that is when we got our first STTR. Then we were given a chance to convert the STTR into an SBIR, and we did because the SBIR offered more advantages.[15] So the STTR/SBIR from NSF created Language Weaver and what we are today. Without that we would have shelved the technology."

Language Weaver has received a total of $150,000 in Phase I SBIR grants and $1,500,000 in Phase II grants. It has received SBIR grants from NSF and the U.S. Army. The amount the company has received in SBIR grants since its founding in 2002 totals $1,150,000. Table App-C-4 summarizes the company's SBIR/STTR grants in number and amount. In addition to its SBIR grants, the company received a multiyear grant from the Advanced Technology Program, for a large scale syntax-based system, expected to bear fruit several years out.

BUSINESS STRATEGY, COMMERCIALIZATION, AND BENEFITS

The company is "a core technology house based on licensing its software." It licenses its product, statistical machine translation software (SMTS) directly to customers, and through partners, such as solution vendors who add multilingual capability to their applications with Language Weaver. Underpinning its ability to license are more than 50 patents pending worldwide on SMTS. These partners are important marketing vehicles. "We have a symbiotic relationship with our partners," Mr. Wong explained. "We promote our products and our partner's product lines that contain our technology; our partners do the same thing."

In its first two years the company focused on heavy government funding. Grants and development contracts comprising 80-90 percent of company revenue were the company's "life's blood." In 2004, the company received more

[15]At the time, research funds were extremely limited. The company found it advantageous to provide the university an ownership share of the company rather than give up the limited STTR research dollars to it.

TABLE App-C-4 Language Weaver: SBIR/STTR Grants from NSF and Other Agencies

NSF Grants	Number	Amount ($)	Other Agency Grants	Number	Amount ($)	Total Number	Total Amount ($)
SBIR Phase I	1	100,000	SBIR Phase I	1	50,000	2	150,000
SBIR Phase II[a]	1	500,000	SBIR Phase II[a]				
SBIR Phase IIB/ Enhancements	1	500,000	SBIR Phase IIB/ Enhancements				
STTR Phase I	[b]		STTR Phase I				
STTR Phase II			STTR Phase II				
STTR Phase IIB/ Enhancements			STTR Phase IIB/ Enhancements				
Totals							

[a]Excludes Phase IIB/Enhancement grants which are listed separately.
[b]An STTR Phase I grant was converted to an SBIR Phase I grant.
SOURCE: Language Weaver.

revenue from licensing, allowing it to get the government share down to 60-70 percent of revenues. According to Mr. Wong, the goal in 2005 is to cut the share of government grants to less than half of company revenue, with the majority coming from licensing.

The company is "not hanging onto research," explained Mr. Wong. "Yes, we will continue to have researchers internally, but the growth vehicle is the market now. We are sticking with the company's core technology and then building supporting technologies around it to make the core more useful." An additional avenue for research support, reminded Mr. Wong, is the fact that the company will continue to benefit from research advances made at the University of Southern California through Language Weaver's rights to future discoveries.

Mr. Wong elaborated on the ongoing relationship between Language Weaver and the university. "Whatever they are working on that improves the quality of automated language translation, Language Weaver will gain the rights to. This arrangement used to be unusual, especially unusual for research university institutions because they are not focused on commercial prospects. We were probably the poster child—actually the second one—to break this path where we not only receive current rights but also future rights (for five years) from the university. Because the university shares in the company's equity, and we are a healthy company, the university benefits from our success. In addition, Language Weaver hires a steady stream of graduates coming out of the university program."

Mr. Wong also provided perspective about the company's view of and use of venture capital and external business management, explaining, "Early on we made the decision that we wouldn't be control freaks, and that we couldn't handle all of the management aspects. We found our CEO—Bryce Benjamin—through the Tech Coast Angels. From then on we just wanted reasonable ownership of the company for the amount of money they were giving us. We decided not to worry about the fact that we are giving up shares. Rather we would worry about how much value is being added. We did not have to give up majority ownership in order to attract funding. We stayed away from people who didn't see the future of the technology, and just seemed out to get majority ownership. To develop a beneficial relationship with venture capitalists, you need to be in that phase when you are moving towards having customers, but need to grow more; in this phase venture capital can help and investors can see potential."

Language Weaver's technology offers societal benefits in several ways: First, it reportedly provides a significantly higher rate of accuracy in translation than counterpart rule-based machine translations, providing customers more value. Second, it is able to provide translation systems for languages for which there is a shortage of available translators and for which there is considerable demand for translations, particularly for defense purposes. Third, it can be a more cost-effective solution for translation of large volumes of information than human translators. Fourth, the technology may offer a faster means to obtain needed translations by its ability to process large volumes of data quickly. For example, it reportedly can process in one minute what a human translator would take several days to produce.

VIEWS ABOUT THE SBIR PROGRAM AND ITS PROCESSES

Because NSF SBIR grants make up most of Language Weaver's funding received through the SBIR program, Mr. Wong's views on the SBIR program, which follow, mainly reflect NSF's program.

Commercialization Assistance Program

Language Weaver chose not to participate because when the opportunity arose, the company had just brought in business management and a salesperson with an extensive directory of contacts. The company found the optional nature of participation to its liking.

NSF's Emphasis on Commercialization

"The NSF provides a lot of encouragement to companies to look more at the commercialization side," stated Mr. Wong. "You've got to find your marketing or sales guy; you've got to find your customers." They push you to ask "What do your customers really want."

Assessing this focus, Mr. Wong stated, "I think it was about the right emphasis in our case, being in the software industry. I can see how it might be too fast for some technology areas such as materials, manufacturing, and chemicals, or pharmaceuticals. The time required to commercialize a technology is definitely industry dependent. In the software area, minimizing the time to market is important. We have to worry that someone in some other place will copy our technology—even though at this point there are only a handful of people who understand statistical machine translation, and they are a very tight group, making it less easy to copy. In any case, speed is of the essence."

NSF's Phase IIB Matching Funds Requirement

Mr. Wong noted, "NSF's Phase IIB was very good; it worked well for Language Weaver." However, he expressed concern that other start-up companies not able to move as fast into commercialization would find it very difficult to meet the matching funds condition of Phase IIB. He explained that getting a Phase I, Phase II, and Phase IIB is not enough to enable a start-up company to bring in a marketing person and is not enough to allow a CEO like Mr. Benjamin to build a business. A start-up company must have already found additional funding in order to position itself to do what is being asked to do at the Phase IIB stage. Thus, the implied sequence of the SBIR phases feeding directly into one another as a tool to launch a business is most cases would be quite problematic. A company will need to go out and find more funding sources and partners very early on. In the words of Mr. Wong, "We were lucky that we were able to do that."

In response to a question about whether the company was able to use its government defense-related contracts as match for its Phase IIB grant, Mr. Wong responded as follows: "We needed commercial contracts. We could not use our DoD contracts as match. We used venture capital funding as our match. We had to show that we had the money in the bank. It meant actually showing bank receipts.

NSF's Flexibility and Empowerment of Program Managers

"NSF's flexibility was extremely helpful to us," noted Mr. Wong, explaining that because the company was just getting started, it needed to make some changes in its research plan in Phase II after getting underway. He observed that NSF empowers its program managers and provides them enough leeway to make decisions that allow changes from the original research plan if it seems warranted—provided the company reports on it and explains why it was beneficial and the results.

Another point Mr. Wong made about the advantages of NSF's flexibility was that the NSF allowed Language Weaver to go at an accelerated pace and finish early—without having to turn back money. The ability to accelerate research was reportedly very important to meeting the rapidly developing market demand for a better machine translation system.

Financing Gap

"A lot of us were willing to work for free during that early time and that helped relieve the financial stress. But I can see how surviving during the start-up phase would be a hardship for some companies," said Mr. Wong.

Opinion about Phase I Grants

In the words of Mr. Wong, "I think requiring Phase I is a good idea. Always a good idea! You don't know if an idea is even interesting to someone and if you can get it together in the beginning. So I totally see why we need to have the Phase I—especially when the follow-on phases are so much bigger. Of course it would be nice to get more money up-front, but the Phase I is a way for the government to take a manageable risk."

NSF Proposal Review Process

In Mr. Wong's opinion, the process seemed fair and the quality of reviews seemed good. "It seemed more academic in nature," he said, "but that was good, because it reinforced that we had something new and interesting. And as a new company we needed that third-party affirmation that our ideas were worthwhile. Because we received the NSF SBIR, and the affirmation it gave us, we were able to do follow-on work."

He noted that there seemed more comments on the business side at the Phase II review. And, the grading seemed pass/no pass. "Is the project sound; is there an inkling of business sense?"

"At the Phase IIB review stage, the emphasis in the review was definitely financial; nuts and bolts. The CEO, Mr. Benjamin and others gave a presentation before a panel for Phase IIB. The presentation was very business oriented, addressing why we have potential as a company."

Idea for "Phase IIC" Grant

"I think commercialization is very hard for people," said Mr. Wong. "If we hadn't been able to recruit our CEO from Tech Coast Angels, we would have had the same problems as everyone else. Making available a Phase IIC grant would be very helpful to extend the time to get to market for longer lead time technologies.

If everything is looking good through the Phase IIB period, a little more money might make a big difference."

SUMMARY

This case study illustrates how an NSF SBIR grant was critical to bootstrapping a technology with national security and economic potential out of a university into use on a fast-track basis. The credibility afforded the technology by the SBIR grant enabled the company to get the management, additional funding, and strategic partners it needed to make a business. Without the NSF SBIR grant, a technology that turned out to be extremely timely would not have been developed in the same time frame. The technology is statistical machine translation that Language Weaver has applied to translating Arabic, Farsi, Chinese, and other languages. At this time, it is being used mainly for military-related purposes such as to create translations of Arabic broadcasts. From its founding in 2002, Language Weaver has moved from being almost entirely dependent on government grants to receiving the majority of its revenue from licensing fees. The case illustrates the speed capabilities of a software company as well as the speed imperatives that often characterize this field. The interview provided many observations that may help improve the SBIR program.

MicroStrain, Inc.[16]

Rosalie Ruegg
TIA Consulting

THE COMPANY

While pursuing a graduate degree in mechanical engineering at the University of Vermont, Steve Arms witnessed an incident that led him to his future business. During a horse vaulting gymnastics competition, a friend flipping off the back of a horse injured the anterior cruciate ligaments in both knees when she landed. That set Steve, an avid sportsman himself, wondering about the amount of strain a human knee can take and how to measure that strain. Soon he was making tiny devices called "sensors" in his dorm room to measure biomechanical strain, and soon afterwards he was making money for graduate school by selling sensors around the world—the first a tiny sensor designed for arthroscopic implantation on human knee ligaments.

In 1985, Steve Arms left graduate school to start his company, MicroStrain, Inc. Its business: sensors. "In many ways," he said, "an excellent time to start a business is when you first leave school and it is easier to take the risk, the opportunity cost is small, and one is used to living on a budget." He operated the business out of his home at first.

The company is not a university spin-off, but the company has a number of academic collaborators. Among them are the University of Vermont, Carnegie Mellon, the University of Arizona, Penn State, and Dartmouth University.

He located the company in Vermont to be close to family and friends, and to continue to enjoy the excellent quality of life offered by that location. In the longer run, the location has proven positive for high employee retention.

From its initial focus on micro sensors with biomechanical applications, MicroStrain moved into producing micro sensors for a variety of applications. Its sensor networks are in defense applications, security systems, assembly line testing, condition-based maintenance, and in applications that increase the smartness of machines, structures, and materials.

[16]The following informational sources informed the case study: interview with Mr. Steve Arms, President of MicroStrain, conducted at the Navy Opportunity Forum, May 2-4, 2005, Reston, VA; the company Web site, <*http://www.microstrain.com*>; company product brochures; a paper, "Power Management for Energy Harvesting Wireless Sensors," by S. W. Arms, C. P. Townsend, D. L. Churchill, J. H. Galbreath, S. W. Mundell, presented at SPHASE IE International Symposium on Smart Structures and Smart Materials, March 9, 2005, San Diego, CA; a book chapter, "Wireless Sensor Networks: Principles and Applications," *Sensor Technology Handbook*, Jon S. Wilson, ed., Elsevier Newnes, 2005, Ch. 22, pp. 575-589; a University of Vermont Alumnus Profile of Mr. Arms; a presentation by Mr. Arms at a DHHS SBIR conference; and a recent Dun & Bradstreet Company Profile Report.

MICROSTRAIN, INC.: COMPANY FACTS AT A GLANCE

- **Address:** 310 Hurricane Lane, Suite 4, Williston, VT 05495-3211
- **Telephone:** 802-862-6629
- **Year Started:** 1985
- **Ownership:** Privately held
- **Revenue:** Approx. $3.0 million in 2004
  - —Revenue share from SBIR/STTR and other government grants: approx. 25 percent
  - —Revenue share from sale of product and contract research: approx. 75 percent
- **Number of Employees:** 22
- **SIC:** **Primary SIC:** 3823 Industrial Instruments for Measurement, Display, and Control of Process Variables, and Related Products

  Secondary SICs:

  3625 Relays and Industrial Controls

  3679 Electronic Components, not elsewhere classified

  3812 Search, Detection, Navigation, Guidance, Aeronautical, and Nautical Systems and Instruments

  3823 8711 Engineering Services
- **Technology Focus:** Wireless sensors and sensor networks for monitoring strain, loads, temperature, and orientation
- **Application Areas:** Condition-based maintenance; smart machines, smart structures, and smart materials; vibration and acoustic noise testing; sports performance and sports medicine analysis; security systems; assembly line testing
- **Funding Sources:** Product sales, contract research, and federal government grants
- **Number of SBIR Grants:**
  - —From NSF: 3 Phase I, 3 Phase II, and 3 Phase IIB
  - —From other agencies: 6 Phase I, 2 Phase II, 1 Phase III

The company has grown to approximately 22 employees, including mechanical and electrical engineers. It occupies 4,200 square feet of industrial space near Burlington, Vermont. Its annual sales revenue was recently reported as $3.0 million in 2004, with revenues growing at about 30 percent per year. Revenues are expected to reach $4.0 million in 2005.

THE TECHNOLOGY AND ITS USES

A "sensor" is a device that detects a change in a physical stimulus, such as sound, electric charge, magnetic flux, optical wave velocity, thermal flux, or mechanical force, and turns it into a signal that can be measured and recorded. Often, a given stimulus may be measured by using different physical phenomena, and, hence, detected by different kinds of sensors. The best sensor depends on the application and consideration of a host of other variables.

MicroStrain focuses on producing smarter and smaller sensors, capable of operating in scaleable networks. Its technology goal is to provide networks of smart wireless sensing nodes that can be used to perform testing and evaluation automatically and autonomously in the field and to report resulting data to decision makers in a timely and convenient manner. The data can be used to monitor structural health and maintenance requirements of such things as bridges, roads, trains, dams, buildings, ground vehicles, aircraft, and watercraft. The resulting reports can alert those responsible to problems before they become serious or even turn into disasters. They can eliminate unnecessary maintenance and improve the safety and reliability of transportation and military system infrastructure, while reducing overall costs.

Among the features that determine how useful sensors will be for the type of system monitoring function described above are the degree to which the sensors are integrated into the structures, machinery, and environments they are to monitor; the degree to which the systems are autonomous, i.e., operate on their own with little need for frequent servicing; and the degree to which they provide efficient and effective delivery of sensed information back to users. MicroStrain's research has focused on improving its technology with respect to each of these performance features.

Another way to look at it is that MicroStrain has addressed barriers that were impeding the wider use of networks of sensors. For example, MicroStrain was one of the first sensor companies to add wireless capability. Wireless technology overcomes the barrier imposed by the long wire bundles that are costly to install, tend to break, have connector failures, and are costly to maintain. A recently passed international standard for wireless sensors (IEEE 802.15-4) is expected to facilitate wider acceptance of wireless networks.

A barrier to the use of wireless sensor networks is the time and cost of changing batteries. MicroStrain is an innovator in making its networks autonomous, without need of battery changes, by pursuing two strategies: First, it has adopted various passive energy harvesting systems to supply power, such as by using piezoelectric materials to convert strain energy from a structure into electrical energy for powering a wireless sensing node, or by harvesting energy from vibrating machinery and rotating structures, or by using solar cells. Second, the company has reduced the need for power consumption by such strategies as using sleep modes for the networks in between data samples.

A recent newsworthy application of MicroStrain's sensors was to assist the

National Park Service move the Liberty Bell into a new museum. The Bell has a hairline fracture that extends from its famous larger crack, making the Bell quite frail. MicroStrain applied its wireless sensors developed as part of an NSF SBIR grant to detect motion in the crack and fracture as small as 1/100th the width of a human hair. During a lifting operation at the end of the move, the sensors detected shearing motions of about 15 microns (roughly half the width of a human hair) at the visible crack with simultaneous strain activity at the hairline crack's tip. MicroStrain's engineers stopped the riggers during this activity, and the sensor readings returned to baseline. Further lifting proceeded very slowly, and no further readings of concern were observed. The Bell was protected by this early warning detection system, which saved it by literally splitting hairs.

Another newsworthy application by the company of a sensor network was to the Ben Franklin Bridge which links Philadelphia and Camden, New Jersey, across the Delaware River. The bridge carries automobile, train, and pedestrian traffic. At issue was the possible need for major and costly structural upgrades to accommodate strains on the bridge from high-speed commuter trains crossing the bridge. MicroStrain placed a wireless network of strain sensors on the tracks of the commuter train to generate the data needed to assess the added strain to the bridge. "For a cost of only about $20,000 for installing the wireless sensor network, millions were saved in unnecessary retrofit costs," explained Mr. Arms.

In the future, military systems will benefit from the cost-saving information from MicroStrain's sensor networks. Current development projects include power-harvesting wireless sensors for use aboard Navy ships, and damage-tracking wireless sensors for use on Navy aircraft. Mr. Arms explained that the data collected in this application is expected to result in recognition that the lives of the aircraft can be safely extended, avoiding billions of dollars of replacement costs.

THE ROLE OF SBIR IN COMPANY FUNDING

Early on, SBIR funding played an important role in supporting company research. While in graduate school at the University of Vermont, Mr. Arms was involved in proposal writing. He also had learned of the SBIR program. "Were it not for this," he said, "the application process may have seemed intimidating." He tapped Vermont's EPSCoR[17] Phase O grants to leverage his ability to gain federal SBIR grants. EPSCoR Phase O grants provide about $10,000 per grant. According to Mr. Arms, these Phase O grants helped the company get preliminary data for convincing results and helped it write competitive proposals. The company has leveraged a total of $40,500 in EPSCoR grants to obtain $3.6 million in SBIR funds.

[17]EPSCoR (Experimental Program to Stimulate Competitive Research Program) is aimed currently at 25 states, Puerto Rico and the U.S. Virgin Islands—jurisdictions that have historically received lesser amounts of federal R&D funding.

According to Mr. Arms, he found the NSF SBIR program with its "more open topics" particularly helpful in the early stages when the company was building capacity. "The open topics allowed the company to pursue the technical development that best fit its know-how," he explained. "Now the company is better able to respond to the solicitations of the Navy and the other agencies that issue very specific topics."

The company regards receipt of an SBIR grant as "a strong positive factor that is helpful in seeking other funding," said Mr. Arms. "It is used not only to fund the development of new products, but as a marketing tool," he continued, pointing out that the company issues a press release whenever it receives an SBIR grant.

MicroStrain has received a total of nine Phase I SBIR grants, five Phase II grants, three Phase IIB supplemental grants, and one Phase III grant. It has received SBIR grants from National Science Foundation (NSF), Navy, Army, and the Department of Health and Human Services. The amount the company has received in SBIR grants since its founding in 1985 totals about $3.6 million. Table App-C-5 summarizes the company's SBIR/STTR grants in number and amount.

According to Mr. Arms, the receipt of additional SBIR grants in the future is hoped for as a means to enable it to continue to innovate and stay at the forefront of its field. The company is targeting about 25 percent of its total funding to come from SBIR grants in coming years.

TABLE App-C-5 MicroStrain, Inc.: SBIR/STTR Grants from NSF and Other Agencies

NSF Grants	Number	Amount ($)	Other Agency Grants	Number	Amount ($)	Total Number	Total Amount ($)
SBIR Phase I	3	224,800	SBIR Phase I	6	445,800	9	670,600
SBIR Phase II[a]	3	1,198,800	SBIR Phase II[a]	2	1,349,100	5	2,547,900
SBIR Phase IIB/ Enhancements	3	346,900	SBIR Phase III	1	63,000	4	409,900
STTR Phase I	0	0	STTR Phase I	0		0	0
STTR Phase II	0	0	STTR Phase II	0		0	0
STTR Phase IIB/ Enhancements	0	0	STTR Phase IIB/ Enhancements	0		0	0
Totals	9	1,770,500		9	1,857,900	18	3,628,400

[a]Excludes Phase IIB/Enhancement grants which are listed separately.
SOURCE: MicroStrain, Inc.

BUSINESS STRATEGY, COMMERCIALIZATION, AND BENEFITS

The company operates at an applied R&D level, and, unlike most R&D-based companies, has had sales from its beginning. Mr. Arms, the company founder and president, emphasized his belief in the need to produce product "to make it real as soon as possible." Continuing, Mr. Arms said, "Having products lets people know you know how to commercialize and that you intend to do it."

Mr. Arms sees the company's main competitive advantage as its role as an integrator of networked sensors. "Our goal is to produce the ideal wireless sensor networks," he explained, "smart, tiny in size, networked and scaleable in number, able to run on very little power, software programmable from a remote site, capable of fast, accurate data delivery over the long run, capable of automated data analysis and reporting, low in cost to purchase and install, and with essentially no maintenance costs." These features are important because they help to overcome the multiple barriers that were impeding the wider acceptance of sensors.

While the company sells its sensors mainly in domestic markets, it has from the beginning shipped sensors to customers around the world. Now the company sees market potential particularly in Japan and China. Patenting is reportedly very important to the company's commercialization strategy.

MicroStrain has received a number of grants in recognition of outstanding new product development in the sensors industry. It has received seven new product grants in the "Best of Sensors Expo" competition. Products that have been recognized by grants include the company's V-Link/G-Link/SG-Link microdatalogging transceivers for high speed sensor datalogging and bidirectional wireless communications; its WWSN wireless Web sensor networks for remote, internet enabled, ad hoc sensor node monitoring; its FAS-G gyro enhanced MEMS based inclinometer; its MG-DVRT microgauging linear displacement sensor; its 3DM-G gyro enhanced MEMS based orientation sensor; its EMBEDSENSE embeddable sensing RFID tag; AGILE-Link frequency agile wireless sensor networks; and INERTIA-LINK wireless inertial sensor.

Society stands to benefit in a variety of ways from improved sensors and networks of sensors. Structures, such as buildings, bridges, and dams, as well as transportation and industrial equipment should have fewer catastrophic failures, because managers will be alerted to emerging problems in time to take preventative action. Homeland security should be enhanced by smarter networks of sensor-based warning systems. Manufacturing productivity may be increased by better planning of required maintenance and avoidance of costly, unplanned downtime. In general, integration of smart sensor networks into civilian and military structures and infrastructure, transportation equipment, machinery, and even the human body can conserve resources, improve performance, and increase safety.

VIEWS ON THE SBIR PROGRAM AND ITS PROCESSES

Mr. Arms made the following several observations about the SBIR program and its processes, some of which focused on the NSF program, some on the Navy program.

Topic Specification

Mr. Arms contrasted the "open topics" of NSF with the "very specific topics" of the Navy and other agencies, noting the former is particularly important to a company when it is "building capacity," while the latter is important when the company is positioned to generate a variety of new products.

Financing Gap

Mr. Arms noted that "early on in the life of the company the funding gap was very difficult, but now the company is able to bridge the gap using its sales revenue."

Value of Keeping Phase I Grants as Prerequisite to Phase II

"Phase I grants are important for getting a reaction to an area; to understanding better a technology's potential," said Mr. Arms. "I would not want to see this phase eliminated or by-passed."

Size of Grants

"It is great that the agencies are beginning to increase the size of their grants," commented Mr. Arms. "I especially like the NSF's Phase IIB match grant; it fits well with my company's commercial emphasis."

Application Process

Mr. Arms finds the Navy's SBIR application process particularly agreeable, calling it "the best!"

Value of Commercialization Assistance

The company has not participated in an NSF-sponsored commercialization assistance program, but it has participated in Navy-sponsored opportunity forums and in NSF conferences. It has found the networking provided by these forums and conferences to be very valuable. In fact, it was at an NSF-sponsored conference that MicroStrain made contact with Caterpillar Company, leading it to become a participant in a joint venture led by Caterpillar and sponsored by the Advanced Technology Program.

Observation about NSF's and Navy's SBIR Program Manager Systems

"The way NSF conferences facilitate face to face meetings between program managers, who have extensive business experience, with budding entrepreneur-scientists is excellent," Mr. Arms said. He expressed special enthusiasm for the Navy program managers, calling them "extremely knowledgeable and focused."

NSF's Student and Teacher Programs (outside SBIR)

Like several of the other companies interviewed, MicroStrain has used the NSF students program, "but, regretfully, not the teacher program." Like the other companies that have used these programs, Mr. Arms said MicroStrain had found the NSF students program valuable. "I think it would be a great thing to expand this idea to the other agencies," he suggested.

SUMMARY

This case shows a still-small company that has emphasized product sales since its inception in 1985. It has leveraged $40,500 of Vermont's EPSCoR "Phase O" grants to obtain $3.6 million in federal SBIR grants. With SBIR support it developed an innovative line of microminiature, digital wireless sensors, which it is manufacturing. These sensors can autonomously and automatically collect and report data in a variety of applications. Unlike most research companies, MicroStrain, started by a graduate student, has emphasized product sales since its inception in 1985. Its sensors have been used to protect the Liberty Bell during a move and to determine the need for major retrofit of a bridge linking Philadelphia and Camden. Current development projects include power-harvesting wireless sensors for use aboard Navy ships, and damage-tracking wireless sensors for use on Navy aircraft. Although annual revenues are relatively small ($3 million in 2004), the company can document many millions of dollars of savings achieved by users of its wireless sensor networks. A little more than a quarter of the company's revenue come from government sources.

National Recovery Technologies, Inc.[18]

Rosalie Ruegg
TIA Consulting

THE COMPANY

At the time National Recovery Technologies (NRT) was founded, the growth potential for municipal solid waste recycling looked promising. To develop municipal recycling technology, Dr. Ed Sommer applied for an SBIR grant from the U.S. Department of Energy (DoE) soon after starting NRT in 1981. NRT applied first to DoE's SBIR program for funding, because of the energy implications of municipal waste recycling. After being granted a Phase II DoE SBIR grant, Dr. Sommer said he was advised by a DoE program manager that further research to develop the plastics sorting technology would better fit the mission of EPA because of the environmental benefits of reducing PVC plastic waste in incineration. According to Dr. Sommer, having a close fit with EPA's mission made it more likely to receive SBIR grants.

Dr. Sommer described his company's location in Tennessee as very positive from a business standpoint. However, he noted that a drawback is the lack of technology infrastructure in the State. In developing proposals to the SBIR, "you are on your own," he said. There are not the incubators and other institutional assistance provided by some of the states that have a stronger technology infrastructure. He noted that NRT is the first- or second-largest recipient of SBIR grants in Tennessee.

NRT developed and commercialized several innovative processes for high-speed, accurate analyzing and sorting of municipal solid waste streams. The initial customer base had its origins in state recycling laws. Demand for the company's initial product was politically driven, not economically driven. In 1991, with venture capital funding, the company installed process lines in large sorting plants located in states with recycling requirements.

An EPA-granted SBIR project was to remove chlorine-bearing PVC from municipal waste streams prior to incineration for the purpose of emissions control. The company was successful in bridging to the new application, and quickly became a world leader in the recycling equipment industry, providing equipment and automated systems for analyzing and sorting plastics and curbside collected materials. It continues to have worldwide equipment sales, mainly in North

[18]The following informational sources informed the case study: interview at the company with company founder, President and CEO, Dr. Ed Sommer, Jr.; company Web site, <*http://www.nrt-inc.com*>; company brochures; and a recent Dun & Bradstreet company profile report.

NATIONAL RECOVERY TECHNOLOGIES, INC.: COMPANY FACTS AT A GLANCE

- **Address:** 566 Mainstream Drive, Suite 300, Nashville, TN 37228
  Telephone: 615-734-6400
- **Year Started:** 1983
- **Ownership:** Private
- **Annual Sales:** ~$4 million
- **Number of Employees:** 14 total; 5 in R&D
- **3-year Sales Growth Rate:** 67 percent
- **SIC:** Primary SIC: 3589, Manufacture Service Industry Machinery
  358890300, Manufacture Sewage and Water Treatment Equipment
  Secondary SIC: 8731, Commercial Physical Research
  87310202, Commercial Research Laboratory
- **Technology Focus:** Initial Focus—Mixed municipal solid waste recycling system. Later focus—Automated process for sorting plastics by type with high throughput and accuracy for cost-effective recycling; electronics-driven metals recycling system; inspection technology for security checking in airports and other security check points; and continuation of the plastics sorting business.
- **Funding Sources:** Internal revenue mainly from sales of plastics sorting equipment, venture capital, SBIR funding, and ATP (as subcontractor pass-through).
- **Number of SBIR Grants:** From NSF, 3 Phase I, 3 Phase II, and 2 Phase IIB; plus additional grants from DoE and EPA.

America, Europe, Japan, and Australia. As a result of this technology development, NRT received EPA's National Small Business of the Year grant in 1992.

In 1994, the U.S. Supreme Court ruled on a case brought by waste haulers that found that a city violated their rights by requiring them to take collected waste to recycling plants. Because it was cheaper to take it to landfills, many haulers stopped taking it to the sorting plants—taking it to landfills instead—causing large numbers of sorting plants to shut down. Plants that employed their own haulers were more likely to survive. At the same time, there was a move for presorting of curbside waste pick-up, requiring less sorting by secondary processors, further reducing demand for the company's equipment.

As a result of these developments, the company was in trouble. It needed to move into others areas or go out of business. As it looked for new ways to leverage its existing intellectual capital and technology capabilities, it identified new areas in which to apply its technical capabilities: metals recycling and airport security. It continued to pursue automated process technology for high throughput sorting of plastics.

Today the company maintains sales of plastics analysis and sorting equipment with annual sales running about $4 million. At the same time, it is developing new lines of business that were not yet generating sales at the time of the interview. The company employs a staff of 14, five of whom are in R&D.

THE TECHNOLOGIES AND THEIR USES

For plastics recycling, NRT used such technologies as IR spectroscopy for polymer identification, machine vision for color sorting, concurrent parallel processing for rapid identification, quick real-time sorting, and precision air jet selection of materials.

An idea that emerged from discussions with potential customers was metals sorting, smelting, and refining. NRT undertook a research effort now in its seventh year to develop metals processing technologies. It is collaborating with another company, wTe Corporation of Bedford, Massachusetts, which has an automobile shredder division, to develop a novel optoelectronic process for sorting metals at ultra-high speeds into pure metals and alloys. It also joined with wTe to form a new company, Spectramet LLC, to serve as the operator of metals reprocessing plants.

An idea for a new technology/business platform that emerged just in the past several years is in the security area. The stream of objects moving along a conveyer belt at an airport security system resembles in many ways a mixed waste stream in the recycling business. NRT's approach combines fast-throughput materials detection technology with data compilation, retrieval, analysis, storage, and reporting to provide an improvement over the current nonautomated, manual inspection system. The Transportation Security Administration (TSA) is evaluating NRT's system, a necessary step in qualifying it for use in airport security. According to Dr. Sommer, NRT anticipates product sales in 2005.

THE ROLE OF SBIR IN COMPANY FUNDING

According to Dr. Sommer, "Without the SBIR program, NRT wouldn't have a business. We couldn't have done the necessary technical development and achieved the internal intellectual growth." The SBIR program was critical, he explained, both in developing NRT's initial technology, and in responding to market forces to develop new technologies after a Supreme Court decision caused many municipal solid waste sorting plants to close and the growth potential of the initial waste recycling technology to decline. "SBIR saved our bacon," said Dr. Sommer. As a result of the intellectual capabilities built within the company through SBIR-funded research, "we continue to be able to contribute."

In 1985, the company had received a Phase II grant from DoE to pursue development of an automated process for sorting municipal solid waste. Subsequently the company received a series of SBIR grants from EPA, including eight completed Phase II grants between 1989 and 1996, aimed at developing and refining plastics recycling technologies. In 1996, NRT received SBIR grants again from DoE for plastics recycling and mixed radioactive waste recycling. Since 1996, the company has received SBIR grants from both EPA and NSF. Its first Phase II grant from NSF was received in 1999, when the company was in its second decade of operation. The funded project was aimed at developing new technologies in scrap metals processing. Later NSF SBIR funding was aimed at developing new technologies in the security area.

From the NSF, the company has received a total of three Phase I grants, totaling $0.3 million; three Phase II grants, totaling $1.4 million; and, reportedly two Phase IIB grants. Altogether, the company has received more than $5 million in SBIR funding combined from DoE, EPA, and NSF over the past 20 years.

As time passed and the growth potential of the plastics recycling business flattened, Dr. Sommer said that he turned to NSF's SBIR program to develop new lines of technology. He said that NSF's SBIR program was particularly appealing because its solicitation is the broadest among the agency programs. NSF's solicitations allowed the company more leeway to propose new technology development projects that it believed would lead to business opportunities with higher growth potential. At the same time, he characterized NSF's SBIR program as "very competitive" and its grants "the hardest to get." With NSF funding, NRT was able to develop metals recycling technology and, more recently, the detection system for airport security.

The "openness" of NSF, Dr. Sommer said, was critical to his business model, which entails first finding a need for a new product, performing market analysis, and then looking for funding to perform research needed to bring a product to the prototype stage. Contrasting NSF's openness with the "narrow" solicitations of the Department of Defense (DoD), Dr. Sommer explained that he did not apply to the DoD SBIR program because "for us it is not conducive to developing products aimed at a general market."

Dr. Sommer related how his company (in a subcontractor position to the wTe Corporation, Bedford, Massachusetts) had subsequently looked to the ATP for larger amounts of funding to help further develop the metals reprocessing technology. In describing the sequence, he said there was a "spring-off from SBIR to ATP."

BUSINESS STRATEGY, COMMERCIALIZATION, AND BENEFITS

While it develops the metals reprocessing and security product lines, NRT has maintained a steady revenue stream on the order of $2 million-4 million/year, primarily from sales of plastics analysis and sorting equipment. Dr. Sommer sees a larger market potential in metals reprocessing—which includes partnerships to operate as well as provide equipment—with projected annual sales revenue in the range of $10-30 million. He sees a larger potential in the security market, with projected sales revenue in the range of $100 million/year. Dr. Sommer expressed his intention to take the company into a faster growth mode with the development of these new lines of business.

Annual sales revenue is currently running approximately $4 million. Revenues reportedly generated as a result of SBIR grants, referred to as "Phase III revenues," totaled approximately $44 million from the start of the company up to November 2003.

At least eight products are on the market derived from DoE and EPA SBIR-funded research, including, for example, the following:

- NRT VinylCycle® system—a grant winning sorting system for separating PVC from a mixed stream of plastic bottles introduced in 1991
- NRT MultiSort®IR System—an advanced plastic bottle sorting system for separating specific polymers from a mixed stream of materials
- NRT Preburn™ Mixed Waste Recycling System—a facility with integrated technologies to provide a system for achieving maximum material recovery from waste streams otherwise slated for landfill

Products funded by the NSF SBIR program are still in the development stage. The metal alloy sorting technology under development with NRT's commercialization partner wTe Corporation is planned to be used by the Spectramet LLC spin-off, jointly owned by NRT and wTe Corp, for processing of metals as opposed to the technology being made available as a commercial equipment product. The advanced third generation of this sorting technology is installed and in initial commercial operation processing selected loads of scrap metals from various suppliers.

Two patents resulting from the NSF funded research have been issued. Four additional patents are pending.

Dr. Sommer identified four types of social benefits that have resulted, or may result, from the SBIR-funded technologies. (1) Knowledge creation and dissemination result from patents the company filed on the intellectual capital coming out of its NSF research, and from presentations. Patents signal the creation of new knowledge and provide a path of dissemination. Company researchers have also presented at conferences in the fields of metallurgy and plastics recycling. (2) Safety effects arise from the automation of sorting machines in recycling plants, which appear to have reduced injuries as compared with conveyer belts using labor-intensive hand-picking techniques that bring the worker in close interface with potentially unhealthy waste streams and possibly injurious equipment. (3) Environmental effects result because NRT's sensing and sorting technologies are based in electronics, not chemicals. Using "dry processes" rather than "wet processes" avoids the runoff of chemicals into waste streams and the associated pollution. Additionally, environmental benefits result directly from recycling plastics and metals into reuse instead of dumping them into landfills. The availability of automated systems that increase the efficiency of the process helps to enable cost-effective recycling of diverse materials around the world. (4) National security benefits may result if NRT is successful in leveraging its automated materials sensing technology into the security arena, improving the efficiency—and more important—the effectiveness of security at airports and other security check points. NRT's technology is currently under evaluation for airport security applications, and not yet in use.

VIEWS ON THE SBIR PROGRAM AND PROCESSES

Submitting Proposals Through NSF's FastLane

Discussing the SBIR application process and how the application process compares among agencies, Dr. Sommer noted that the answer is very time dependent given that the agencies have recently developed more computerized applications processes. He noted that NSF's FastLane system is "very slick." It is also very complex, he said, with many modules, which make navigating around the system hard on a newcomer. However, once one becomes familiar with it, it becomes more useful, he concluded.

NSF's Review Process

Dr. Sommer's view was that NSF does a "fabulous job" with its review of SBIR proposals. He noted that earlier there was an issue—"too heavy a reliance on university reviewers"—but believes that now there is more use of reviewers who come from the commercial sector who are better able to assess proposals for technology development with commercial potential. He noted that he was so impressed that he wanted to give back to the system, and volunteered to serve on a review panel for NSF. The experience, he said, gave him confidence in the process as being fair. He also saw the experience as a good way to learn the ins and outs of preparing higher quality proposals.

NSF's Feedback on Reviews

Dr. Sommer also found useful NSF's feedback system to give applicants information from review results. He said his company had resubmitted a rejected proposal, taking into account feedback received, and was successful with the resubmitted proposal. Asked if he felt his company's proprietary ideas had ever been threatened during the proposal and review process, he responded with a definite no with respect to DoE, EPA, and NSF.

NSF Program Managers

In speaking of NSF program managers, Dr. Sommer praised those with whom he had direct experience as "extremely dedicated."

Grant Size

When asked if he thought the size of SBIR grants should be increased, possibly in trade-off to a decrease in the number of grants, Dr. Sommer responded that he thought the size of Phase I grants is about right, and noted that "it is good to spread around the funding," rather than concentrate it in fewer grants. He said that he would, however, like to see somewhat larger Phase II grants. In this regard, he characterized NSF's Phase IIB grant as "a very good tool, providing a boost to finding other dollars." He noted that the Phase IIB requirements fit well with his business model. He also reiterated that it is good that the ATP is available to provide larger research grants.

Funding Gap

Dr. Sommer noted that there often is a lag—a funding gap—between Phase I and Phase II grants that can "put the brakes on research." He explained that he is fortunate in having an ongoing business with a revenue stream that can help him bridge the gap with internal funding, rather than shut down the research as he would otherwise have to do.

Phase I as a Prerequisite to Phase II

When asked if he would like the opportunity to bypass the Phase I grant and go directly to Phase II, Dr. Sommer responded that often research funded in-house positions him to have the ability to apply directly for a Phase II grant. "Phase I," he said, "makes you do your feasibility analysis more thoroughly than you might otherwise do, but this can slow you down." He saw both pros and cons to keeping Phase I as a prerequisite; his response was inconclusive.

Commercialization Assistance Program

A company founder and CEO, Dr. Sommer, holds a doctorate degree in physics from Vanderbilt University. He went into business soon after receiving the degree. According to Dr. Sommer, he earlier resisted participating in the commercialization assistance programs sponsored by SBIR programs, and dropped out of a program that he had begun. He thought that the time requirements were excessive, and he resisted a diversion from his focus on technical issues. In 2000, however, he enrolled for a second time in the commercialization assistance program provided by Dawnbreaker Company. This time he completed the program, which he described as highly beneficial, providing him with insights, vision, know-how, and tools to more aggressively pursue business opportunities. He said he needed the training.

NSF's Solicitation Topics

Dr. Sommer emphasized that he would like to see preserved the "openness" of NSF's solicitation, which he called "critical" to maintain.

NSF's Emphasis on Commercialization

Another important feature of the NSF program that Dr. Sommer thought should be kept is the emphasis of the program on commercialization.

NSF's Review Process

Dr. Sommer expressed the view that it is very important that NSF achieve a balance in the use of university reviewers and those knowledgeable and experienced in business.

SUMMARY

This case study shows how SBIR is used not only by a start-up company to help it establish a technology platform from which it can launch a business, but also by a mature company that needs to rejuvenate its technology platform in order to meet changing markets. The case study company, NRT, used SBIR grants initially to support R&D underlying its first line of business. In its second decade, it used SBIR grants to leverage its existing technological base in a directional change that offers potential for future robust growth.

The NSF SBIR grant solicitation with its broad topic areas and emphasis on commercialization fits particularly well the company's business strategy of first identifying a potential market opportunity, developing a research plan to bring a product/process to the prototype stage, and then looking for early stage research funding to make it happen.

This case also illustrates a business trajectory that, although up to this time, is not dramatic in terms of its growth, is significant when considered as one of many such companies enabled by the SBIR program. The company has survived for over 20 years, remaining small but steadily employing approximately 10-15 people, generating on the order of $4 million per year, meeting niche market needs that have energy and environmental implications, and poised to make further, potentially substantial technological contributions and to achieve growth. It may be argued that this company represents a component of the R&D landscape, which in its aggregate is an extremely important part of the nation's capacity to innovate.

NVE Corporation[19]

Rosalie Ruegg
TIA Consulting

THE COMPANY

This small company traces its origins to a very large company, Honeywell, Inc. The company founder, Dr. James Daughton, co-invented "Magnetoresistive" Random Assess Memory (MRAM) while at Honeywell. On retiring from Honeywell, he licensed the technology for pursuit of civilian development and applications, and, in 1989, founded Nonvolatile Electronics, Inc. (NVE).[20] Other former Honeywell employees joined NVE following a downsizing at Honeywell and continue with NVE today.

Initially the company operated out of the founder's home in a Minneapolis suburb, but in the early 1990s after receiving research funding from the SBIR program and from the Advanced Technology Program (ATP), it found space in a nearby Eden Prairie, Minnesota, industrial park. Today it leases a facility of approximately 21,000 square feet, which includes offices and a clean-room for research and fabrication of semiconductor devices. It employs a staff of 70, including 24 in R&D—12 at the PhD level, and the rest in product development, manufacturing, and administration.

The company has licensing arrangements with other companies, among them Honeywell, Motorola, Cypress Semiconductor, and Agilent Technologies. It also has affiliations with a number of universities, including the University of Minnesota, Iowa State University, and the University of Alabama, and it has sponsored a university student through an NSF program.

THE TECHNOLOGY AND ITS USE

Since its founding, the company has continued development of MRAM, a revolutionary technology that fabricates memory with nanotechnology and uses electron spin to store data. MRAM computer chips could prevent accidental losses

[19]The following informational sources informed the case study: interview at the company with Robert Schneider, Director of Marketing, Richard George, Chief Financial Officer, and John Myers, Vice President of Development; information from the company Web site, <*http://www.nve.com*>; the company's Annual Report for 2004; company press releases and selected press clippings; an article about the company's technology in Sensors, 21(3), March 2004; Dun & Bradstreet company profile report; earlier interview results compiled by Ritchie Coryell, NSF (retired); and an earlier "status report" developed by ATP.

[20]In 2000 through a reverse merger, the company officially took the corporate name of NVE Corp. It trades on the Nasdaq SmallCap Market under the symbol "NVEC."

NVE CORPORATION: COMPANY FACTS AT A GLANCE

- **Address:** 11409 Valley View Road, Eden Prairie, MN 55344
- **Telephone:** 952-829-9217
- **Year Started:** 1989 (Incorporated 1989)
- **Ownership:** Traded on the NASDAQ Small Cap Market as "NVEC"
- **Revenue:** $12 million annually
  - —Share from SBIR/STTR grants: 35 percent
  - —Share from product sales, R&D contracts, and licensing: 65 percent
- **Number of Employees:** 70
- **SIC:** Primary SIC 3674
- **Technology Focus:** Spintronics-based semiconductors, a nanotechnology which utilizes electron spin rather than electron charge to acquire, store and transmit information
- **Application Areas:** Electronic memory, sensors, and isolated data couplers
- **Facilities:** Leases 21,362 sq. ft.
- **Funding Sources:** Commercial sales, government sales, licensing fees, federal government grants & contracts, stock issue, private investment, venture capital funding, and reinvestment of retained earnings.
- **Issued Patent Portfolio:** 34 issued U.S.; more than 100 patents worldwide issued, pending, or licensed from others
- **Number of SBIR grants:**
  - —From NSF: 31
  - —From other agencies: 90

of information when the power is interrupted, extend battery life, and replace essentially all RAM technology in use today. A Web site devoted to MRAM news (*<http://mram-info.com>*) calls the technology the "holy-grail" of memory, and states that it "promises to provide non-volatile, low-power, high-speed and low-cost memory." MRAM is based on effects named "Giant MagnetoResistance" (GMR) and "spin-dependent tunneling," whereby a sandwich of metals shows a substantially greater change in resistance than a single metal of the same size when exposed to a magnetic field. These effects enabled researchers to increase signal strength while increasing density and decreasing size.

The technology has proven a challenging pursuit indeed. A decade later MRAM remains largely a promise still falling short of the realization of its huge commercial potential. However, NVE has developed significant intellectual property in MRAM that it has licensed to both Motorola and USTC for initial license fees and future royalty payments when it is put into commercial use. Other companies, including IBM Corporation, Hewlett-Packard Company, Infineon Technologies AG, NEC Corporation, Fujitsu Limited, Sony Corporation, and Samsung Electronics are also seeking to develop MRAM chips.

It should be noted that there are currently available nonvolatile memories, such as "flash" memories and ferroelectric random access memories (FRAMs), and there are also emerging technologies that are expected to compete with MRAM, such as polymeric ferroelectric random access memory (PFRAM), ovonic unified memory (OUM), and carbon nanotubes. However, according to its developers, MRAM offers advantages over the existing nonvolatile memories in terms of speed, lower power use, longer life expectancy, and freedom from other limitations, and also advantages over the emerging technologies, including being closer to market.

As NVE pursued MRAM development, it saw other potential commercial applications in the GMR effect. By the mid-1990s, NVE was making and selling GMR-based sensing products for such diverse applications as automotive braking systems, medical devices, and portable traffic monitoring instruments.

NVE now describes its technical focus as "spintronics," a nanotechnology based on MRAM research, which, like MRAM, takes advantage of the property of electron spin. The technology combines quantum mechanical tunneling with magnetic scattering from the spin of electrons, resulting in a new phenomenon called "spin-dependent tunneling (SDT)" or "magnetic tunnel junctions (MTJ)."(*SENSORS*, March 2004). The targeted application area is magnetic field sensing for which very small, inexpensive sensors with high sensitivity to small changes in the magnetic field are required. Standard silicon microprocessing methods can be used to fabricate SDT devices. The company expects to enter the market within several years with SDT sensors in complex magnetometer systems, in small simple event detectors, in arrays for perimeter security, vehicle detection, and other security systems, and also for detection of deep cracks, corrosion, and other deeply buried flaws. Further in the future, applications are anticipated for physiological monitoring, advanced magnetic imaging, and other areas not yet identified.

THE ROLE OF SBIR AND OTHER GOVERNMENT R&D FUNDING

Federal grants for R&D—both from SBIR and ATP—have played an essential role in the company's start-up, survival, and growth. During its early days, the company's founder credited government R&D funding with preventing the company from failing and improving its ability to attract capital from other sources.

More recently, NVE's vice president called the SBIR program "the mother of invention." The company currently derives approximately half of its funding from government funding, including SBIRs and BAAs (Broad Area Announcements that federal agencies may use to solicit contract work), and the remaining half from commercial sales, up-front license fees and royalties, stockholder investment, and retained earnings. It views SBIR and other government R&D funding programs as essential to being able to perform the advanced R&D that has allowed the company subsequently to produce products for sale and to license intellectual property.

Table App-C-6 summarizes the company's SBIR/STTR Grants from the National Science Foundation (NSF) and other agencies. The 121 SBIR and STTR grants received have totaled $34.3 million in R&D funding. Approximately a fourth of the 121 grants have come from NSF.

BUSINESS STRATEGY, COMMERCIALIZATION, AND BENEFITS

NVE President and CEO, Daniel Baker, has been recently quoted as saying, "We believe that NVE is well-positioned with critical intellectual property covering a broad range of near-term and long-term MRAM designs. Our MRAM strategy, therefore, will be to focus on an intellectual property business model, providing technology to enable revolutionary memory designs rather than both providing technology and selling devices." (NVE Press Release, April 19, 2005)

TABLE App-C-6 NVE Corporation: SBIR/STTR Grants from NSF and Other Agencies

NSF Grants	Number	Amount ($)	Other Agency Grants	Number	Amount ($)	Total Number	Total Amount ($)
SBIR Phase I	19	1,487,000	SBIR Phase I	51	4,512,000	70	5,999,000
SBIR Phase II[a]	9	3,540,000	SBIR Phase II[a]	34	22,428,000	43	25,968,000
SBIR Phase IIB/ Enhancements	2	582,000	SBIR Phase IIB/ Enhancements	2	488,000	4	1,070,000
STTR Phase I	1	100,000	STTR Phase I	1	70,000	2	170,000
STTR Phase II	0	0	STTR Phase II	2	1,120,000	2	1,120,000
STTR Phase IIB/ Enhancements	0	0	STTR Phase IIB/ Enhancements	0	0	0	0
Totals	31	5,709,000		90	28,618,000	121	34,327,000

[a]Excludes Phase IIB/Enhancement grants which are listed separately.
SOURCE: NVE Corporation.

Regarding NVE's strategy regarding sensors and signal couplers, it appears that NVE will continue to build its intellectual property base in SDT and GMR sensors, as well as continue to design, fabricate, and sell a variety of sensor and signal coupler devices for both commercial and defense applications.

Product revenue from sales of spintronic sensors and couplers has steadily increased, reaching $5.4 million in FY2004, and the company projects its commercial product revenues to continue to grow. R&D revenue for FY2004 rose to $6.6 million, as government contract revenue increased. Total revenue for FY2004 totaled $12 million. As of the end of FY2004, NVE had been profitable for two years, and the firm was projecting continued profitability into FY2005.

The spintronic sensors and couplers sold by NVE offer value-added benefits to users in terms of accuracy and data rates. The firm's unique components provide up to 3 to 4 times the accuracy and twice the data rate of conventional electronics, allowing users to make better products at lower costs.

NVE has stated recently that the commercial viability of MRAM technology is now more assured. It expects FY2005 to be pivotal for MRAM commercialization. In additional to nonvolatility, MRAM's potential benefits to users are high speed, small size, increased life expectancy, lower power use, and scalability. Nevertheless, as indicated earlier, the competition among firms and among technologies is intense.

VIEWS ON THE SBIR PROGRAM

The interviewees strongly supported the SBIR program in general. They noted that it has fostered the development of a large number of small R&D companies like itself that "collectively comprise the modern-day equivalents of the Bell Labs of the past." They emphasized the advantages of performing R&D in a small-company environment where "there is much more freedom to innovate" and "R&D is not viewed as an unwelcome tax on what is considered the productive part of the corporation." They expressed the hope that program officials and policy makers will not underrate the importance of this collective group of small firms by assuming that only if they individually grow into large companies are they worthwhile from the standpoint of the economy and its innovative capacity.

At the interviewer's request, the three NVE officials shifted their focus to NSF's SBIR program, beginning with a positive comment and then turning to areas for potential improvement:

Praise for NSF Portfolio Mangers

NVE officials emphasized the high quality of NSF's program managers. They identified several of the program managers by name, calling them as a group "a class act."

Concern about an Unofficial Limit on Number of Grants to a Firm

The NVE team voiced concern that NSF (NIST was also mentioned) appears to be imposing an unofficial criterion on top of the official proposal eligibility criteria, in the form of a limit on the number of grants a given company can receive. According to the company representatives, the company is sometimes informed that it must choose a subset of the total number of grants for which it has been deemed eligible, as evaluated against the published criteria.

Moreover, since the company is aware of firms that it believes have received many more SBIR grants than NVE, it has the impression that the agency may be applying an unofficial limit on grants unevenly among applicant firms. At the same time, other agencies do not appear to have such a criterion—either officially or unofficially.

NVE's position is that if the company has submitted multiple proposals that meet an agency's published SBIR eligibility requirements, it should be able to receive the grants for which it is eligible without limit, given there is no overlap among them and adequate funding. If there is a limit on the number of grants that a company can receive, NVE believes this limit should be made official, explicit, and be evenly applied so that companies can know up-front exactly what rules apply before they incur the considerable costs of proposing.

Concern about Limit on Number of Proposals Allowed

NVE also noted a limit on the number of proposals a company can submit to NSF in a given year. If topical solicitations are spread out over the year, as they are for NSF, this limit means that a company must make decisions about how it will spend its proposal "quota" among the topic areas before it may be ready to do so strategically.

Need to Overcome a Timing Problem

NVE noted that NSF's timing of its solicitations mean that if you miss a key date, you miss a whole grant cycle. Extending the proposal submittal window or opening it several times a year would help to overcome this timing problem.

Comments on Commercialization Assistance

According NVE's Director of Marketing, participation in the Foresight activity was of greater value to the company than was the Dawnbreaker activity. A reason given was that the Foresight staff appeared to have more industry experience.

Lack of NSF Travel Funds

NVE officials expressed concern that NSF program managers are not able to conduct company site visits. "On-site visits would be good." At the same time, they noted that the annual SBIR/STTR Phase II meeting provides them the opportunity to discuss their projects with the NSF program managers.

SUMMARY

This case study shows how a company, which traces its origins to a large company, used SBIR and other federal grants to help launch the company, to keep it from failing, and to improve its ability to attract capital from other sources. Since its founding, the company has pursued development of MRAM technology that uses electron spin to store data and promises nonvolatile, low-power, high-speed, small-size, extended-life, and low-cost computer memory. As NVE pursued MRAM development, it saw related potential applications such as magnetic field sensors. NVE has developed substantial intellectual property in MRAM technology. The company has licensing arrangements with a number of other companies. Approximately half of the company's funding comes from government funding, and the remainder from commercial sales of magnetic field sensors, up-front license fees, and royalties. The company is now traded on the NASDAQ Small Cap Market.

Physical Sciences, Inc.

Irwin Feller
American Association for the Advancement of Science

THE COMPANY

Physical Sciences, Inc. (PSI) was established in 1973 by Robert Weiss, Kurt Wray, Michael Finson, George Caledonia, and other colleagues at the Avco-Everett Research Laboratory, a Massachusetts-based, DoD-oriented research firm. The founders left Avco-Everett to start their own firm in part because they sought a smaller firm research and working environment than was possible at Avco-Everett, which at the time of their leaving had grown to about 900 employees.

PSI is located in Massachusetts, and has retained its major laboratories and corporate headquarters there because it is where its founders live. It has additional operations in Bedford, Massachusetts, Princeton, New Jersey, Lanham, Maryland and San Ramon, California.

The firm's growth was modest at first. Its initial contracts were with the Air Force and the Department of Energy. By the early 1980s, it had approximately $10m in revenues and a staff of 35-50. Sizeable reductions in DoE's R&D budget in the early 1980s caused its contract revenues to fall by approximately one-third. The firm recovered after that period, in part by diversifying the range of its Federal customers, such as participation in NSF's Research Applied to National Needs (RANN) program. As the breadth of its technical competencies kept pace with rapidly changing advances in laser and optics technology, and as it become more actively involved in the SBIR program, it was able to expand its range of technological expertise as well as of Federal governmental and private sector customers. For FY2005-2006, its revenues are estimated at $35m, and its employment level at 175. (These figures include sales and employment levels at its wholly owned subsidiaries Q-Peak and Research Support Instruments, Inc., but exclude employment at Confluent Photonics Corp., a commercial spinout.) Approximately 40 percent of estimated FY2005 revenues are derived from SBIR awards. SBIR awards have contributed a diminishing portion of firm revenues, falling from a peak of about 60 percent in the late 1990s to a projected 35 percent in FY2006.

The founding vision for the firm was to do world-class basic and applied research and prototype product development under contracts from government and private-sector sponsors. It has grown primarily by self-financing and employee stock ownership. This strategy rather, than one based on pursuit of external angel or venture capital, has been adopted in order to avoid dilution of owner/employee direction of the firm. Related to this vision of being a premier research organiza-

PHYSICAL SCIENCES, INC.:
COMPANY FACTS AT A GLANCE

- **Address:** 20 New England Business Center
  Andover, MA 01810
- **Telephone:** 978-689-0003
- **Year Started:** 1973
- **Ownership:** Employee Stock Ownership Trust
- **Revenue:** $35 million (estimated FY2005)
- **Total Number of Employees:** 175 (Physical Sciences Inc and its two subsidiaries, Q-Peak, Inc. and Research Support Instruments, Inc.)
- **Number of Patent disclosures:** 100 since 1992; approximately 12 per year since 2000
- **Number of Patents Issued:** 39 U.S., 54 foreign patents (24 issued to PSI, or pending, and five issued to Q-Peak, a PSI subsidiary, are directly related to SBIR/STTR programs)
- **SIC:** Primary (8731)
  Secondary (none)
- **Technology Focus:** Optical sensors, contaminant monitors, aerospace materials, weapons of mass destruction detectors, power sources, signal processing, system modeling, weapons testing
- **Funding Sources:** Federal government R&D contracts and services (80 percent); sales to the private sector, domestic and international (20 percent)
- **Number of SBIR Awards:** Total
  Phase I—435
  Phase II—176
- **DoD SBIR Awards:** Phase I—337
  Phase II—98
- **Number of STTR Awards:** (included in the data above)
- **Publications:** Total: Over 1200 to date, probably 50 percent of which are SBIR/STTR-related (an accurate number has not been determined)

tion was the expectation that the firm's staff would publish research findings in the open literature. These foundational principles have continued in force to the present, accounting for the firm's emphasis on R&D and prototype development rather than manufacturing, which would require additional external capital.

The firm's initial financing came from the assets of its founders, including mortgages on their homes, and funds from family and friends. The firm has drawn little support and seen few benefits in the various technology development programs operated by the Commonwealth of Massachusetts. Massachusetts is seen as lagging behind other states in the scale and flexibility of programs targeted at fostering the establishment and growth of small, high-technology firms.

In keeping with its pursuit of autonomy and a concentration on contract R&D, PSI has limited its involvement with venture capital firms. Its engagement with them has generally involved the launching of spin-off firms to commercialize products derived from PSI's R&D technological developments, all of which flowed from SBIR funding. To date, this involvement has been infrequent, and the economic record has been mixed. One such firm in the area of medical instrumentation failed when it couldn't raise sufficient (3rd stage) venture capital funding to complete clinical trials and obtain FDA approval. Another spin-off firm underwritten by venture capital funds did succeed, and was acquired by a strategic partner. A more recent venture in the area of optical communications is currently manufacturing components for the telecommunications and cable television industries.

From its inception, the firm's business strategy has been to specialize in the performance of contract research and development and prototype development. In terms of DoD's categorization of R&D, the firm sees itself as oriented toward 6.2 and 6.3 projects. In the terminology derived from Donald Stoke's classic work, "Pasteur's Quadrant," it has strategically chosen to position itself in Pasteur's Quadrant, that is as a performer of R&D characterized by the pursuit of both fundamental understanding and utility.

PSI's initial research expertise was based upon and has continued to center on the development and application of laser and optics technology. Reflecting the experience of its founders, the firm initially targeted the aerospace industry as its primary customer. As optical technology has evolved to an ever-wider set of applications, the firm's technological and market bases have widened to encompass applied R&D, production operations, and bundling of "hands-on" service delivery with the application of newly developed products, especially in the areas of instrument development, diagnostics and monitoring.

Over time, PSI has applied its core research expertise to a widening, more diversified set of technological applications and for an increasingly diversified set of government and private sector clients, both in the U.S. and internationally. It has strategically positioned itself in an R&D market niche defined by multidisciplinary expertise and research infrastructure in specialized high-tech areas too small to attract major investments by large DoD prime contractors, while at the same time too mission-driven to elicit competition from universities. Its interdisciplinary orientation reflects its origin: Its founding partners represented expertise in aeronautical and mechanical engineering, physical chemistry, and physics. Its current R&D projects encompass optical sensors, laser systems, space hardware, contaminant monitors, aerospace materials, weapons of mass destruction detectors, power sources, signal processing, system modeling, and weapons testing. Reflecting the breadth and interdisciplinary nature of this R&D portfolio, its research staff has R&D expertise in fields extending from astrophysics to zoology.

Given this breadth of activity, the firm operates on a matrix model; it has multiple divisions, arrayed across general topical areas. Its research staff, however, operate across divisions, employing their specific expertise to multiple projects It also employs cross-division review procedures to set priorities, sift prospective responses to DoD solicitations for Phase I proposals, and provide critical technical evaluation of work in process.

The firm's primary customer is the Department of Defense. DoD R&D contracts drawn from across several services account for an estimated 70 percent of firm revenues. In recounting the impacts of its R&D endeavors, PSI emphasize the application of its technologies and the beneficial impacts these applications have had on the ability of DoD sponsors to achieve mission objectives. Given this emphasis, it sees concepts and related measures of technology transfer, applications, contributions to mission needs, and impact as more important indicators of the quality of the work it performs under Phase I and Phase II SBIR awards than more commonly used measures of commercial impact.

Some of the firm's contracts with DoD involve development of specialized, one of a kind, technologies, that are seen as significantly contributing to the service's mission, but for which the total market, public or private sector as measured by sales volume, is quite small or nonexistent. Other DoD contracts lead to the development of technologies, mainly in the area of instruments, that the firm seeks to market to the private sector. For example, PSI's development under SBIR awards of sensor technology to detect methane gas leaks has been sold to gas companies. In general, sales to the private sector are largely based on technologies developed for DoD under SBIR awards.

PSI's strategy of emphasizing contractual R&D has led it to purposefully limit the degree to which it seeks to move beyond bench and field prototypes, especially in the scale of manufacturing activities. Thus, it engages in limited production for specific instruments for DoD and other federal agencies. When its technological developments lead to commercially viable products, PSI follows a mixed strategy. One strategy is to form new firms, with new, independent management, that operate as partially owned spinouts. Shaping this business decision is the firm's view that the "cultures" and operational needs of contract R&D and manufacturing firms differ sufficiently that it is more efficient to operate them as separate entities rather than attempt to combine them into one larger firm. Conversely, wholly owned subsidiaries, which are focused on R&D activities, have become eligible for SBIR competition on their own. Since 1990, PSI has formed four new product companies employing technologies developed by their R&D activities. In general, though, PSI sees itself as operating in technologically sophisticated areas whose commercial markets are too small to attract the interest of venture capital firms.

Another strategy is to sell directly to a customer. This strategy is followed for products where the scale of production is low and does not require extensive capital investment in new plant and equipment. In cases where the product has larger market potential but primarily as a subcomponent in a larger complex technological product, PSI has licensed the (patented) technology to an industry leader.

PSI also notes that the gestation period of the technological advances contained in several of its DoD R&D-funded projects is often quite lengthy, with the implication that attempts to measure the commercial import within short periods of time, say 3 years, can be misleading. Its experiences with DoD also have demonstrated the multiple but at times different uses to which a technology has been applied rather than that projected in an initial proposal. PSI also conducts classified research, one effect of which is to limit public disclosure of the technological or economic impact of some of its activities. Its 30-year history also highlights cases in which a different service than the one that supported the initial Phase I award has made beneficial use of the resulting research findings or technology.

EXPERIENCES WITH SBIR

PSI received its first SBIR award in 1983, 10 years after its founding. The SBIR program however is credited by the firm for contributing significantly to growth and diversification since then. As stated in its corporate material. "The Small Business Innovation (SBIR) program has played a pivotal role in PSI's technical and commercial success, and has been responsible for a family of intelligent instrumentation products based on proprietary electro-optical, and electromechanical technologies."

Since its first award, PSI has been a highly successful competitor for Phase I and Phase II SBIR awards. As of 2005, summed across all federal agencies, it had received 435 Phase I and 176 Phase II awards, placing it among the top five recipients of SBIR awardees. PSI has received SBIR awards from multiple agencies, including several DoD services, NIH, NSF, NASA, DoE, NIST and EPA.

Acknowledging its distinctive performance in SBIR competitions, PSI, however, rejects the label that it an "SBIR mill." Rather, it sees itself as winning SBIR awards because it provides valuable R&D services to its (repeat) federal agency customers, who have limited discretionary resources other than SBIR.

SBIR awards are seen as an especially flexible mechanism by which DoD can contract for the development and application of advanced instrumentation for monitoring and testing. SBIR Phase I and Phase II awards are seen as an especially effective and appropriate mechanism to further DoD's 6.2 R&D objectives, especially in advancing technologies to the stage of a bench prototype. It notes that the Phase II award frequently culminates at that stage; additional R&D support is seen as needed to move the technology through the stages of field prototypes, engineering prototypes, and eventually to manufacturing prototypes, with each of these stages being necessarily preludes to the commercial introduction of a new product.

Addressing the delays between Phase I and Phase II awards, PSI considers it prudent to avoid spending money on Phase II awards until it receives formal notice that its proposal has been successful. However, at times, it will allocate company funds to bridge the gap between awards in order to keep an R&D project going. Since this support invariably involves closing down or deferring other R&D projects, at times those being conducted by other divisions, decisions about the use of internal funds involve extensive consultation with R&D staff. PSI's current size and matrix organization are seen as enabling it to somewhat buffer these delays, an advantage it now sees itself as having as compared with smaller firms or those with limited SBIR award portfolios. It will shift staff among projects, as needed, to minimize interruptions in the course of work on projects deemed likely to win Phase II competitions. (The Navy is singled out for commendation on its ability to compress the time between Phase I and Phase II awards).

RECOMMENDATIONS FOR SBIR

PSI believes that SBIR needs to maintain and indeed increase its emphasis on breakthrough technologies. It is concerned that the increasing emphasis being placed on and within SBIR towards commercialization will cause it to "die by incrementalism." Commercialization is conventionally measured by sales, at times with the implication that only those to the private sector "count." In the view of the firm, this narrowing of the objectives of the SBIR program omits or obscures the contributions that the "application" of the outputs of specific SBIR projects can make to the mission requirements of DoD. As stated by PSI representatives, a root cause of this problem is the failure at times to recognize that the legislative intent of SBIR is both to meet the mission-oriented needs of the sponsoring agencies *and* to produce commercial spin-off, wherever possible. Over time, the two objectives have incorrectly been interpreted as one, with the latter one becoming the exclusive criterion for evaluating the aggregate performance of SBIR awardees, and the program itself.

PSI also sees an increase in the dollar size of Phase I and Phase II awards as needed, even at the trade-off of DoD and other federal agencies thus being able to make fewer awards. Administrative expenses chargeable to the SBIR program also are seen as needed to reduce the unduly lengthy review processes for both types of award and to shorten the time between Phase I and Phase II awards. PSI recognizes that its views on administrative costs differ from that of most other participants in the SBIR program, who see such charges as subtracting from the amount available for awards to firms.

SBIR's award processes are described as fair, but not necessarily competent. Acknowledging that agencies may encounter difficulties in recruiting competent reviewers who do not have conflicts-of-interest, PSI's reaction to some reviews of its proposals is that some reviewers are flat-out incompetent. Among federal agencies with SBIR programs, DoD is viewed to have the most efficiently run program, with the Navy being deemed the best of all services. One reason is the DoD culture that encourages one-on-one conversations with program managers and cutting edge technology. Similarly, NIH is held to have a highly effective SBIR program. It is seen as truly viewing small firms as contributing to technological innovation, and as understanding that multiple Phase II awards are frequently necessary to convert findings generated from Phase I awards into marketable products and processes. NIH also is commended for the breadth of its outreach activities; these include meetings between the firm and NIH program managers, and opportunities at larger forums for small firms to interact with university researchers and other, larger firms. NIH review procedures though are criticized for the propensity of some reviewers to confuse SBIR proposals with RO1 submissions. While the proposals are arguably of equal quality, the scoring system used for SBIRS is different from that used to evaluate an RO1.

At the other end of the distribution, NSF's SBIR program is said to be the worst among federal agencies, both because of its protracted review and award processes and the confusing commercial emphasis of its (mostly academic) reviewers. It is also the only agency that restricts companies to four proposals per year. DoE is seen as having very smart personnel, but lacking respect for the R&D capabilities of small businesses. Instead, in its operation of the SBIR program, DoE sees small businesses mainly as vendors of new products, particularly instruments, that are to be used in national laboratories. NASA has mission objectives similar to the DoD, but needs to improve on communicating its goals and requirements through program manager-to-company interactions.

SAM Technologies

Robin Gaster
North Atlantic Research

BACKGROUND AND HISTORY

Dr. Gevins founded SAM in 1986, at about the same time that he founded its sister nonprofit research organization, the San Francisco Brain Research Institute, founded by Gevins in 1980 and previously part of the University of California School of Medicine in San Francisco. Dr. Gevins was focused on a project he had conceived while a freshman at MIT, to build a technology that could measure the intensity of mental work in the brain—reflecting in real time the concentration and attention capacity of the user.

Since 1986, SAM has consistently pursued this single goal, using all its SBIR and other awards to help build a prototype to measure signals in the brain that reflect attention and memory. This is, in short, a case study in how multiple SBIR and other awards can help to support a visionary and very high risk project in long-term biomedical research.

Dr. Gevins had received RO1 grants at UCSF, where he was offered a tenured position in the psychology department. However, RO1 reviewers were not in the mid-1980s friendly to technology-oriented projects, and Dr. Gevins found that SBIR was a better channel for his engineering activities.

Over the past 30 years, Dr. Gevins has received continuous federal support from the Air Force, the Navy, DARPA, NASA, NSF, and seven NIH institutes. These awards have been used to maintain a core staff working on the central project of the company. To fund such a complex and long-term project, Dr. Gevins systematically divided it into essential individual subprojects, and sought funding for them through unsolicited federal basic research and SBIR awards. This minimized overall risk, and SAM's work has been supported by many SBIR awards from many agencies, as the project covers many possible applications of the technology. For example, SAM has received significant support from support from the National Institute on Ageing, because SAM's assessment and analysis technology could have a very large impact on seniors facing performance deficits of many kinds.

SAM has developed both the hardware that measures brain signals and transmits that signal to a processing device (now a PC), as well as the software used to integrate different kinds of brain stimulation signals. One early SBIR was designed to build the meters necessary to capture the EEG signals SAM intended to work with, as these meters were not then available elsewhere.

In 2005-2006, SAM completed the first commercial product in the MM line, the world's first medical test that directly measures brain signals regulating attention and memory, the SAM Test (Sustained Attention & Memory Test). The SAM Test is covered by four U.S. patents and by a number of trade secrets. The test is designed to fill an urgent need for an objective measure of how a patient's cognitive brain functioning is affected by a disease, injury, or treatment in a wide range of areas including head injuries, sleep disorders, mild cognitive impairment of aging, attention deficit hyperactivity disorder, epilepsy, and depression.

The next proposed product, the Online Mental Meter, is designed for widespread use beyond medical care as a computer peripheral that provides continuous information about the user's state of alertness and mental overload or under load, by measuring mental activity in real time while people perform everyday tasks at a computer. The Online MM constitutes a substantial technical leap from the SAM Test, which requires that a subject perform a standardized repetitive psychometric test of sustained focused attention and memory. SBIR projects 17 and 18 (see Table App-C-7) have paved the way for this advance.

Continuous real-time measurement of mental effort could become a key enabling technology for a wide variety of advanced adaptive systems that will vary the sharing of tasks between a human and a computer in an optimal manner depending on the user's cognitive state. SAM believes that such systems may well be ubiquitous in the future.

SAM aims to become the gold standard for medical testing in neurology. Currently, the brain measurement component of psychological testing requires a PET or MRI scene, which is inconvenient and very costly ($4,000 or more each). In addition, existing performance-based tests can be misleading, as they fail to measure brain activity directly. For example, early Alzheimer's patients often produce acceptable memory and brain performance, because these patients are able to compensate for their initial problems. Direct brain measurement would reveal what performance analysis obscures—the actual problem at the neuron level.

Federal funding has allowed SAM to reject overtures from venture capital companies. According to Dr. Gevins, venture companies have "a different agenda, timescale, and process." In contrast, SBIR supports a transition from basic research to the next step." Dr. Gevins observed that venture capital companies in general have declining interest in truly innovative work, because such work often takes too long to get to market for venture capital timescales.

SAM is currently working with consultants under the LARTA commercialization support program to develop a strategic alliance with a large corporation in order to make the SAM Test commercially available as a fee-for-service medical test. The partner will need to undertake independent clinical trials, FDA registration, approval for third-party reimbursement and a major marketing and sales campaign, activities that could take at least 3 years and cost in excess of $15 million.

Staff

SAM has 13 scientists, engineers, and associates, and several outside consultants, covering a range of disciplines. The eight most senior staff members have been with SAM an average of 11 years. Collaborations with scientists and doctors at universities, medical schools, and government labs are used to leverage internal research efforts, and SAM has made distribution agreements with medical device companies which account for most product sales. SAM is an FDA registered Medical Device Manufacturer.

OUTCOMES

Commercial Products

Six of the SBIR-funded projects (#3, 7, 9-12) have to date resulted in two commercial products.

Image Vue™

Image Vue™ is a software package for visualizing brain function and structure by fusing EEG data with MRI (Magnetic Resonance Images), using patented algorithms to integrate functional and structural information about the brain, in order to localize epileptic seizures in a patient's brain. A wizard-driven software system running under Windows XP, it co-registers EEGs with MRIs, performing patented DEBLURRING™ spatial enhancement and several types of source localization analysis, and provides interactive 3-D graphics visualization. The patented XCALIPER™ hardware and associated software facilitates rapid measurement of EEG electrode positions needed for co-registration with MRIs.

The product is used primarily to visualize and localize the origin and spread of epileptic seizures in the human brain in planning neurosurgical treatment of complex partial seizure disorders that are refractory to treatment with antiepileptic drugs.

Image Vue™ is FDA-registered and is sold by Nicolet Biomedical, Inc. (a subsidiary of Viasys Healthcare, Inc), the world's largest supplier to the clinical neurology market. Nicolet has purchased approximately 100 systems from SAM to date, from which they have generated about $2,000,000 in revenues. A number of competing products worldwide have been modeled on Image Vue™

MANSCAN®

MANSCAN® evolved from basic research completed under prior NIH R01s, which with the aid of SBIR awards has been turned into robust algorithms embodied in a convenient, integrated system to enable research on human brain function that would not otherwise be commercially available.

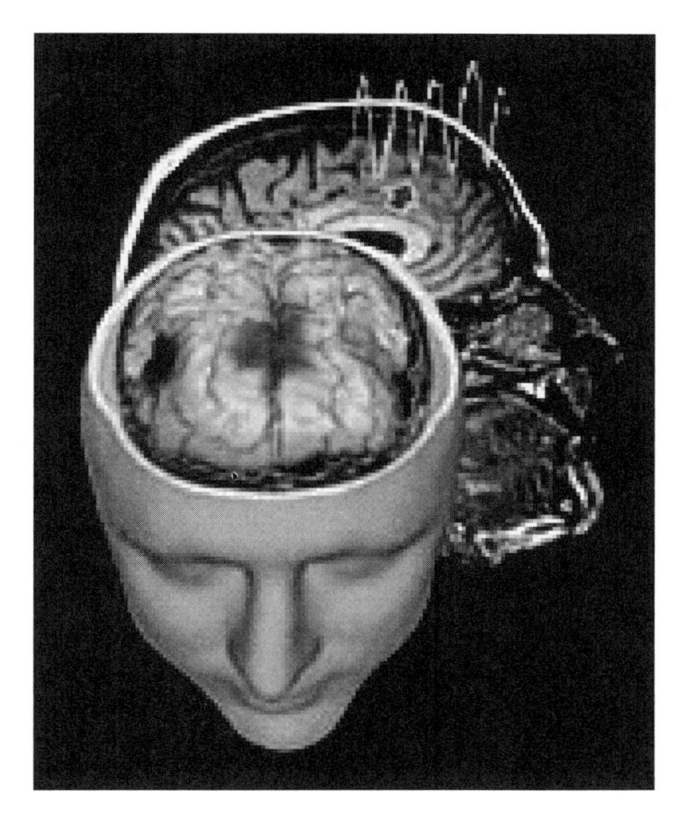

FIGURE App-C-1 MANSCAN®.
SOURCE: SAM Technologies.

MANSCAN® is an integrated software and hardware system for performing brain function research via high-resolution EEG and event-related potential (ERP) studies, and for integrating the results with magnetic resonance images. MANSCAN® It was the first system to integrate the high time resolution of EEG with the high anatomical resolution of MRI, and the first to allow subsecond measurement of rapidly shifting functional cortical networks. It enabled a new generation of research, and a number of significant advances in understanding attention, memory and other basic cognitive brain functions have been made with it.

Results of these studies provide unique views of structural and functional neuroanatomy. MANSCAN® analysis and visualization functions quickly and easily quantify features from EEGs and ERPs, leading neuroscience toward the goal of uniting brain electrical activity with brain anatomy.

MANSCAN®'s hardware includes quick application electrode caps, an efficient device called XCALIPER™ for measuring electrode positions, and an advanced digital amplifier called MICROAMPS™. MANSCAN® software is fully integrated with the Microsoft NT/W2000/XP operating system running on a PC.

Thirty MANSCAN® systems have been sold to qualified scientists at U.S. universities, medical schools and government labs, where it is helping to perform advanced research. MANSCAN® has generated approximately $650,000 of revenue. Several competing products worldwide have been modeled on MANSCAN®. MANSCAN® is also specifically designed as a step toward the MM.

Knowledge Effects

SAM aims to produce commercial products—clearly, its entire mission is focused on commercial outcomes in the long run. However, there have been significant knowledge effect benefits during the course of this high-risk research. Nine of the projects listed below led to over 50 peer-reviewed scientific and engineering publications. Thirteen of the projects led to 18 U.S. patents. More widely, SAM staff have published more than 150 peer-reviewed publications including five papers in *Science*.

SBIR Issues and Concerns

The Selection Process

In recent years, SAM has been criticized as being "insufficiently innovative," possibly because the field is catching up with SAM. Dr. Gevins notes that the recent drive for closer attention to commercialization is impacting SBIR reviews at NIH, but also that reviewers from academia have a better understanding of more basic research and are likely to be somewhat biased toward it. He also notes that academic reviewers are not themselves unbiased, in that they tend to focus on whether outputs from the project in question will be useful in their own research. Review quality and outcomes also vary very substantially by study section.

Conflict of Interest

Dr. Gevins is very concerned about potentially *major* conflicts of interest stemming from the use of industry participants on study sections. He believes procedures for addressing such conflicts are "pathetic": section members are handed a written conflict of interest description immediately before the panel meets, and are then on the honor system to disqualify themselves. No NDA is signed, and little attention is paid to the process. Dr. Gevins believes that the current approach is just designed to protect NIH from awkward questions, rather than to provide real protections to applicants.

SAM always reviews membership of review panels, and not infrequently requests that the SRA exclude a panel member from reviewing a SAM proposal. This veto process mostly works, according to Dr. Gevins, but is not foolproof. There is clearly some element of risk involved in releasing internal plans to outsiders (Dr. Gevins pointed out that this risk is endemic to the funding process—and that venture capitalists never sign NDAs, so there are also risks involved in working with venture capitalists).

In short, Dr. Gevins argues that while formal protections are in place, no effort is made at NIH to define or verify the absence of conflicts of interest, even though SRAs are in general honest and conscientious.

Recommendation: NIH must take the conflict of interest problem much more seriously. It should implement its own conflict of interest policies more effectively, and should consider mechanisms for auditing reviewer activities, at least on a random basis.

Expertise

Dr. Gevins sees a disconnect between the SRA running the selection process, and the IC which will eventually fund the project, and which has technical expertise in the subject area. This contrasts with funding at DoD, NSF, where a single point of contact essentially determines funding and manages the award.

Recommendation: It should be mandatory that the primary reviewer have technical competence in the field covered by the proposal.

Commercial Review

Dr. Gevins sees substantial room for improvement in addressing commercializing concerns. He does not support the commercialization index in use at DoD, which he regards as highly oversimplified and biased toward short-cycle projects. He did agree that a commercialization review could be useful, but though there might be helpful ways to separate out technical/scientific review from commercialization, which could be addressed by a separate perhaps permanent panel of experts, and where problems could be addressed within a single funding cycle rather than requiring full resubmission, which means at least one and possibly two funding cycles delay.

Reviewer Evaluation

CSR manages this function and does so fairly effectively. Dr. Gevins believes that the key motivation for reviewer participation is to follow activity near the cutting edge in a particular field.

Overly Random Scoring

Like many interviewees, Dr. Gevins noted the substantial random element in the review process. In particular, he believed that that there are quite substantial differences in scoring tendency between different review panels.

Recommendation: Scores should be normalized across study sections, just as they are for RO1's. Otherwise it is perfectly possible—indeed likely—that one study section will tend to systematically provide higher scores than another. As all scores are integrated into a single priority score list for a given IC, this would inevitably generate a bias toward projects that were reviewed by the higher scoring study section.

Funding Issues

Size of Awards

Dr. Gevins said that in his experience, over-limit applications were always discussed beforehand with the program manager. SAM makes a point of mentioning this in the application, to ensure that reviewers know that the relevant program manager is in the loop. Extra-large awards can sometimes be held up for one or more funding cycles by the program manager, even if they are technically inside the payline. This is a less formal procedure, but similar to that in place at NIH for RO1 awards.

Recommendation: SAM supports increasing the size of Phase I awards and reducing the number of awards. Some program announcements already call for Phase Is in the vicinity of $500,000. SAM also supports increasing the size of Phase II awards and reducing the number. SAM would recommend a three-year Phase II award for $1 million, possibly requiring prior approval from the program manager.

Commercialization Support

SAM has actively participated in the new LARTA-led program. It sees the program as useful, particularly because it forces companies to focus on commercialization. However, at the time of the interview, SAM had received almost no useful time from the consultants, who appeared to have too many companies in their portfolios (20 or more each).

In general, the basic outline of the LARTA support program met SAM's needs, which focused not on preparing for public presentations, but on moving steadily through the steps of developing a good commercialization plan, focused on strategic alliances. SAM had received a steady flow of reminder/check-up calls, pushing the company to focus on the commercial element of the business.

Recommendation: NIH should consider allowing LARTA to focus its resources more tightly on fewer companies, providing them with more resources.

Program Managers

Dr. Gevins wondered what role SBIR program managers or liaisons play at the various institutes. As they did not appear to manage the financial and reporting aspects of individual grants, and did not substantially influence selection, he was unclear as to their role? He saw program managers as providing no value added for recipients, and added that in some cases they could be highly destructive (a point also made by other interviewees). He believed that some managers clearly had their own research agendas, and sought to impose these on the SBIR application process. Dr. Gevins also noted that being a program manager with responsibilities for many SBIR awards was not a plum job at NIH; as a result, it was often handed off to the least senior staff member.

Recommendation: SAM suggested adding a program manager review section to the final report for each project, which would allow NIH to gather better feedback about program manager performance.

Funding Gap Issues

SAM handles the gap by operating with multiple overlapping project. Current work has taken 6 years on a specific stage of the overall brain measurement project.

Dr. Gevins also noted that IC's do not always fund projects immediately—the latter can be delayed for one or more funding cycle. SAM believes that April applications are likely to be funded fastest, because they show up at the beginning of the fiscal year and require least juggling from the IC. For example, SAM had a project that was approved during a September Council meeting, but had still not been funded by the IC as of the following March.

Other Concerns and Recommendations

SAM offered a range of other concerns and suggestions:

- Believes awards are too short. Six months for Phase I is "a joke," as Phase I research always takes about a year. SAM has never completed a two-year Phase II award in the standard two years.
- Does not support FastTrack, partly because reviewers tend to split FastTrack applications in two anyway, and also because there are too few advantages in this process for companies, in comparison to the additional uncertainty.
- Supports direct access to Phase II, without prior Phase I. SAM believes that the time required to apply for and complete a Phase I is the problem, as this can add a year or more to a project.
- Supports the view that drug development funding could be distorting the overall shape of the SBIR program.
- Supports competing continuation awards.

TABLE App-C-7 SBIR Awards to SAM Technologies

	Awarding Agency	Year	Award Number	Technology Description
1	NIH	1988	R44-RR03553	Removal of Distortion from Magnetic Resonance Images
2	NIMH	1989	R44-MH42725	Active Electrode Hat for EEG Imaging of Schizophrenics
3	AFOSR	1989	F49620-89-C-0049	Software Tools for Signal Identification using Neural Networks
4	AFRL	1989	F33615-89-C-0605	Flight Helmet EEG System
5	NIMH	1991	R44-MH43075	EEG ArtifactDetection
6	AFOSR	1991	F49620-92-C-0013	Physiological Indices of Mental Workload
7	NINDS	1992	R44-NS27392	Neurofunctional Research Workstation
8	NIAAA	1993	R44-AA08680	Cognitive Performance Assessment for Alcohol Intoxication
9	NIMH	1993	N44-MH30023	128 Channel Automated EEG Recording System
10	NINDS	1994	R44-NS28623	Multimodality Workstation for Seizure Localization
11	NINDS	1995	R44-NS32241	Functional Brain Imaging of Unconstrained Subjects
12	NASA	1995	NAS9-19333	Spacecrew Testing and Recording System
13	AFRL	1995	F41624-95-C-6000	Decontamination of Physiological Signals of Mental Effort
14	NIMH	1996	N44-MH60027	Neurocognitive Experiment Authoring Tool
15	NIAAA	1997	R44-AA11702	Attention & Alertness Neurometer
16	NINDS	1998	N44-NS-0-2394	Assessment of Alertness in Patients with Sleep Disorders
17	AFOSR	1998	F49620-98-C-0049	Sustained Attention Meter for Monitoring Cognitive Load
18	AFOSR	1998	F49620-96-C-0021	Brain Automatization Monitor
19	AFOSR	1998	F41624-98-C-6007	Operator State Classifier Developer's Toolkit
20	AFRL	1998	F41624-97-C-6030	Rapid Application Cutaneous Electrode (RACE) System
21	AFRL	1999	F41624-99-C-6007	An Ambulatory Neurophysiological Monitoring System
22	NIMH	2001	R44-MH60053	NeuroCognitive Assessment Meter For Psychiatric Drugs
23	NIDA	2001	R44-DA12840	Neurocognitive Index of Cannabis Effects
24	NIA	2002	R44-AG17397	Neurocognitive Assessment of the Elderly
25	NICHD	2002	R44-HD37728	
26	NINDS	2003	R44-NS42992	
27	DARPA	2003	DAAH01-03-C-R292	Multicompartment Neuroworkload Monitor
28	NHLBI	2004	R44-HL065265	
29	ONR	2004	N00014-04-C-0431	Multitasking Personnel Selection Test

SOURCE: SAM Technologies.

Savi Technology

Irwin Feller
American Association for the Advancement of Science

FIRM HISTORY

Savi Technology was established in 1989 by Robert Reis, a Stanford University engineering graduate and serial entrepreneur, about the core technological concept of installing radio frequency emitters, or tags, in products as a means of identifying their location. Based on an experience in which Reis had difficulty locating his young son in a store, the original market concept was to install the technology in children's shoes as a way for parents to monitor their whereabouts. This concept quickly proved technically and commercially unworkable.

The value of integrating radio frequency identification devices (RIFD) with the Internet for purposes of supply chain management soon became evident, and it is along these lines that Savi has developed, becoming an international leader. From the late 1980s through the early 1990s, the firm experienced modest growth. Its growth has increased rapidly since then, especially after adoption of its technology by the Army, where it is credited with greatly improving the efficiency and effectiveness of DoD's logistic management capabilities.

By the mid-1990s, Savi made a major business decision to systematically focus on the defense market. This strategy has led to a sequential extension of its

SAVI TECHNOLOGY: COMPANY FACTS AT A GLANCE

- **Address:** 615 Tasman Drive
  Sunnyvale, California
- **Phone:** 408-743-8000
- **Fax:** 408-543-8650
- **Revenues:** $89 million (2004)
- **Employment:** 250
- **Number of SBIR Awards:** 4
  Phase I: 3
  Phase II: 1
- **Number of Patents:** 13; 20 pending
- **Selected Awards:** 1994 National Small Business Innovation Research Company of the Year
  1996 Tibbetts Award

technology to a widening set of DoD requirements, international defense customers, homeland security, and asset security management.

Savi is now the major supplier of RFID technology to the Department of Defense, and one of the key technologies provides to its Global Total Asset Visibility Network. It has developed a strong international presence, being a major supplier of RIFD and related technologies to the United Kingdom's Ministry of Defense, NATO, NATO member nations, and Australia. It also is increasingly engaged in the development of globally interoperable logistics monitoring systems with major international ports. It has also built a steadily increasingly commercial business, especially among multinational firms.

If its technological path has been relatively straightforward, consisting of a continuing stream of improvements and widened applications of its radio frequent identification technology, Savi's history as a firm has been circuitous. As a relatively small firm, with about 40 employees and $10 million in sales but with limited markets, Savi was sold in 1990 to Texas Instruments (TI) for $40 million, which at that time was following a diversification strategy. Soon after however, following the death of TI's chief executive office, Savi's place within TI became unclear. In 1997, TI sold Savi to Raytheon, which was in the process of acquiring several firms as part of a diversification strategy. Raytheon's business strategy soon gave way to one of concentrating on core businesses, with Savi at the margins of Raytheon's operations. In 1999, Savi's management entered into a buyout agreement, purchasing the firm from Raytheon for $10 million.

CAPITAL FORMATION

Begun as a start-up operation, augmented with an infusion of angel capital in 1992, Savi remains a privately held firm, albeit with several rounds of venture capital since its management buyout. Being sold twice and then regaining its autonomy via a management buyout, although at one level detracting from Savi's ability to articulate and operate a focused technology development and business strategy, has over time proven beneficial to the firm. TI and Raytheon are estimated to have invested $50 million in Savi's R&D. As a consequence, when Savi regained its independence, it was on stronger technological and production basis that when it was first sold. As described by Vikram Verman, Savi's CEO, the firm's history thus resembles the story of Jonah, albeit with a positive outcome: As a fledging firm, it was swallowed up by a whale—actually two of them, nurtured inside their bellies, and then disgorged as a stronger unit, better able to fend for itself.

Savi has had four rounds of venture capital funding since 1999, raising a total of $150 million. Among these investors are Accel, UPS Strategic Investment Fund, Mohr Davidow, Temasek, an investment holding company for the Port of Singapore and Neptune Orient Lines, Hutchison Whampoa and Mitusi, among others.

EXPERIENCES UNDER SBIR

The SBIR program provided two key inputs into Savi's long-term growth. First, a combination of DARPA and Navy SBIR awards to the firm in 1989 and 1990 provided it with the seed capital that enabled it to refine the initial technological concept of radio frequency identification tags, such that it performed as needed when its initial market opportunity surfaced during the First Gulf War. Second, it provided the funding that led to the employment of Vikram Verman, then a Stanford University PhD student in engineering. Verma was born in India, moving to the United States at age 18 to study electrical engineering. Upon completing his undergraduate degree at Florida Institute of Technology, he moved first to the University of Michigan and then to Stanford University for graduate work. He joined Savi in 1990, advancing steadily from staff engineer, to vice president for engineering, to his current position as CEO.

In total, Savi received 38 SBIR awards, with the last of these awards being received in 1992. Savi credits its subsequent technological and business success to these early SBIR awards, even though they did not commercialize the technology identified in the initial SBIR projects. Rather, in its formative period, essentially between 1989-1991, SBIR awards were critical in helping Savi build the organizational infrastructure, engineering teams and knowledge base that undergirded its subsequent growth. Furthermore, although successful in competing for SBIR R&D awards, Savi never saw itself as a government contract R&D firm. Rather, its founding and continuing objective has been to be a commercially oriented, product-based firm. R&D is a means to obtain a competitive advantage for its products and related services.

Savi sees its greatest asset to be the know-how and organizational infrastructure gained from multiple R&D projects. These internal, often intangible, assets permit it to deal from a position of strength when it negotiates with external investors, such as venture capital funds. Verma estimates that Savi received a total of $3 million in SBIR awards for the development of RFID. For its part, the firm invested about $150 million in development, and that development took approximately 10 years.

INTELLECTUAL PROPERTY

Intellectual property protection in the form of patents is seen as of modest importance to Savi. It files patent, primarily to defend it technology and market position, not as a means though of securing license revenues or of entering into cross-licensing agreement. As viewed by the firm, its know-how and organizational capacity to assemble high-performing engineering teams are the main sources of its continuing technological innovativeness.

INVOLVEMENT IN STATE GOVERNMENT PROGRAMS

Savi has not participated in any State of California high-tech or economic development programs.

RECOMMENDATIONS

Savi's participation in the SBIR program ended by the mid-1990s. Accordingly, its assessments of the SBIR program and recommendations for its improvement relate more to the general place of the program within the U.S. national innovation system than to its specific programmatic details. In its view, the SBIR program accords with the federal government's role of financing R&D before a technology is commercially viable, and before a fledging firm can attract external capital. SBIR should be viewed and used as a source of seed capital. It is a means to an end, the end being the development and production of a usable and competitively marketable product. Firms that make a habit of living off the SBIR program are misusing the program. To guard against this practice, selection criteria should include requirements that firms detail how they plan to commercialize a product. Also, the importance and contributions of the SBIR program need to be more fully and effectively communicated.

SUMMARY

The SBIR program served both as a source of seed capital for Savi's early R&D on RFID devices that have been the base of its employment and revenue growth and as the foundation for the knowledge and organizational infrastructure that have made it an internationally prominent firm in supply chain management technology. Although it no longer participates in the program, Savi views SBIR as an essential element—a national treasure in America's long-term capacity to compete internationally on the basis of technological innovation.

Sociometrics Corporation

Robin Gaster
North Atlantic Research

EXECUTIVE SUMMARY

Sociometrics is a woman- and minority-owned company of approximately 16 employees that develops commercial products and services based on state-of-the-art behavioral and social research. Located in California's Bay Area R&D hub, it has received funding from diverse government and private sources, generating revenues of more than $28 million since its foundation in 1983. By introducing the concept of packaged replication programs to the practice of social behavior change, Sociometrics has changed the nature of the field.

Sociometrics illustrates well one type of successful SBIR company. Winner of an NIH SBIR award during the first year of the program in 1984, Sociometrics has gone on to continue winning awards with consistency. Between 1992 and 2002, the company received 19 Phase I awards from NIH (see annex to this case study); 18 of these have become Phase II projects, a very high 94 percent conversion rate. Since 2003, six new Phase I projects have been awarded; two additional Phase I applications have received priority scores at the fundable level. Sociometrics will be submitting eight Phase II applications from these projects at the appropriate time.

Consistent with the goals of the SBIR program, Sociometrics has become a product-oriented company. Every Phase II it has conducted has generated a highly-marketable, commercial product. Sociometrics has now developed several product lines, secured a distribution agreement with a major electronic publishing house, and had products chosen by the Centers for Disease Control and Prevention (CDC) for its public health initiatives. In 2002, the majority of the firm's profits shifted from contract and grant research fees to product sales, a testament to the firm's SBIR-related success. This success has been due in part to the cumulative and strategic nature of Sociometrics' efforts. New projects and products build on previous ones, adding one or more innovations in the process. The company has also leveraged the ubiquity of the World Wide Web to increase its products' reach and public access, designing most products for download or interactive use on the Internet.

Despite Sociometrics' own success, its market niche—the development of behavioral and social science-based commercial products—remains under-resourced with private funds. In part, this situation obtains because typical customers for such products are nonprofit organizations that appreciate and use the resources but cannot afford to pay very much for them. Responding to this important need, Sociometrics has kept its product pricing at close to production cost, leveraging instead good business practice with a sense of public service. For these reasons the company continues to rely on SBIR funding to provide the necessary development support to create new products and expand its range of services. While SBIR grants remain an important revenue stream for many small companies in the program, Sociometrics' involvement has distinguished itself with: a) a wide set of well-regarded and widely-used products; b) a profitable business model, distinguishing it from many other behavioral and social science-focused SBIR firms; and c) Phase II funding as sufficient support to bring the company's products to market, in contrast to many biotech- and pharmaceutical-oriented companies.

Motivated once again by both business and public service concerns, Sociometrics staff members have published extensively in peer-reviewed journals. The company does not develop patentable products.

PRIMARY OUTCOMES

- Four major product lines, with a fifth under development;
- Consistent profitability from inception;
- Industry standard for topically focused social science data and program archives;
- Distribution agreement for data products with world-leading provider of authoritative reference information solutions;
- CDC adoption of program archive products for nationwide distribution;
- Product line with social impact: Effective program replication kits have changed the way behavioral practitioners operate in community settings; and
- More than 60 peer-reviewed publications based on SBIR-funded projects

KEY SBIR ISSUES

Sociometrics has found the SBIR program a very productive platform for its work. The program has allowed Sociometrics to: a) take state-of-the-art social and behavioral research in its topical areas of expertise (reproductive health, HIV/AIDS, drug abuse, and mental health); b) use scientist expert panels to assess the research and identify the best available data, practices, and knowledge; and c) develop commercial products and services based on the panels' selections. Sociometrics' products are aimed at a diverse set of target audiences. For example, its data archives and evaluation instruments are meant for use by researchers, faculty members, and students. Its effective program replication kits, evaluation publications, and program development and evaluation training workshops are intended for use by health practitioners in schools, clinics and community-based organizations. Its forthcoming Web-based behavioral and social science information resources, summarizing state-of-the-art research knowledge in select topical areas in nonscientist language, will be aimed at both academic and practitioner audiences.

BACKGROUND

Sociometrics Corporation was established in September 1983 as a corporation in the State of California. Within a year of its founding, Sociometrics had applied for its first SBIR project. Dr. Josefina Card, Sociometrics founder and CEO, was encouraged to found the company as a for-profit organization by Dr. Wendy Baldwin, then a program manager at NIH. (Dr. Baldwin later went on to become Deputy Director of NIH for Extramural Research.) Dr. Baldwin suggested that Sociometrics incorporate as a for-profit entity in order to benefit from the newly created SBIR program without precluding the possibility of obtaining basic and applied research grants.

Sociometrics' goals are:

- to conduct applied behavioral and social research to further our understanding of contemporary health and social problems;
- to promote evidence-based policymaking and intervention program development;
- to conduct evaluation research to assess the effectiveness of health-related prevention and treatment programs;
- to facilitate data sharing among social scientists as well as public access to exemplary behavioral and social data; and
- to help nonexperts utilize and benefit from social science and related technologies and tools.

In carrying out its mission, several areas of corporate expertise have been developed:

- the design and operation of machine-readable, topically focused data archives;
- the development of powerful, yet user-friendly, software for search and retrieval of information in health and social science databases;
- the harnessing of state-of-the-art developments in computer hardware and software to facilitate access to, and use of, the best data in a given research area;
- primary and secondary analysis of computer data bases using a variety of commercially available statistical packages as well as custom-designed software;
- the design, execution, and analysis of program evaluations;
- the design, execution, and analysis of health and social surveys;

- the collection and analysis of social and psychological data using a variety of modes (mail; telephone; focus groups; in person interviews);
- the collection and dissemination of social intervention programs with demonstrated promise of effectiveness; and
- the provision of training and technical assistance on all the above topics.

Sociometrics currently has 16 employees, six of whom have PhDs and five of whom have Masters degrees. Expertise of its staff spans a diverse set of behavioral and social science fields, including sociology, social psychology, clinical psychology, demography, linguistics, education, and public administration. In June 2005, a seventh PhD, specializing in political science and international relations, will be joining the firm.

Sociometrics has been the recipient of several awards including:

- a Medsite award for "quality and useful health-related information on the Internet";
- a U.S. Small Business Administration Administrator's Award for Excellence "in recognition of outstanding contribution and service to the nation by a small business in satisfying the needs of the Federal procurement system"; and
- a certificate of recognition for Project HOT (Housing Options for Teachers) by the California State Senate, the California State Assembly, and the Palo Alto Council of PTAs "in appreciation for service supporting Palo Alto Unified School District's teachers and staff."

PRODUCTS

Sociometrics currently provides four research-based product lines:

- data archives and analysis tools;
- replication kits for effective social and behavioral intervention programs;
- evaluation research; and
- training and technical assistance services.

Sociometrics staff members are currently developing a fifth product line, online behavioral and social science-based information resources, to facilitate "distance learning."

Data Archives and Analysis Tools

The Sociometrics data archives are collections of primary research data. Each collection is focused on a topic of central interest to an NIH Institute or Center (IC). The data sets comprising each collection are selected and vetted by high-level scientist advisory boards to ensure that each collection is best-of-breed. Data sets are acquired from their holders, packaged and documented in standard fashion, and then made publicly available both online and on CD-ROM. Each successive data archive has leveraged features of previous archives that promote ease of use and has then developed new features of its own. This cumulative development effort has resulted in a digital library consisting of several hundred topically-focused datasets that are easy to use, even by novices such as students and early-career researchers. Data in Sociometrics' archives are accompanied by standard documentation, SPSS and SAS analytic program statements, and the company's proprietary search and retrieval tools. By providing high-quality data resources, and adding features facilitating appropriate and easy use, Sociometrics has created a niche of standardized, quality data products that complement the larger, though not universally standardized, data resources offered by other major data providers such as the University of Michigan. Currently, Sociometrics publishes nine data archives (see Box App-1), with a tenth data archive on childhood problem behaviors under development. The nine collections are disseminated both as single data sets (by Sociometrics) as well as via institutional

BOX App-1
The Sociometrics Data Archives: Topical Foci and Scope

- AIDS/STD. Nineteen studies comprising 30 data sets with over 18,000 variables
- ADOLESCENT PREGNANCY & PREGNANCY PREVENTION. One hundred fifty-six studies comprising 260 data sets with over 60,000 variables
- AGING. Three studies comprising 22 data sets with over 19,000 variables
- AMERICAN FAMILY. Twenty studies comprising 122 data sets with over 70,000 variables
- CHILD WELL-BEING and POVERTY Eleven studies comprising 35 data sets with over 20,000 variables
- COMPLEMENTARY AND ALTERNATIVE MEDICINE. Eight studies comprising 17 data sets with over 10,000 variables
- CONTEXTUAL DATA ARCHIVE. Thirteen data sets compiled from over 29 sources with over 20,000 variables
- DISABILITY. Nineteen studies comprising 40 data sets with over 23,000 variables
- MATERNAL DRUG ABUSE. Seven studies comprising 13 data sets with over 5,000 variables

subscriptions to the entire collection known as the *Social Science Electronic Data Library* (SSEDL).

Sociometrics' first data archive on adolescent pregnancy and pregnancy prevention was funded as part of the very first cohort of SBIR awards at NIH. More than 20 years after its initial release, this archive continues to be highly relevant, utilized, and regularly updated by Sociometrics. The archive was originally published on mainframe tapes, and has since been delivered to its customers using ever-changing computer data storage technologies including 5 ¼ and 3 ½ inch floppy disks and CD-ROM. It is now available 24/7 on the World Wide Web, where it has been accessible for the past 8 years.

A year ago, Sociometrics entered into a five-year distribution agreement for the *Social Science Electronic Data Library* (SSEDL) with Thomson Gale, a world-leading provider of authoritative reference information solutions. The agreement calls for Thomson Gale to market SSEDL via subscription to its wide range of academic and research library customers, while allowing Sociometrics to continue selling individual datasets from its own Web site. Sociometrics receives a portion of the Thomson Gale subscription sales in royalties. Several thousand SSEDL data sets have been downloaded from the Sociometrics Web site over the last three years, some by pay-as-you-go customers who execute a secure credit card transaction, others by faculty members and students able to download the data sets at no charge because their university is an SSEDL subscriber. Sales of the data archives and associated products have yielded approximately $125,000 in profits over the last three years. This figure does not yet include royalties from the Thomson Gale agreement which came into effect at the close of the 2003-2004 fiscal year, the latest date for which figures are available.

Competition

The Inter-University Consortium for Political and Social Research (ICPSR) data archive at the University of Michigan provides the main competition for Sociometrics' data archives. Despite ICPSR's much larger collection of data sets, Sociometrics has maintained a specialized niche, leveraging organization around selected health-related topics, careful selection of exemplary data by Scientist Expert Panels standardized documentation, ease of use of data sets, and value-added SPSS and SAS data analytic statements into a specialized collection tailored to data novices such as students and early-career researchers, that complements the ICPSR collection.

Replication Kits

Having developed the Data Archives, Sociometrics realized in 1992 that additional public service would occur if it extended and adapted its work beyond selection, packaging, and distribution of exemplary data to selection, packaging, and distribution of practices shown by such data to be effective in changing unhealthy or problem behaviors. Diverse health issues with important behavioral determinants (adolescent pregnancy, STD/HIV/AIDS, and substance abuse) were selected to showcase the new product line. Prior to 1992, information on effective programs was limited to brief descriptions in scientific journals often not read by health practitioners, a serious barrier to their widespread use. Using SBIR funding, Sociometrics sought to overcome this barrier by adapting to this new product line the time-tested methods it used to establish its data archives.

The company again worked with Scientist Expert Panels to identify and select effective programs based on their empirical support, collaborated with developers of selected programs to create replication kits for Panel-selected behavior change interventions, and partnered with networks of health professionals to disseminate these kits to schools, clinics, and community-based organizations. Replication kits were conceptualized as boxes containing all the materials required to reimplement the effective intervention. Typical replication kits contain a user's guide to the program, a teacher's or facilitator's manual, a student or participant workbook, one or more videos, and forms for "homework" assignments or group exercises.

The first effective program collection, the Program Archive on Sexuality, Health & Adolescence (PASHA) now comprises 29 replication kits. The newer HIV/AIDS Prevention Program Archive (HAPPA) and the Youth Substance Abuse Prevention Program Archive (YSAPPA) encompass 11 and 12 replication kits, respectively. Despite considerable initial skepticism from academics and some practitioners, the program-in-a-box approach has been received with considerable enthusiasm. PASHA, HAPPA, and YSAPPA have all proven to be social successes, with their programs being implemented in hundreds of schools, clinics, and communities across the country. They have also proven to be commercial successes, generating profits totaling over half a million dollars in the last three years.

Like the Sociometrics Data Archives, the Sociometrics replication kits (known collectively as the Sociometrics Program Archives) are topically focused, best-of-class collections, selected using clearly defined effectiveness criteria by Scientist Expert Panels, and sold with free technical support for purchasers. This complimentary technical assistance has been lauded as an extremely valuable service by Sociometrics' customers and the company's reputation follows, in part, from the excellent product support it provides.

The replication kits have been sold individually from the Sociometrics Web site to such customers as schools, community health and service organizations, and medical clinics. They have also been displayed at exhibit booths at annual meetings of health practitioner professional organizations. Their dissemination is further supported by a company newsletter published three times annually and reaching 30,000 recipients. In 2003 CDC became an important customer for several replication kits, providing "train the trainer" workshops in Atlanta for hundreds of practitioners in use of selected kits. These new trainers have in turn returned to their hometowns and home organizations to train more staff, resulting in further sales of the replication kits and dissemination of important prevention programs.

Competition

CDC has provided the impetus for the only real competition to Sociometrics in the field of replication kit development for effective teen pregnancy and HIV/AIDS prevention programs. The CDC initiatives have been based on a decentralized distribution model, with CDC funding program developers to publish their programs themselves or to seek out their own commercial distributors. In contrast, the Sociometrics distribution model is centralized, with Sociometrics' Web site serving as a one-stop-shopping-point for highly effective programs in the areas in which the company operates.

The inevitable delays in implementing a new initiative, plus changes in policy at CDC and substantial budget cuts, have put one of the CDC's two development programs on hold (the one on teen pregnancy prevention), leaving the Program Archive on Sexuality, Health, and Adolescence without significant current competition. The other CDC program on HIV/AIDS prevention, a competitor to Sociometrics' HIV/AIDS Prevention Program Archive, is using replication kits developed at Sociometrics for some of its selected programs, boosting sales of the Sociometrics HIV/AIDS Prevention Program Archive by an order of magnitude. In this manner Sociometrics Program Archives have complemented the larger CDC efforts, just as the Sociometrics Data Archives have complemented the larger University of Michigan efforts.

Longer-Term Challenges and Opportunities

Over the longer term it is possible that commercial challenges may arise from changes in the academic world, where more and more developers of effective programs are deciding to publish their work themselves, releasing kits or parts of kits through their own Web sites or negotiating other arrangements with commercial publishers. Sociometrics is not overly concerned by these developments as it regards its work as complementary to, and supportive of, developers' efforts to get their effective programs in the public domain. Recent history is supportive. During the initial establishment of Sociometrics' HIV/AIDS Prevention Program Archive (HAPPA), the advisory panel recommended 18 effective programs for inclusion in the archive; of these one was withdrawn as "obsolete" by its original developer, seven developers had previously decided to use a commercial publisher, and ten were made available through HAPPA. Thus with the help of Sociometrics' efforts complementing existing efforts, replication kits for almost all effective programs are now publicly available to community-based organizations striving to prevent HIV. This constitutes an important public service in terms of: (1) packaging the most promising interventions to enhance their usability; (2) facilitating low-cost access to, and widespread awareness of, these interventions; (3) encouraging additional rigorous tests of the interventions' effectiveness in a variety of populations; and (4) demonstrating the value of, and providing a model for, the research-to-practice feedback loop.

Further opportunities for enhanced product dissemination arise from Sociometrics' collaboration with other organizations besides CDC. In particular, large nonprofit networks provide many opportunities for partnership. For example, on the teen pregnancy prevention program archive, Sociometrics has worked with the National Campaign to Prevent Teen Pregnancy, Advocates for Youth, and the National Organization for Adolescent Pregnancy, Prevention, and Parenting, Inc. These organizations have become bulk purchasers of replication kits. They have provided other marketing support as well; for example, Advocates for Youth placed a link to Sociometrics on its Web site, and marketed Sociometrics' kits to its constituency from there.

Evaluation Research

Sociometrics has considerable expertise in program evaluation research and technical assistance. Over the last 15 years, the company has conducted many studies and provided technical assistance to many nonprofit organizations to determine whether a particular social intervention program was able to meet its short-term goals and long-term objectives. While most of the company's work developing its data and program archives has been funded by the SBIR program, Sociometrics' evaluation work has been funded primarily by state governments (such as California, Minnesota, and Wisconsin), local governments (such as Santa Clara County), private sources, especially foundations seeking an evaluation of the efforts of their grantees (the Packard Foundation, the Mott Foundation, the Northwest Area Foundation, and the Kaiser Family Foundation), and nonprofits seeking an evaluation of the effectiveness of their work. Sociometrics publishes a number of books and resource materials on program evaluation (e.g., *Data Management: An Introductory Workbook for Teen Pregnancy Program Evaluators*). It also offers at low cost (15 cents per page) evaluation research instruments that have been used in national surveys or in successfully implemented and published evaluation efforts.

Training Services

Sociometrics conducts workshops and courses to familiarize practitioners with the tools and benefits of social science and related technologies. These courses have recently been put online to increase their reach while lowering access costs. Training is offered in a variety of social science areas, particularly in effective program selection, development, and implementation; program evaluation concepts, design, and execution; and data collection, management, and analysis.

Science-Based Information Modules

These new products, still under development, will integrate the research literature in a given topical area, describe what science says in language and format easily understood by nonscientists (eighth grade reading level), and disseminate the information online via the Internet for easy "distance learning" access by all.

Into the Future

Most of Sociometrics' products are available for 24/7 download (with payment by credit card) on its award-winning Web site at *<http://www.socio.com>*. Its data archives, collectively known as *The Social Science Electronic Data Library* (SSEDL), are also available via institutional subscriptions marketed to universities and research libraries by Sociometrics' dissemination partner Thomson Gale. Sociometrics will continue its development of its Web site as a major product platform. In 2003, this Web site received over 1.7 million hits resulting in 29,729 product downloads. The company will also continue to develop additional subscription products. For example, Sociometrics plans to bundle its HIV and teen pregnancy replication kits, evaluation resources, training courses, and information module products and disseminate these bundled products to academics and health practitioners via online subscriptions. Eventually, mental health resources will be added to the Data Library, HIV, and teen pregnancy subscription resources as a fourth subscription line. Two current and two forthcoming SBIR Phase I grants support expansion into this important topical focus of mental health.

Profits and Revenues

Sociometrics' gross annual revenues are approximately $2.3 million with approximately 22 percent of this amount being profit (Table App-C-8). Profits from product sales are now substantially larger than profits from SBIR project fees, a testament to the success of Sociometrics as an SBIR firm. However, the profit stream is still insufficient to replace SBIR as the primary funding engine for future development efforts. Sociometrics does not market price its products, as many of its customers are small, community-based nonprofits that cannot afford products fully priced to market. Rather Sociometrics' products are priced at the cost of production with a small profit mark-up equivalent to a technical assistance retainer. Sociometrics sees this focus on widespread dissemination

TABLE App-C-8 Sources of Revenue and Profit, Sociometrics Corporation, 1984 to 2004

		Breakdown of Gross Revenue Given in Column 2 ($)		Breakdown of Profit Given in Column 4 ($)		
Fiscal Year	Gross Revenue ($)	Project Base (Direct Costs +Overhead)	Profit	Project Fees	Product Sales	Other Income
84/85	295,073	279,802	15,271	12,539	977	1,754
85/86	346,502	322,799	23,703	14,873	4,958	3,871
86/87	628,932	593,309	35,622	24,529	5,303	5,789
87/88	597,225	553,748	43,476	17,696	11,739	14,039
88/89	484,871	412,906	71,965	42,308	20,467	9,189
89/90	764,349	632,519	131,829	108,271	12,682	10,875
90/91	728,580	654,255	74,325	45,536	14,409	14,379
91/92	1,225,910	1,087,726	138,184	103,862	9,603	24,717
92/93	1,216,676	1,105,338	111,338	69,787	17,644	23,906
93/94	1,174,597	1,014,256	160,340	106,888	38,053	15,398
94/95	1,328,429	1,152,854	175,575	119,909	33,655	22,010
95/96	1,632,612	1,435,536	197,075	112,533	49,244	35,298
96/97	2,012,046	1,648,160	363,886	264,688	49,063	50,134
97/98	1,929,552	1,646,048	283,503	181,654	64,470	37,378
98/99	2,018,483	1,756,005	262,477	145,565	77,034	39,877
99/00	2,412,523	2,078,184	334,339	154,454	122,231	57,653
00/01	2,861,756	2,448,389	413,367	193,661	128,257	91,448
01/02	2,329,169	1,968,113	361,056	155,909	145,066	60,080
02/03	2,285,660	1,770,709	514,951	207,126	270,178	37,646
03/04	2,322,086	1,814,404	507,682	146,021	333,105	28,556

NOTE: In the last three fiscal years the contribution to profit of the various product lines was as follows: data archives, 18 percent; program archives, 67 percent; evaluation and training, 15 percent. The distribution agreement entered into with Thomson Gale to market institutional subscriptions to the *Social Science Electronic Data Library* (SSEDL) occurred late in fiscal year 2003-2004; therefore royalties from this agreement are not yet included in this table.
SOURCE: Sociometrics Corporation.

and use (as opposed to single-minded emphasis on profits alone) as part of its important public service.

REVIEW PROCESS

Sociometrics is generally satisfied with the SBIR proposal review process. It believes that it has "learned to compete successfully on paper." The company takes a very pragmatic approach to review. It understands that there is a substantial random element in the process (a study conducted by NSF in the early 1990s found that the chance effect for whether a journal article or proposal is accepted by peers is approximately 50 percent). Therefore it believes that the best approach to unfunded proposals is to consider all reviewer comments seriously and resubmit the proposal whenever these comments can be addressed. Another source of variability is the frequent change in study section make-up from one review to the next. As a result, comments made by one Panel may be negated by the next Panel who have new considerations and concerns. Nevertheless, Sociometrics believes that with tenacity its good proposals will eventually be approved through the current SBIR peer review mechanism. The company estimates that 70 percent of its applications for SBIR Phase I support and 95 percent of its applications for SBIR Phase II support are eventually successful. Forty percent of Phase I applications and 75 percent of Phase II applications are successful at first submission; the other applications require one or two resubmissions before they are eventually funded.

Commercialization has always been a strength for Sociometrics. The company has been pleased that this success criterion has received increased emphasis in recent application guidelines. Consistent with this, Sociometrics staff have observed that reviewer comments have recently praised the company's strength in this area.

Sociometrics makes sure that its SBIR applications highlight its sales track record as well as its sales and marketing expertise. The company notes that this is very different from R01 research grant applications, for which these capacities are essentially irrelevant.

There needs to be ongoing evaluation of reviewers serving on study sections, providing some accountability. Sociometrics supports the concept that "bad reviewers" should be eliminated, but also understands that it is hard to find reviewers. Related to this problem is the concern that reviewers have appropriate expertise for the proposals they evaluate, which is a challenge when study sections cover quite broad areas.

OTHER OUTCOMES

Other Funding

Sociometrics has been funded by many NIH agencies and by government agencies outside NIH. Initial SBIR funding came from the Office of Population Affairs under the Deputy Assistant Secretary for Population Affairs, Department of Health and Human Services. Other projects have been funded by the National Science Foundation (NSF), the Centers for Disease Control and prevention (CDC), the Veterans Administration (VA), the National Center for Health Statistics (NCHS), private companies, nonprofit organizations, state and local governments, and private foundations. Sociometrics has never sought venture capital funding because its profits, while impressive for a behavioral and social science firm, are not large enough to make the company sustainable without SBIR funding, a requirement for venture capital funding.

Changing the Field

NIH and CDC officials, among others, regard Sociometrics' effective-program replication kits as an important innovation helping to bridge the gap between health-related research and practice. The program-in-a-box opened the door for researchers to generate something more than an article or book as an output from their studies, and many researchers were especially pleased to find a way to connect their work to the improvement of practice.

Publications

Sociometrics' staff members have more than 250 peer-reviewed publications; approximately 60 of these are based on the company's SBIR work.

Training Effects

Sociometrics is poised for expansion now that younger staff are becoming qualified as PIs in their own right, which relieves some of the PI burden from the two senior managers, who were until recently PIs on all projects.

NIH Institutes and Centers (IC's)

Sociometrics has had the longest relationship with—and is closest to—the National Institute on Child Health and Human Development (NICHD). Sociometrics' Founder and CEO, Dr. Josefina Card, has served on several NICHD study sections, and has also been on the NICHD National Advisory Council.

SUPPLEMENTAL FUNDING

Supplemental funding procedures vary substantially by IC. Typically, small supplement requests—up to 25 percent of the annual award amount at NIMH—are available at the discretion of the program officer (depending on funding availability). These are referred to as "non-competing administrative supplements." Large supplementary funding requests must compete with other similar requests, seeking a "competing supplementary award." Sociometrics has obtained a few non-competing administrative supplements. It has also obtained three larger supplement awards by expanding the scope of the funded Phase II grant in a way deemed "high priority" by the funding agency or by competing successfully via another funding mechanism (such as an RFA) with the funding agency then deciding, for administrative simplicity reasons, to add monies to the SBIR grant instead of issuing a new grant award. Examples include:

- **Teen pregnancy prevention program replication kits.** Originally funded by NICHD, Sociometrics sought a third year of Phase II support through a supplement to expand the scope of the Program Archive on Sexuality, Health & Adolescence (PASHA) from teen pregnancy prevention alone to teen STD/HIV/AIDS prevention as well. This expansion had been recommended by the PASHA Scientist Expert Panel, in light of the national spotlight on HIV/AIDS and the similar sexual-risk behaviors underlying both unintended pregnancy and STD/HIV/AIDS. The supplement request was forwarded by NICHD to the Deputy Assistant Secretary of Population Affairs who serves simultaneously as Director of the Office of Population Affairs. This political appointee interviewed the Sociometrics PI, and then personally approved the requested additional $750,000 in Phase II funding, transferring the monies to the NICHD grant.

- **Program Archive for HIV/AIDS in adults.** Initially funded by the National Institute of Allergy and Infectious Diseases (NIAID), supplementary funding was requested for the HIV/AIDS Prevention Program Archive (HAPPA) to expand the project to include programs targeted directly at minorities. In this case, the request was for approximately $575,000 over three years. However, an end-of-year budget under-run at NIAID resulted in the full requested funding being provided over one year, instead the requested three years.
- **Complementary and Alternative Medicine Data Archive.** Sociometrics had Phase II funding from the National Center on Alternative Medicine (NCAM) to establish the Complementary and Alternative Medicine Data Archive (CAMDA) when it responded to an RFA issued by NCAM encouraging research on minorities and CAM. Sociometrics responded to the RFA by proposing to expand CAMDA to include data sets especially focused on minority populations. Its proposal received a high priority score and NCAM decided to fund the project via an administrative supplement to the Phase II project rather than via a new grant award.

RECOMMENDATIONS

Sociometrics believes that the SBIR program provides an essential resource for generating innovative and effective research-based products in efficient fashion. In response to questions about its support for various issues and trends in the program, Sociometrics makes the following recommendations:

- **Normalization of scores.** Scores should be normalized across SBIR study sections.
- **Award size and duration.** Phase I duration should be one year, and additional funding (beyond $100,000) should be available with justification. Phase II size and duration limits could remain as they are ($750,000 over two years); Sociometrics has always found it possible to split larger projects into two or more ideas qualifying for separate SBIR funding. While

Sociometrics has no *a priori* objection to "supersized" Phase II awards (awards exceeding the Phase II guidelines of $750,000), it recommends that if such awards are indeed becoming common, then information about them should be fully communicated to applicants and transparency increased. The increasing prevalence of larger Phase II awards might tend to benefit well-established companies and could result in fewer SBIR grants being made. These consequences should be taken into account in approving very large Phase II awards.

- **Direct to Phase II.** Phase II competition should be open to all applicants meeting small business qualifications, permitting bypass of Phase I awards (though not of the need to show equivalent results). This might also, however, tend to benefit well-established companies.
- **Resubmission.** The one-page Phase I proposal limit for summarizing applicants' responses to reviewer comments is insufficient. The limit should be increased to two pages or even three, as is the case for Phase II proposals.
- **Evaluation.** NIH should develop a program to evaluate the health, social, and economic impact of SBIR projects. Sociometrics would very much like to undertake evaluations either of its own SBIR projects, or of a group of projects that would include some of its own.
- **Chartered study sections for SBIR.** Given the now-permanent character of the program, NIH should consider asking Congress to charter what are currently Special Emphasis Panels (SEPs), or should consider changing its guidelines for SEPs to mimic those for chartered study sections. In this manner, the composition of review panels would be more stable from one review round to the next, resulting in better reviews.

Sociometrics—Annex: Sociometrics' SBIR Awards, NIH-Sponsored Phase I Projects Started in FY1992-2002

Fiscal Year	Phase Type	Award Amount ($)	Project Title	IC
1994	Phase I	80,991	AMERICAN FAMILY DATA CENTER	HD
1995	Phase II	216,369	AMERICAN FAMILY DATA CENTER	HD
1996	Phase II	222,496	AMERICAN FAMILY DATA CENTER	HD
1998	Phase II	114,998	AMERICAN FAMILY DATA CENTER	HD
1997	Phase I	99,817	ARCHIVE—EFFECTIVE YOUTH DRUG ABUSE PREVENTION PROGRAMS	DA
1998	Phase II	343,632	ARCHIVE—EFFECTIVE YOUTH DRUG ABUSE PREVENTION PROGRAMS	DA
1999	Phase II	396,232	ARCHIVE—EFFECTIVE YOUTH DRUG ABUSE PREVENTION PROGRAMS	DA
2001	Phase II	24,999	ARCHIVE—EFFECTIVE YOUTH DRUG ABUSE PREVENTION PROGRAMS	DA
1999	Phase I	95,825	CHILD WELL-BEING & POVERTY: STATISTICAL ABSTRACT & DATA	HD
2001	Phase II	377,391	CHILD WELL-BEING & POVERTY: STATISTICAL ABSTRACT & DATA	HD
2002	Phase II	371,768	CHILD WELL-BEING & POVERTY: STATISTICAL ABSTRACT & DATA	HD
1999	Phase I	94,833	COMPLEMENTARY AND ALTERNATIVE MEDICINE DATA ARCHIVE	AT
2002	Phase II	375,345	COMPLEMENTARY AND ALTERNATIVE MEDICINE DATA ARCHIVE	AT
2003	Phase II	374,587	COMPLEMENTARY AND ALTERNATIVE MEDICINE DATA ARCHIVE	AT
2004	Phase II	215,070	COMPLEMENTARY AND ALTERNATIVE MEDICINE DATA ARCHIVE	AT
2005	Phase II	258,084	COMPLEMENTARY AND ALTERNATIVE MEDICINE DATA ARCHIVE	AT
1992	Phase II	162,492	DATA ARCHIVE ON MATERNAL DRUG ABUSE	DA
1993	Phase II	174,720	DATA ARCHIVE ON MATERNAL DRUG ABUSE	DA
1993	Phase II	18,336	DATA ARCHIVE ON MATERNAL DRUG ABUSE	DA
1997	Phase I	99,834	DATASET DEVELOPMENT SOFTWARE & FAMILY RESEARCH ITEM BANK	HD
1998	Phase II	382,614	DATASET DEVELOPMENT SOFTWARE & FAMILY RESEARCH ITEM BANK	HD
1999	Phase II	356,870	DATASET DEVELOPMENT SOFTWARE & FAMILY RESEARCH ITEM BANK	HD

Sociometrics—Annex: Sociometrics' SBIR Awards, NIH-Sponsored Phase I Projects Started in FY1992-2002 Continued

Fiscal Year	Phase Type	Award Amount ($)	Project Title	IC
1995	Phase I	70,770	ESTABLISHING A CONTEXTUAL DATA ARCHIVE	HD
1996	Phase II	338,926	ESTABLISHING A CONTEXTUAL DATA ARCHIVE	HD
1997	Phase II	409,298	ESTABLISHING A CONTEXTUAL DATA ARCHIVE	HD
1993	Phase I	49,975	ESTABLISHMENT OF A RESEARCH ARCHIVE ON DISABILITY	HD
1994	Phase II	236,145	ESTABLISHMENT OF A RESEARCH ARCHIVE ON DISABILITY	HD
1995	Phase II	230,731	ESTABLISHMENT OF A RESEARCH ARCHIVE ON DISABILITY	HD
1996	Phase II	31,288	ESTABLISHMENT OF A RESEARCH ARCHIVE ON DISABILITY	HD
1997	Phase II	149,709	ESTABLISHMENT OF A RESEARCH ARCHIVE ON DISABILITY	HD
1992	Phase II	220,955	ESTABLISHMENT OF AN AIDS/STD DATA ARCHIVE	HD
1993	Phase II	236,301	ESTABLISHMENT OF AN AIDS/STD DATA ARCHIVE	HD
1994	Phase II	42,070	ESTABLISHMENT OF AN AIDS/STD DATA ARCHIVE	HD
1995	Phase II	10,000	ESTABLISHMENT OF AN AIDS/STD DATA ARCHIVE	HD
1998	Phase II	350,655	HIV/AIDS PREVENTION PROGRAM ARCHIVE	AI
1999	Phase II	397,449	HIV/AIDS PREVENTION PROGRAM ARCHIVE	AI
2000	Phase II	574,670	HIV/AIDS PREVENTION PROGRAM ARCHIVE	AI
1997	Phase I	99,221	INSTITUTE FOR PROGRAM DEVELOPMENT AND EVALUATION	HD
1999	Phase II	337,465	INSTITUTE FOR PROGRAM DEVELOPMENT AND EVALUATION	HD
2000	Phase II	412,385	INSTITUTE FOR PROGRAM DEVELOPMENT AND EVALUATION	HD
2001	Phase II	49,987	INSTITUTE FOR PROGRAM DEVELOPMENT AND EVALUATION	HD
1993	Phase I	49,494	INSTRUMENT ARCHIVE OF SOCIAL RESEARCH ON AGING	AG
1992	Phase II	223,017	MICROCOMPUTER DATA ARCHIVE OF SOCIAL RESEARCH ON AGING	AG
1993	Phase II	19,106	MICROCOMPUTER DATA ARCHIVE OF SOCIAL RESEARCH ON AGING	AG
2001	Phase I	197,562	PROMOTING EVALUATION/TEACHING/RESEARCH ON AIDS (PETRA)	MH
2003	Phase II	420,281	PROMOTING EVALUATION/TEACHING/RESEARCH ON AIDS (PETRA)	MH

continued

Sociometrics—Annex: Sociometrics' SBIR Awards, NIH-Sponsored Phase I Projects Started in FY1992-2002 Continued

Fiscal Year	Phase Type	Award Amount ($)	Project Title	IC
2004	Phase II	329,507	PROMOTING EVALUATION/TEACHING/RESEARCH ON AIDS (PETRA)	MH
1994	Phase I	80,991	SOCIONET—ONLINE ACCESS TO SOCIAL SCIENCE DATA	HD
1995	Phase II	446,682	SOCIONET: ONLINE ACCESS TO SOCIAL SCIENCE DATA	HD
1996	Phase II	303,033	SOCIONET: ONLINE ACCESS TO SOCIAL SCIENCE DATA	HD
1998	Phase I	99,340	STATISTICS ON DEMAND: DATA ANALYSIS OVER THE INTERNET	HD
2000	Phase II	381,525	STATISTICS USING MIDAS: DATA ANALYSIS OVER THE INTERNET	HD
1999	Phase II	367,726	STATISTICS USING MIDAS: DATA ANALYSIS OVER THE INTERNET	HD
2002	Phase I	99,011	VIRTUAL PROGRAM EVALUATION CONSULTANT (VPEC)	HD
2003	Phase II	383,357	VIRTUAL PRACTIONER EVALUATION CONSULTANT (VPEC)	HD
2004	Phase II	363,923	VIRTUAL PRACTIONER EVALUATION CONSULTANT (VPEC)	
1993	Phase I	50,000	ARCHIVE OF TEEN PREGNANCY PREVENTION PROGRAMS	HD
1995	Phase II	408,644	PROGRAM ARCHIVE ON SEXUALITY, HEALTH, & ADOLESCENCE	HD
1996	Phase II	987,378	PROGRAM ARCHIVE ON SEXUALITY, HEALTH, & ADOLESCENCE	HD
1997	Phase II	45,000	PROGRAM ARCHIVE ON SEXUALITY, HEALTH, & ADOLESCENCE	OPA
1994	Phase I	79,383	ESTABLISHING A STROKE DATA ARCHIVE	NS
2002	Phase I	199,922	PROMOTING CULTURALLY COMPETENT/EFFECTIVE HIV/AIDS PREVENTION PROGRAMS	AI
2004	Phase II	361,223	PROMOTING CULTURALLY COMPETENT/EFFECTIVE HIV/AIDS PREVENTION PROGRAMS	AI

SOURCE: Sociometrics Corporation.

Appendix D

Bibliography

Acs, Z. and D. Audretsch. 1988. "Innovation in Large and Small Firms: An Empirical Analysis." *The American Economic Review.* 78(4):678-690.

Acs, Z., and D. Audretsch. 1990. *Innovation and Small Firms.* Cambridge, MA: MIT Press.

Advanced Technology Program. 2001. *Performance of 50 Completed ATP Projects, Status Report 2.* National Institute of Standards and Technology Special Publication 950-2. Washington, DC: Advanced Technology Program/National Institute of Standards and Technology/U.S. Department of Commerce.

Alic, John A., Lewis Branscomb, Harvey Brooks, Ashton B. Carter, and Gerald L. Epstein. 1992. *Beyond Spinoff: Military and Commercial Technologies in a Changing World.* Boston, MA: Harvard Business School Press.

American Association for the Advancement of Science. "R&D Funding Update on NSF in the FY2007." Available online at *<http://www.aaas.org/spp/rd/nsf07hf1.pdf>*.

American Psychological Association. 2002. "Criteria for Evaluating Treatment Guidelines." *American Psychologist.* 57(12):1052-1059.

Archibald, R., and D. Finifter. 2000. "Evaluation of the Department of Defense Small Business Innovation Research Program and the Fast Track Initiative: A Balanced Approach." In National Research Council. *The Small Business Innovation Research Program: An Assessment of the Department of Defense Fast Track Initiative.* Charles W. Wessner, ed. Washington, DC: National Academy Press.

Arrow, Kenneth. 1962. "Economic welfare and the allocation of resources for Invention." Pp. 609-625 in *The Rate and Direction of Inventive Activity: Economic and Social Factors.* Princeton, NJ: Princeton University Press.

Arrow, Kenneth. 1973. "The theory of discrimination." Pp. 3-31 in *Discrimination in Labor Market.* Orley Ashenfelter and Albert Rees, eds. Princeton, NJ: Princeton University Press.

Audretsch, David B. 1995. *Innovation and Industry Evolution.* Cambridge, MA: MIT Press.

Audretsch, David B., and Maryann P. Feldman. 1996. "R&D spillovers and the geography of innovation and production." *American Economic Review* 86(3):630-640.

Audretsch, David B., and Paula E. Stephan. 1996. "Company-scientist locational links: The case of biotechnology." *American Economic Review* 86(3):641-642.

Audretsch, D., and R. Thurik. 1999. *Innovation, Industry Evolution, and Employment.* Cambridge, MA: MIT Press.

Baker, Alan. No date. "Commercialization Support at NSF." Draft.

Barfield, C., and W. Schambra, eds. 1986. *The Politics of Industrial Policy.* Washington, DC: American Enterprise Institute for Public Policy Research.

Baron, Jonathan. 1998. "DoD SBIR/STTR Program Manager." Comments at the Methodology Workshop on the Assessment of Current SBIR Program Initiatives, Washington, DC, October.

Barry, C. B. 1994. "New directions in research on venture capital finance." *Financial Management* 23 (Autumn):3-15.

Bator, Francis. 1958. "The anatomy of market failure." *Quarterly Journal of Economics* 72: 351-379.

Bingham, R. 1998. *Industrial Policy American Style: From Hamilton to HDTV.* New York: M.E. Sharpe.

Birch, D. 1981. "Who Creates Jobs." *The Public Interest* 65 (Fall):3-14.

Branscomb, Lewis M., Kennth P. Morse, Michael J. Roberts, and Darin Boville. 2000. *Managing Technical Risk: Understanding Private Sector Decision Making on Early Stage Technology Based Projects.* Washington, DC: Department of Commerce/National Institute of Standards and Technology.

Branscomb, Lewis M., and Philip E. Auerswald. 2001. *Taking Technical Risks: How Innovators, Managers, and Investors Manage Risk in High-Tech Innovations*, Cambridge, MA: MIT Press.

Branscomb, L. M. and P. E. Auerswald. 2002. *Between Invention and Innovation: An Analysis of Funding for Early-Stage Technology Development.* Gaithersburg, MD: National Institute of Standards and Technology.

Branscomb, L. M. and P. E. Auerswald. 2003. "Valleys of Death and Darwinian Seas: Financing the Invention to Innovation Transition in the United States." *The Journal of Technology Transfer* 28(3-4).

Branscomb, Lewis M., and J. Keller. 1998. *Investing in Innovation: Creating a Research and Innovation Policy.* Cambridge, MA: MIT Press.

Brav, A., and P. A. Gompers. 1997. "Myth or reality?: Long-run underperformance of initial public offerings; Evidence from venture capital and nonventure capital-backed IPOs." *Journal of Finance* 52:1791-1821.

Brodd, R. J. 2005. *Factors Affecting U.S. Production Decisions: Why Are There No Volume Lithium-Ion Battery Manufacturers in the United States?* ATP Working Paper No. 05-01, June 2005.

Brown, G., and Turner J. 1999. "Reworking the Federal Role in Small Business Research." *Issues in Science and Technology* XV, no. 4 (Summer).

Bush, Vannevar. 1946. *Science—the Endless Frontier.* Republished in 1960 by U.S. National Science Foundation, Washington, DC.

Carden, S. D., and O. Darragh. 2004. "A Halo for Angel Investors." *The McKinsey Quarterly* 1.

Cassell, G. 2004. "Setting Realistic Expectations for Success." In National Research Council. *SBIR: Program Diversity and Assessment Challenges.* Charles W. Wessner, ed. Washington, DC: The National Academies Press.

Caves, Richard E. 1998. "Industrial organization and new findings on the turnover and mobility of firms." *Journal of Economic Literature* 36(4):1947-1982.

Christensen, C. 1997. *The Innovator's Dilemma.* Boston, MA: Harvard Business School Press.

Clinton, William Jefferson. 1994. *Economic Report of the President.* Washington, DC: U.S. Government Printing Office.

Clinton, William Jefferson. 1994. *The State of Small Business.* Washington, DC: U.S. Government Printing Office.

Coburn, C., and D. Bergland. 1995. *Partnerships: A Compendium of State and Federal Cooperative Technology Programs.* Columbus, OH: Battelle.

Cochrane, J. H. 2005. "The Risk and Return of Venture Capital." *Journal of Financial Economics* 75(1):3-52.

Cohen, L. R., and R. G. Noll. 1991. *The Technology Pork Barrel.* Washington, DC: The Brookings Institution.

Congressional Commission on the Advancement of Women and Minorities in Science, Engineering, and Technology Development. 2000. *Land of Plenty: Diversity as America's Competitive Edge in Science, Engineering and Technology.* Washington, DC: National Science Foundation/U.S. Government Printing Office.

Cooper, R. G. 2001. *Winning at New Products: Accelerating the process from idea to launch.* In Dawnbreaker, Inc. 2005. "The Phase III Challenge: Commercialization Assistance Programs 1990–2005." White paper. July 15.

Council of Economic Advisers. 1995. *Supporting Research and Development to Promote Economic Growth: The Federal Government's Role.* Washington, DC.

Council on Competitiveness. 2005. *Innovate America: Thriving in a World of Challenge and Change.* Washington, DC: Council on Competitiveness.

Cramer, Reid. 2000. "Patterns of Firm Participation in the Small Business Innovation Research Program in Southwestern and Mountain States." In National Research Council. 2000. *The Small Business Innovation Research Program: An Assessment of the Department of Defense Fast Track Initiative.* Charles W. Wessner, ed. Washington, DC: National Academy Press.

David, P. A., B. H. Hall, and A. A. Tool. 1999. "Is Public R&D a Complement or Substitute for Private R&D? A Review of the Econometric Evidence." NBER Working Paper 7373. October.

Davidsson, P. 1996. "Methodological Concerns in the Estimation of Job Creation in Different Firm Size Classes." Working Paper. Jönköping International Business School.

Davis, S. J., J. Haltiwanger, and S. Schuh. 1994. "Small Business and Job Creation: Dissecting the Myth and Reassessing the Facts," *Business Economics* 29(3):113-122.

Dawnbreaker, Inc. 2005. "The Phase III Challenge: Commercialization Assistance Programs 1990–2005." White paper. July 15.

Dertouzos. 1989. *Made in America: The MIT Commission on Industrial Productivity.* Cambridge, MA: MIT Press.

Dess, G. G. and D. W. Beard. 1984. "Dimensions of Organizational Task Environments." *Administrative Science Quarterly* 29:52-73.

Devenow, A. and I. Welch. 1996. "Rational Herding in Financial Economics. *European Economic Review* 40(April):603-615.

DoE Opportunity Forum. 2005. "Partnering and Investment Opportunities for the Future." Tysons Corner, VA. October 24-25.

Eckstein, Otto. 1984. *DRI Report on U.S. Manufacturing Industries.* New York: McGraw Hill.

Eisinger, P. K. 1988. *The Rise of the Entrepreneurial State: State and Local Economic Development Policy in the United State.* Madison, WI: University of Wisconsin Press.

Feldman, Maryann P. 1994. *The Geography of Knowledge.* Boston, MA: Kluwer Academic.

Feldman, Maryann P. 1994. "Knowledge complementarity and innovation." *Small Business Economics* 6(5):363-372.

Feldman, M. P., and M. R. Kelley. 2001. "Leveraging Research and Development: The Impact of the Advanced Technology Program." In National Research Council. *The Advanced Technology Program.* Charles W. Wessner, ed. Washington, DC: National Academy Press.

Feldman, M. P., and M. R. Kelley. 2001. *Winning an Award from the Advanced Technology Program: Pursuing R&D Strategies in the Public Interest and Benefiting from a Halo Effect.* NISTIR 6577. Washington, DC: Advanced Technology Program/National Institute of Standards and Technology/U.S. Department of Commerce.

Fenn, G. W., N. Liang, and S. Prowse. 1995. *The Economics of the Private Equity Market.* Washington, DC: Board of Governors of the Federal Reserve System.

Flamm, K. 1988. *Creating the Computer.* Washington, DC: The Brookings Institution.

Flender, J. O., and R. S. Morse. 1975. *The Role of New Technical Enterprise in the U.S. Economy.* Cambridge, MA: MIT Development Foundation.

Freear, J., and W. E. Wetzel Jr. 1990. "Who bankrolls high-tech entrepreneurs?" *Journal of Business Venturing* 5:77-89.

Freeman, Chris, and Luc Soete. 1997. *The Economics of Industrial Innovation.* Cambridge, MA: MIT Press.

Galbraith, J. K. 1957. *The New Industrial State.* Boston: Houghton Mifflin.

Geroski, Paul A. 1995. "What do we know about entry?" *International Journal of Industrial Organization* 13(4):421-440.

Geshwiler, J., J. May, and M. Hudson. 2006. "State of Angel Groups." Kansas City, MO: Kauffman Foundation.

Gompers, P. A. and J. Lerner. 1977. "Risk and Reward in Private Equity Investments: The Challenge of Performance Assessment." *Journal of Private Equity* 1:5-12.

Gompers, P. A. 1995. "Optimal investment, monitoring, and the staging of venture capital." *Journal of Finance* 50:1461-1489.

Gompers, P. A., and J. Lerner. 1996. "The use of covenants: An empirical analysis of venture partnership agreements." *Journal of Law and Economics* 39:463-498.

Gompers, P. A., and J. Lerner. 1998. "Capital formation and investment in venture markets: A report to the NBER and the Advanced Technology Program." Unpublished working paper. Harvard University.

Gompers, P. A., and J. Lerner. 1998. "What drives venture capital fund-raising?" Unpublished working paper. Harvard University.

Gompers, P. A., and J. Lerner. 1999. "An analysis of compensation in the U.S. venture capital partnership." *Journal of Financial Economics* 51(1):3-7.

Gompers, P. A., and J. Lerner. 1999. *The Venture Cycle.* Cambridge, MA: MIT Press.

Good, M. L. 1995. Prepared testimony before the Senate Commerce, Science, and Transportation Committee, Subcommittee on Science, Technology, and Space (photocopy, U.S. Department of Commerce).

Goodnight, J. 2003. Presentation at National Research Council Symposium. "The Small Business Innovation Research Program: Identifying Best Practice." Washington, DC May 28.

Graham, O. L. 1992. *Losing Time: The Industrial Policy Debate.* Cambridge, MA: Harvard University Press.

Greenwald, B. C., J. E. Stiglitz, and A. Weiss. 1984. "Information imperfections in the capital market and macroeconomic fluctuations." *American Economic Review Papers and Proceedings* 74:194-199.

Griliches, Z. 1990. *The Search for R&D Spillovers.* Cambridge, MA: Harvard University Press.

Groves, R. M., F. J. Fowler, Jr., M. P. Couper, J. M. Lepkowski, E. Singer, and R. Tourangeau. 2004. *Survey Methodology.* Hoboken, NJ: John Wiley & Sons, Inc.

Hall, Bronwyn H. 1992. "Investment and research and development: Does the source of financing matter?" Working Paper No. 92-194, Department of Economics/University of California at Berkeley.

Hall, Bronwyn H. 1993. "Industrial research during the 1980s: Did the rate of return fall?" Brookings Papers: *Microeconomics* 2:289-343.

Haltiwanger, J. and C. J. Krizan. 1999. "Small Businesses and Job Creation in the United States: The Role of New and Young Businesses" in *Are Small Firms Important? Their Role and Impact*, Zoltan J. Acs, ed., Dordrecht: Kluwer.

Hamberg, Dan. 1963. "Invention in the industrial research laboratory." *Journal of Political Economy* (April):95-115.

Hao, K. Y., and A. B. Jaffe. 1993. "Effect of liquidity on firms' R&D spending." *Economics of Innovation and New Technology* 2:275-282.

Hebert, Robert F., and Albert N. Link. 1989. "In search of the meaning of entrepreneurship." *Small Business Economics* 1(1):39-49.

Heilman, C. 2005. "Partnering for Vaccines: The NIAID Perspective" in Charles W. Wessner, ed. *Partnering Against Terrorism: Summary of a Workshop.* Washington, DC: The National Academies Press.

Held, B., T. Edison, S. L. Pfleeger, P. Anton, and J. Clancy. 2006. *Evaluation and Recommendations for Improvement of the Department of Defense Small Business Innovation Research (SBIR) Program.* Arlington, VA: RAND National Defense Research Institute.

Holland, C. 2007. "Meeting Mission Needs." In National Research Council. *SBIR and the Phase III Challenge of Commercialization.* Charles W. Wessner, ed. Washington, DC: The National Academies Press.

Himmelberg, C. P., and B. C. Petersen. 1994. "R&D and internal finance: A panel study of small firms in high-tech industries." *Review of Economics and Statistics* 76:38-51.

Hubbard, R. G. 1998. "Capital-market imperfections and investment." *Journal of Economic Literature* 36:193-225.

Huntsman, B., and J. P. Hoban Jr. 1980. "Investment in new enterprise: Some empirical observations on risk, return, and market structure." *Financial Management* 9 (Summer):44-51.

Institute of Medicine. 1998. "The Urgent Need to Improve Health Care Quality." National Roundtable on Health Care Quality. *Journal of the American Medical Association* 280(11):1003, September 16.

Jacobs, T. 2002. "Biotech Follows Dot.com Boom and Bust." *Nature* 20(10):973.

Jaffe, A. B. 1996. "Economic Analysis of Research Spillovers: Implications for the Advanced Technology Program." Washington, DC: Advanced Technology Program/National Institute of Standards and Technology/U.S. Department of Commerce).

Jaffe, A. B. 1998. "Economic Analysis of Research Spillovers: Implications for the Advanced Technology Program." Washington, DC: Advanced Technology Program/National Institute of Standards and Technology/U.S. Department of Commerce.

Jaffe, A. B. 1998. "The importance of 'spillovers' in the policy mission of the Advanced Technology Program." *Journal of Technology Transfer* (Summer).

Jewkes, J., D. Sawers, and R. Stillerman. 1958. *The Sources of Invention.* New York: St. Martin's Press.

Jarboe, K. P. and R. D. Atkinson. 1998. "The Case for Technology in the Knowledge Economy; R&D, Economic Growth and the Role of Government." Washington, DC: Progressive Policy Institute. Available online at <*http://www.ppionline.org/documents/CaseforTech.pdf*>.

Johnson, M. 2004. "SBIR at the Department of Energy: Achievements, Opportunities, and Challenges." In National Research Council. *SBIR: Program Diversity and Assessment Challenges.* Charles W. Wessner, ed. Washington, DC: The National Academies Press, 2004.

Kauffman Foundation. About the Foundation. Available online at <*http://www.kauffman.org/foundation.cfm*>.

Kleinman, D. L. 1995. *Politics on the Endless Frontier: Postwar Research Policy in the United States.* Durham, NC: Duke University Press.

Kortum, Samuel, and Josh Lerner. 1998. "Does Venture Capital Spur Innovation?" NBER Working Paper No. 6846, National Bureau of Economic Research.

Krugman, P. 1990. *Rethinking International Trade.* Cambridge, MA: MIT Press.

Krugman, P. 1991. *Geography and Trade.* Cambridge, MA: MIT Press.

Langlois, Richard N., and Paul L. Robertson. 1996. "Stop Crying over Spilt Knowledge: A Critical Look at the Theory of Spillovers and Technical Change." Paper prepared for the MERIT Conference on Innovation, Evolution, and Technology. Maastricht, Netherlands, August 25-27.

Langlois, R. N. 2001. "Knowledge, Consumption, and Endogenous Growth." *Journal of Evolutionary Economics* 11:77-93.

Lebow, I. 1995. *Information Highways and Byways: From the Telegraph to the 21st Century.* New York: Institute of Electrical and Electronic Engineering.

Lerner, J. 1994. "The syndication of venture capital investments." *Financial Management* 23 (Autumn):16-27.

Lerner, J. 1995. "Venture capital and the oversight of private firms." *Journal of Finance* 50: 301-318.

Lerner, J. 1996. "The government as venture capitalist: The long-run effects of the SBIR program." Working Paper No. 5753, National Bureau of Economic Research.

Lerner, J. 1998. "Angel financing and public policy: An overview." *Journal of Banking and Finance* 22(6–8):773-784.

Lerner, J. 1999. "The government as venture capitalist: The long-run effects of the SBIR program." *Journal of Business* 72(3):285-297.

Lerner, J. 1999. "Public venture capital: Rationales and evaluation." In *The SBIR Program: Challenges and Opportunities*. Washington, DC: National Academy Press.

Levy, D. M., and N. Terleckyk. 1983. "Effects of government R&D on private R&D investment and productivity: A macroeconomic analysis." *Bell Journal of Economics* 14:551-561.

Liles, P. 1977. *Sustaining the Venture Capital Firm.* Cambridge, MA: Management Analysis Center.

Link, Albert N. 1998. "Public/Private Partnerships as a Tool to Support Industrial R&D: Experiences in the United States." Paper prepared for the working group on Innovation and Technology Policy of the OECD Committee for Science and Technology Policy, Paris.

Link, Albert N., and John Rees. 1990. "Firm size, university based research and the returns to R&D." *Small Business Economics* 2(1):25-32.

Link, Albert N., and John T. Scott. 1998. "Assessing the infrastructural needs of a technology-based service sector: A new approach to technology policy planning." *STI Review* 22:171-207.

Link, Albert N., and John T. Scott. 1998. *Overcoming Market Failure: A Case Study of the ATP Focused Program on Technologies for the Integration of Manufacturing Applications (TIMA).* Draft final report submitted to the Advanced Technology Program. Gaithersburg, MD: National Institute of Technology. October.

Link, Albert N., and John T. Scott. 1998. *Public Accountability: Evaluating Technology-Based Institutions.* Norwell, MA: Kluwer Academic.

Link, A. N. and J. T. Scott. 2005. *Evaluating Public Research Institutions: The U.S. Advanced Technology Program's Intramural Research Initiative.* London: Routledge.

Longini, P. 2003. "Hot buttons for NSF SBIR Research Funds," Pittsburgh Technology Council, *TechyVent*. November 27.

Malone, T. 1995. *The Microprocessor: A Biography.* Hamburg, Germany: Springer Verlag/Telos.

Mansfield, E. 1985. "How Fast Does New Industrial Technology Leak Out?" *Journal of Industrial Economics* 34(2).

Mansfield, E. 1996. *Estimating Social and Private Returns from Innovations Based on the Advanced Technology Program: Problems and Opportunities.* Unpublished report.

Mansfield, E., J. Rapoport, A. Romeo, S. Wagner, and G. Beardsley. 1977. "Social and private rates of return from industrial innovations." *Quarterly Journal of Economics* 91:221-240.

Martin, Justin. 2002. "David Birch." *Fortune Small Business* (December 1).

McCraw, T. 1986. "Mercantilism and the Market: Antecedents of American Industrial Policy." In C. Barfield and W. Schambra, eds. *The Politics of Industrial Policy.* Washington, DC: American Enterprise Institute for Public Policy Research.

Mervis, Jeffrey D. 1996. "A $1 Billion 'Tax' on R&D Funds." *Science* 272:942–944.

Moore, D. 2004. "Turning Failure into Success." In National Research Council. *The Small Business Innovation Research Program: Program Diversity and Assessment Challenges.* Charles W. Wessner, ed. Washington, DC: The National Academies Press.

Morgenthaler, D. 2000. "Assessing Technical Risk," in L. M. Branscomb, K. P. Morse, and M. J. Roberts, eds. *Managing Technical Risk: Understanding Private Sector Decision Making on Early Stage Technology-Based Project.* Gaithersburg, MD: National Institute of Standards and Technology.

Mowery, D. 1998. "Collaborative R&D: how effective is it?" *Issues in Science and Technology* (Fall):37-44.

Mowery, D., and N. Rosenberg. 1989. *Technology and the Pursuit of Economic Growth.* New York: Cambridge University Press.

Mowery, D., and N. Rosenberg. 1998. *Paths of Innovation: Technological Change in 20th Century America.* New York: Cambridge University Press.

Murphy, L. M. and P. L. Edwards. 2003. *Bridging the Valley of Death—Transitioning from Public to Private Sector Financing.* Golden, CO: National Renewable Energy Laboratory. May.

Myers, S., R. L. Stern, and M. L. Rorke. 1983. *A Study of the Small Business Innovation Research Program.* Lake Forest, IL: Mohawk Research Corporation.

Myers, S. C., and N. Majluf. 1984. "Corporate financing and investment decisions when firms have information that investors do not have." *Journal of Financial Economics* 13:187-221.

National Aeronautics and Space Administration. 2002. "Small Business/SBIR: NICMOS Cryocooler—Reactivating a Hubble Instrument." *Aerospace Technology Innovation* 10(4):19-21.

National Aeronautics and Space Administration. 2005. "The NASA SBIR and STTR Programs Participation Guide." Available online at *<http://sbir.gsfc.nasa.gov/SBIR/zips/guide.pdf>*

National Institutes of Health. 2003. Road Map for Medical Research. Available online at *<http://nihroadmap.nih.gov/>*.

National Institutes of Health. 2005. *Report on the Second of the 2005 Measures Updates: NIH SBIR Performance Outcomes Data System (PODS).*

National Research Council. 1986. *The Positive Sum Strategy: Harnessing Technology for Economic Growth.* Washington, DC: National Academy Press.

National Research Council. 1987. *Semiconductor Industry and the National Laboratories: Part of a National Strategy.* Washington, DC: National Academy Press.

National Research Council. 1991. *Mathematical Sciences, Technology, and Economic Competitiveness.* James G. Glimm, ed. Washington, DC: National Academy Press.

National Research Council. 1992. *The Government Role in Civilian Technology: Building a New Alliance.* Washington, DC: National Academy Press.

National Research Council. 1995. *Allocating Federal Funds for R&D.* Washington, DC: National Academy Press.

National Research Council. 1996. *Conflict and Cooperation in National Competition for High-*Technology *Industry.* Washington, DC: National Academy Press.

National Research Council. 1997. *Review of the Research Program of the Partnership for a New Generation of Vehicles: Third Report.* Washington, DC: National Academy Press.

National Research Council. 1999. *The Advanced Technology Program: Challenges and Opportunities.* Charles W. Wessner, ed. Washington, DC: National Academy Press.

National Research Council. 1999. *Funding a Revolution: Government Support for Computing Research.* Washington, DC: National Academy Press.

National Research Council. 1999. *Industry-Laboratory Partnerships: A Review of the Sandia Science and Technology Park Initiative.* Charles W. Wessner, ed. Washington, DC: National Academy Press.

National Research Council. 1999. *New Vistas in Transatlantic Science and Technology Cooperation.* Charles W. Wessner, ed. Washington, DC: National Academy Press.

National Research Council. 1999. *The Small Business Innovation Research Program: Challenges and Opportunities.* Charles W. Wessner, ed. Washington, DC: National Academy Press.

National Research Council. 2000. *The Small Business Innovation Research Program: An Assessment of the Department of Defense Fast Track Initiative.* Charles W. Wessner, ed. Washington, DC: National Academy Press.

National Research Council. 2000. *U.S. Industry in 2000: Studies in Competitive Performance.* Washington, DC: National Academy Press.

National Research Council. 2001. *The Advanced Technology Program: Assessing Outcomes.* Charles W. Wessner, ed. Washington, DC: National Academy Press.

National Research Council. 2001. *Attracting Science and Mathematics Ph.Ds to Secondary School Education.* Washington, DC: National Academy Press.

National Research Council. 2001. *Building a Workforce for the Information Economy.* Washington, DC: National Academy Press.

National Research Council. 2001. *Capitalizing on New Needs and New Opportunities: Government-Industry Partnerships in Biotechnology and Information Technologies.* Charles W. Wessner, ed. Washington, DC: National Academy Press.

National Research Council. 2001. *A Review of the New Initiatives at the NASA Ames Research Center.* Charles W. Wessner, ed. Washington, DC: National Academy Press.

National Research Council. 2001. *Trends in Federal Support of Research and Graduate Education.* Washington, DC: National Academy Press.

National Research Council. 2002. *Government-Industry Partnerships for the Development of New Technologies: Summary Report.* Charles W. Wessner, ed. Washington, DC: The National Academies Press.

National Research Council. 2002. *Making the Nation Safer: The Role of Science and Technology in Countering Terrorism.* Washington, DC: The National Academies Press.

National Research Council. 2002. *Measuring and Sustaining the New Economy.* Dale W. Jorgenson and Charles W. Wessner, eds. Washington, DC: National Academy Press.

National Research Council. 2002. *Partnerships for Solid-State Lighting.* Charles W. Wessner, ed. Washington, DC: The National Academies Press.

National Research Council. 2004. *An Assessment of the Small Business Innovation Research Program: Project Methodology.* Washington, DC: The National Academies Press.

National Research Council. 2004. Capitalizing on Science, Technology, and Innovation: An Assessment of the Small Business Innovation Research Program/Program Manager Survey. Completed by Dr. Joseph Hennessey.

National Research Council. 2004. *Productivity and Cyclicality in Semiconductors: Trends, Implications, and Questions.* Dale W. Jorgenson and Charles W. Wessner, eds. Washington, DC: The National Academies Press.

National Research Council. 2004. *The Small Business Innovation Research Program: Program Diversity and Assessment Challenges.* Charles W. Wessner, ed. Washington, DC: The National Academies Press.

National Research Council. 2006. *Beyond Bias and Barriers: Fulfilling the Potential of Women in Academic Science and Engineering.*

National Research Council. 2006. *Deconstructing the Computer.* Dale W. Jorgenson and Charles W. Wessner, eds. Washington, DC: The National Academies Press.

National Research Council. 2006. *Software, Growth, and the Future of the U.S. Economy.* Dale W. Jorgenson and Charles W. Wessner, eds. Washington, DC: The National Academies Press.

National Research Council. 2006. *The Telecommunications Challenge: Changing Technologies and Evolving Policies.* Dale W. Jorgenson and Charles W. Wessner, eds. Washington, DC: The National Academies Press.

National Research Council. 2007. *Enhancing Productivity Growth in the Information Age: Measuring and Sustaining the New Economy.* Dale W. Jorgenson and Charles W. Wessner, eds. Washington, DC: The National Academies Press.

National Research Council. 2007. *India's Changing Innovation System: Achievements, Challenges, and Opportunities for Cooperation.* Charles W. Wessner and Sujai J. Shivakumar, eds. Washington, DC: The National Academies Press.

National Research Council. 2007. *Innovation Policies for the 21st Century.* Charles W. Wessner, ed. Washington, DC: The National Academies Press.

National Research Council. 2007. *SBIR and the Phase III Challenge of Commercialization.* Charles W. Wessner, ed. Washington, DC: The National Academies Press.

National Research Council. 2008. *An Assessment of the SBIR Program at the Department of Energy.* Charles W. Wessner, ed. Washington, DC: The National Academies Press.

National Research Council. 2008. *An Assessment of the SBIR Program at the National Science Foundation.* Charles W. Wessner, ed. Washington, DC: The National Academies Press.

National Research Council. 2009. *An Assessment of the SBIR Program at the Department of Defense.* Charles W. Wessner, ed. Washington, DC: The National Academies Press.

National Research Council. 2009. *An Assessment of the SBIR Program at the National Aeronautics and Space Administration.* Charles W. Wessner, ed. Washington, DC: The National Academies Press.

National Research Council. 2009. *An Assessment of the SBIR Program at the National Institutes of Health.* Charles W. Wessner, ed. Washington, DC: The National Academies Press.

National Science Board. 2005. *Science and Engineering Indicators 2005.* Arlington, VA: National Science Foundation.

National Science Board. 2006. *Science and Engineering Indicators 2006.* Arlington, VA: National Science Foundation.

National Science Foundation. Committee of Visitors Reports and Annual Updates. Available online at *<http://www.nsf.gov/eng/general/cov/>*.

National Science Foundation. Emerging Technologies. Available online at *<http://www.nsf.gov/eng/sbir/eo.jsp>*.

National Science Foundation. Guidance for Reviewers. Available online at *<http://www.eng.nsf.gov/sbir/peer_review.htm>*.

National Science Foundation. National Science Foundation at a Glance. Available online at *<http://www.nsf.gov/about>*.

National Science Foundation. National Science Foundation Manual 14, *NSF Conflicts of Interest and Standards of Ethical Conduct*. Available online at *<http://www.eng.nsf.gov/sbir/COI_Form.doc>*.

National Science Foundation. The Phase IIB Option. Available online at *<http://www.nsf.gov/eng/sbir/phase_IIB.jsp#ELIGIBILITY>*.

National Science Foundation. Proposal and Grant Manual. Available online at *<http://www.inside.nsf.gov/pubs/2002/pam/pamdec02.6html>*.

National Science Foundation. 2005. Synopsis of SBIR/STTR Program. Available online at *<http://www.nsf.gov/funding/pgm_summ.jsp?Phase Ims_id=13371&org=DMII>*.

National Science Foundation. 2006. "News items from the past year." Press Release. April 10.

National Science Foundation, Office of Industrial Innovation. 2006. "SBIR/STTR Phase II Grantee Conference, Book of Abstracts." Louisville, Kentucky. May 18-20, 2006.

National Science Foundation, Office of Industrial Innovation. Draft Strategic Plan, June 2, 2005.

National Science Foundation, Office of Legislative and Public Affairs. 2003. SBIR Success Story from News Tip. Web's "Best Meta-Search Engine," March 20.

National Science Foundation, Office of Legislative and Public Affairs. 2004. SBIR Success Story: GPRA Fiscal Year 2004 "Nugget." Retrospective Nugget–AuxiGro Crop Yield Enhancers.

Nelson, R. R. 1982. *Government and Technological Progress.* New York: Pergamon.

Nelson, R. R. 1986. "Institutions supporting technical advances in industry." *American Economic Review, Papers and Proceedings* 76(2):188.

Nelson, R. R., ed. 1993. *National Innovation System: A Comparative Study.* New York: Oxford University Press.

Office of Management and Budget. 1996. "Economic analysis of federal regulations under Executive Order 12866."

Office of Management and Budget. 2004. *"What Constitutes Strong Evidence of Program Effectiveness."* Available online at *<http://www.whitehouse.gov/omb/part/2004_program_eval.pdf>*

Office of the President. 1990. *U.S. Technology Policy.* Washington, DC: Executive Office of the President.

Organization for Economic Cooperation and Development. 1982. *Innovation in Small and Medium Firms.* Paris: Organization for Economic Cooperation and Development.

Organization for Economic Cooperation and Development. 1995. *Venture Capital in OECD Countries.* Paris: Organization for Economic Cooperation and Development.

Organization for Economic Cooperation and Development. 1997. *Small Business Job Creation and Growth: Facts, Obstacles, and Best Practices.* Paris: Organization for Economic Cooperation and Development.

Organization for Economic Cooperation and Development. 1998. *Technology, Productivity and Job Creation: Toward Best Policy Practice.* Paris: Organization for Economic Cooperation and Development.

Organization for Economic Cooperation and Development. 2006. "Evaluation of SME Policies and Programs: Draft OECD Handbook." *OECD Handbook.* CFE/SME 17. Paris: Organization for Economic Cooperation and Development.

Pacific Northwest National Laboratory. SBIR Alerting Service. Available online at *<http://www.pnl.gov/edo/sbir>*.

Perko, J. S., and F. Narin. 1997. "The Transfer of Public Science to Patented Technology: A Case Study in Agricultural Science." *Journal of Technology Transfer* 22(3):65-72.

Perret, G. 1989. *A Country Made by War: From the Revolution to Vietnam—The Story of America's Rise to Power.* New York: Random House.

Powell, J. W. 1999. *Business Planning and Progress of Small Firms Engaged in Technology Development through the Advanced Technology Program.* NISTIR 6375. National Institute of Standards and Technology/U.S. Department of Commerce.

Powell, Walter W., and Peter Brantley. 1992. "Competitive cooperation in biotechnology: Learning through networks?" In N. Nohria and R. G. Eccles, eds. *Networks and Organizations: Structure, Form and Action.* Boston, MA: Harvard Business School Press. Pp. 366-394.

Price Waterhouse. 1985. *Survey of Small High-tech Businesses Shows Federal SBIR Awards Spurring Job Growth, Commercial Sales.* Washington, DC: Small Business High Technology Institute.

Roberts, Edward B. 1968. "Entrepreneurship and technology." *Research Management* (July): 249–266.

Romer, P. 1990. "Endogenous technological change." *Journal of Political Economy* 98:71-102.

Rosa, Peter, and Allison Dawson. 2006. "Gender and the commercialization of university science: Academic founders of spinout companies." *Entrepreneurship & Regional Development* 18(4):341-366. July.

Rosenbloom, R., and Spencer, W. 1996. *Engines of Innovation: U.S. Industrial Research at the End of an Era.* Boston, MA: Harvard Business School Press.

Rubenstein, A. H. 1958. *Problems Financing New Research-Based Enterprises in New England.* Boston, MA: Federal Reserve Bank.

Ruegg, Rosalie, and Irwin Feller. 2003. *A Toolkit for Evaluating Public R&D Investment Models, Methods, and Findings from ATP's First Decade.* NIST GCR 03-857.

Ruegg, Rosalie, and Patrick Thomas. 2007. *Linkages from DoE's Vehicle Technologies R&D in Advanced Energy Storage to Hybrid Electric Vehicles, Plug-in Hybrid Electric Vehicles, and Electric Vehicles.* U.S. Department of Energy/Office of Energy Efficiency and Renewable Energy.

Sahlman, W. A. 1990. "The structure and governance of venture capital organizations." *Journal of Financial Economics* 27:473-521.

Saxenian, Annalee. 1994. *Regional Advantage: Culture and Competition in Silicon Valley and Route 128.* Cambridge, MA: Harvard University Press.

SBIR World. SBIR World: A World of Opportunities. Available online at *<http://www.sbirworld.com>*.

Scherer, F. M. 1970. *Industrial Market Structure and Economic Performance.* New York: Rand McNally College Publishing.

Schumpeter, J. 1950. *Capitalism, Socialism, and Democracy.* New York: Harper and Row.

Scotchmer, S. 2004. *Innovation and Incentives.* Cambridge MA: The MIT Press.

Scott, John T. 1998. "Financing and leveraging public/private partnerships: The hurdle-lowering auction." *STI Review* 23:67-84.

Siegel, D., D. Waldman, and A. Link. 2004. "Toward a Model of the Effective Transfer of Scientific Knowledge from Academicians to Practitioners: Qualitative Evidence from the Commercialization of University Technologies." *Journal of Engineering and Technology Management* 21(1-2).

Small Business Administration. 1992. *Results of Three-Year Commercialization Study of the SBIR Program.* Washington, DC: U.S. Government Printing Office.

Small Business Administration. 1994. *Small Business Innovation Development Act: Tenth-Year Results.* Washington, DC: U.S. Government Printing Office (and earlier years).

Society for Prevention Research. 2004. *Standards of Evidence: Criteria for Efficacy, Effectiveness and Dissemination.* Available online at *<http://www.preventionresearch.org/softext.php>*.

Sohl, Jeffrey. 1999. *Venture Capital* 1(2).

Sohl, Jeffery, John Freear, and W.E. Wetzel Jr. 2002. "Angles on Angels: Financing Technology-Based Ventures—An Historical Perspective." *Venture Capital: An International Journal of Entrepreneurial Finance* 4 (4).

Solow, R. S. 1957. "Technical Change and the Aggregate Production Function." *Review of Economics and Statistics* 39:312-320.

Stiglitz, J. E., and A. Weiss. 1981. "Credit rationing in markets with incomplete information." *American Economic Review* 71:393-409.

Stowsky, J. 1996. "Politics and Policy: The Technology Reinvestment Program and the Dilemmas of Dual Use." Mimeo. University of California.

Tassey, Gregory. 1997. *The Economics of R&D Policy.* Westport, CT: Quorum Books.

Tibbetts, R. 1997. "The Role of Small Firms in Developing and Commercializing New Scientific Instrumentation: Lessons from the U.S. Small Business Innovation Research Program," in J. Irvine, B. Martin, D. Griffiths, and R. Gathier, eds. *Equipping Science for the 21st Century.* Cheltenham UK: Edward Elgar Press.

Tirman, John. 1984. *The Militarization of High Technology.* Cambridge, MA: Ballinger.

Tyson, Laura, Tea Petrin, and Halsey Rogers. 1994. "Promoting entrepreneurship in Eastern Europe." *Small Business Economics* 6:165-184.

University of New Hampshire Center for Venture Research. 2007. *The Angel Market in 2006.* Available online at *<http://wsbe2.unh.edu/files/Full%20Year%202006%20Analysis%20Report%20-%20March%202007.pdf>*

U.S. Congress, House Committee on Science, Space, and Technology. 1992. *SBIR and Commercialization: Hearing Before the Subcommittee on Technology and Competitiveness of the House Committee on Science, Space, and Technology, on the Small Business Innovation Research [SBIR] Program.* Testimony of James A. Block, President of Creare, Inc. Pp. 356-361.

U.S. Congress. House Committee on Science, Space, and Technology. 1998. *Unlocking Our Future: Toward a New National Science Policy: A Report to Congress by the House Committee on Science, Space, and Technology.* Washington, DC: Government Printing Office. Available online at *<http://www.access.gpo.gov/congress/house/science/cp105-b/science105b.pdf>*.

U.S. Congress. House Committee on Small Business. Subcommittee on Workforce, Empowerment, and Government Programs. 2005. *The Small Business Innovation Research Program: Opening Doors to New Technology.* Testimony by Joseph Hennessey. 109th Cong., 1st sess., November 8.

U.S. Congress. House Committee on Science, Space, and Technology. Subcommittee on Technology and Innovation. 2007. Hearing on "Small Business Innovation Research Authorization on the 25th Program Anniversary." Testimony by Robert Schmidt. April 26.

U.S. Congress. Senate Committee on Small Business. 1981. Small Business Research Act of 1981. S.R. 194, 97th Congress.

U.S. Congress. Senate Committee on Small Business. 1999. Senate Report 106-330. *Small Business Innovation Research (SBIR) Program.* August 4. Washington, DC: U.S. Government Printing Office.

U.S. Congress. Senate Committee on Small Business. 2006. *Strengthening the Participation of Small Businesses in Federal Contracting and Innovation Research Programs.* Testimony by Michael Squillante. 109th Cong., 2nd sess., July 12.

U.S. Congressional Budget Office. 1985. *Federal financial support for high-technology industries.* Washington, DC: U.S. Congressional Budget Office.

U.S. Department of Education. 2005. "Scientifically-Based Evaluation Methods: Notice of Final Priority." *Federal Register.* 70(15):3586-3589. January 25.

U.S. Food and Drug Administration. 1981. Protecting Human Subjects: Untrue Statements in Application. 21 C.F.R. §314.12

U.S. Food and Drug Administration. *Critical Path Initiative.* Available online at *<http://www.fda.gov/oc/initiatives/criticalpath/>*

U.S. General Accounting Office. 1987. *Federal research: Small Business Innovation Research participants give program high marks.* Washington, DC: U.S. General Accounting Office.

U.S. General Accounting Office. 1989. *Federal Research: Assessment of Small Business Innovation Research Program.* Washington, DC: U.S. General Accounting Office.

U.S. General Accounting Office. 1992. *Federal Research: Small Business Innovation Research Program Shows Success but Can Be Strengthened.* RCED–92–32. Washington, DC: U.S. General Accounting Office.

U.S. General Accounting Office. 1997. *Federal Research: DoD's Small Business Innovation Research Program.* RCED–97–122, Washington, DC: U.S. General Accounting Office.

U. S. General Accounting Office. 1998. *Federal Research: Observations on the Small Business Innovation Research Program.* RCED–98–132. Washington, DC: U.S. General Accounting Office.

U.S. General Accounting Office. 1999. *Federal Research: Evaluations of Small Business Innovation Research Can Be Strengthened.* RCED–99–198, Washington, DC: U.S. General Accounting Office.

U.S. Government Accountability Office. 2006. *Small Business Innovation Research: Agencies Need to Strengthen Efforts to Improve the Completeness, Consistency, and Accuracy of Awards Data*, GAO-07-38, Washington, DC: U.S. Government Accountability Office.

U.S. Government Accountability Office. 2006. *Small Business Innovation Research: Information on Awards made by NIH and DoD in Fiscal years 2001-2004.* GAO-06-565. Washington, DC: U.S. Government Accountability Office.

U.S. Public Law 106-554, Appendix I–H.R. 5667, Section 108.

U.S. Small Business Administration. 1992. *Results of Three-Year Commercialization Study of the SBIR Program.* Washington, DC: U.S. Government Printing Office.

U.S. Small Business Administration. 1994. *Small Business Innovation Development Act: Tenth-Year Results.* Washington, DC: U.S. Government Printing Office.

U.S. Small Business Administration. 1998. "An Analysis of the Distribution of SBIR Awards by States, 1983-1996." Washington, DC: Small Business Administration.

U.S. Small Business Administration. 2006. *Frequently Asked Questions*, June 2006. Available online at *<http://www.sba.gov/advo/stats/sbfaq.pdf>*.

U.S. Small Business Administration. 2006. "Small Business by the Numbers." SBA Office of Advocacy. May.

Venture Economics. 1988. *Exiting Venture Capital Investments.* Wellesley, MA: Venture Economics.

Venture Economics. 1996. "Special Report: Rose-colored asset class." *Venture Capital Journal* 36 (July):32-34 (and earlier years).

VentureOne. 1997. National Venture Capital Association 1996 annual report. San Francisco: VentureOne.

Wallsten, S. J. 1996. The Small Business Innovation Research Program: Encouraging Technological Innovation and Commercialization in Small Firms. Unpublished working paper. Stanford University.

Wallsten, S. J. 1998. "Rethinking the Small Business Innovation Research Program," in *Investing In Innovation.* L. M. Branscomb and J. Keller, eds., Cambridge, MA: The MIT Press.

Weiss, S. 2006. "The Private Equity Continuum." Presentation at the Executive Seminar on Angel Funding, University of California at Riverside, December 8-9, Palm Springs, CA.

Wessner, Charles W. 2004. *Partnering Against Terrorism.* Washington, DC: The National Academies Press.